PROGRESS IN COLLOID & POLYMER SCIENCE

Editors: F. Kremer (Leipzig) and G. Lagaly (Kiel)

Volume 104 (1997)

Optical Methods and Physics of Colloidal Dispersions

Guest Editors:

T. Palberg (Mainz) and
M. Ballauff (Karlsruhe)

ISBN 978-3-662-15702-2
ISSN 0340-255 X

Die Deutsche Bibliothek –
CIP-Einheitsaufnahme

Optical methods and physics of colloidal dispersions / guest ed.: T. Palberg ;
M. Ballauf.
  (Progress in colloid & polymer science ;
  Vol. 104)
  ISBN 978-3-662-15702-2
  ISBN 978-3-7985-1661-8 (eBook)
  DOI 10.1007/978-3-7985-1661-8

© 1997 by Springer-Verlag Berlin Heidelberg
Originally published by Dr. Dietrich Steinkopff
Verlag GmbH & Co. KG, Darmstadt in 1997
Softcover reprint of the hardcover 1st edition 1997

Chemistry editor:
Dr. Maria Magdalene Nabbe;
Production: Holger Frey, Bärbel Flauaus.

Typesetting and Copy Editing:
Macmillan Ltd., Bangalore, India

Progr Colloid Polym Sci (1997) V
© Steinkopff Verlag 1997

This special issue contains the proceedings of the "International Workshop on Optical Methods and the Physics of Colloidal Dispersions" held in memory of late Prof. Dr. Klaus Schätzel on September 30th and October 1st of 1996 at the Johannes Gutenberg Universität, Mainz, Germany. The meeting focused on two special aspects of colloidal science, namely novel optical methods and the physics of colloidal dispersions. It had been these two areas where Klaus Schätzel added much to the development of the whole field and the workshop participants witnessed much of his personal spirit living on in many exciting research activities. The "Correlator Museum" and a special session on the first day were to commemorate Klaus Schätzel and his œvre. It is a great honor for the guest editors to dedicate this volume to Klaus Schätzel.

Although having a long tradition, colloidal physics have recieved increasing interest over the last two decades for several reasons. Advanced instrumental techniques have allowed for many new and exciting experiments. Novel light-scattering equipment gives access to ultraslow and very fast dynamics with applications to gelation or vitrification and to strongly turbid media. Special interest was given to cross correlation schemes to investigate the dynamics in concentrated systems and to study, for example, the influence of hydrodynamic interactions or correlations between microscopic motions and macroscopic viscosity. Small-angle techniques found interesting application to molecular fluids under non-equilibrium near critical conditions, they were also applied to complement rheological experiments allowing to simultaneously study the system structure. Turbidity measurements extended known principles from light scattering to x-rays to access attractive interactions in opaque media. Microscopic methods on the other hand yield important complementary information in real space, e.g. on the epitaxial growth of single hard sphere crystals. Real space correlation methods allow to study α- and β-relaxation processes of colloidal glasses.

Furthermore, the tailoring of particles with particular surface and optical properties has led to well characterized model systems of precisely controlled interactions. Thus rotational diffusion of spherical but optically anisotropic particles or the glass transition in "squishy" hard spheres were reported, entropic attractions were thouroughly discussed and novel developments concerning charge renormalization were reviewed on the basis of recent high precision experiments. In particular, the contributed papers showed that the increase in the quality and number of suitable model systems significantly enhanced our knowledge of colloidal properties, structures and dynamics. Throughout a strong overlap to other fields of colloidal sciences was felt addressing, for example, biological, polymeric, polyelectrolytic systems.

Last but not least statistical mechanics and computer simulations have very much deepened our understanding of structure formation and colloidal dynamics on various time scales in strongly interacting systems. Numerical calculations along integral theories including hydrodynamic interactions predict interesting qualitative differences between hard and soft sphere dynamics, some already supported by experiment. Computer simulations address phase transitions with peculiar pair potentials found in some of the above-mentioned experimental model systems and in restricted geometries.

The meeting provided a platform to both review these recent developments and to exchange new results and ideas. It was attended by more than 160 researchers from 11 countries and offered 17 oral presentations and 48 poster contributions of outstanding quality. On behalf of the family of Klaus Schätzel and the organizing committee we sincerely thank all participants for their contributions to this successfull commemorative meeting. About one-third of attendants came from the city of Mainz and, in particular, from the SFB 262 "Glassy State and Glass Transition of Non-Metallic Amourphous Substances" hosting the workshop, a second lot came from Germany, and more than one-third from abroad. The organizing committee very gratefully acknowledges a generous grant by the Deutsche Forschungsgemeinschaft, which made it possible to invite leading experts and engaged young researchers from all over the world. We also thank the State of Rheinland-Pfalz, the City of Mainz, Johannes Gutenberg University, Daimler-Benz AG, Schott Glaswerke, BASF AG and Hoechst AG which contributed financially and in various other ways. We are indebted to the members of the SFB research groups for their help in preparing and running the workshop. Finally it is a particular pleasure to thank the authors of this volume for their excellent presentations, as well as the referees for their cooperation in realising this volume.

Thomas Palberg
Matthias Ballauff

Progr Colloid Polym Sci (1997) VII
© Steinkopff Verlag 1997

CONTENTS

Progr Colloid Polym Sci (1997) 104:1–3
© Steinkopff Verlag 1997

E.O. Schulz-Du Bois

In memoriam Klaus Schätzel

born 12 November 1952, deceased 4 October 1994

Prof. Dr. Klaus Schätzel 1993 at the age of 40

Prof. Dr. E.O. Schulz-Du Bois (✉)
Brunnenweg 3b
24211 Preetz, Germany

To begin with, let me say a word of thanks to the organizers of this workshop and of its proceedings volume for providing a few moments and a few pages devoted to the memory of Klaus Schätzel. He was professor of experimental physics at Mainz and a member of this Sonderforschungsbereich for little more than a year, yet he left a lasting impact on both organizations and more so on the experimental techniques in colloid sciences everywhere in the world.

Let me recount a few facts about Klaus' life and work. His father, Heinz, a young man at the end of World War II, had to leave his native province of Silesia which subsequently became a part of Poland. As a refugee Heinz settled in Holstein close to the Baltic Sea and married Christa, a local girl. Heinz went into business as a graphic designer. The son, Klaus, also had a considerable talent in graphical design, interestingly he could draw equally well with both hands. At high school, he developed unusual abilities in mathematics and physics, he won a first prize in an international contest "Mathematik Olympiade" and fixed radios and television sets. From 1971 he served two years in the army and became reserve lieutenant in the signal corps. While in the army, he had visited Professor Raether at Hamburg

University, who excited him about a career in physics. Subsequently, he began to study physics on his own. The money he received upon his discharge helped to pay for his education at Christian-Albrechts-Universität in Kiel.

At Kiel, he made rapid progress, passed the Vordiplom in 1975 and a Fulbright scholarship enabled him to study at Virginia Polytechnic Institute for the year 1976/77. There he was introduced to modern coherent optics by Prof. William Dallas, himself a student of Adolf Lohmann, one of the fathers of holography. It is reported that Dallas was uncertain what to do with "this crazy German student" who was bored by the standard lessons. Lohmanns adviced: Let him work on research projects. He liked that and these studies set the direction of his future work.

In 1977 he passed the oral Diploma exam and entered my research group, in which he remained until 1993, with some interruptions. A story is told about Klaus' first day in the group. Other graduate students discussed a chess problem: How can one place 5 queens on a 5 by 5 chess board so that none beats another. While the others talked and scribbled on the blackboard, Klaus did not say a word, but after lunch he presented a computer program including graphics in which the problem was completely solved. Perhaps I may explain that Klaus' decision to work with me, was really a break for me. When called to Kiel in 1974, I had to realize that it was not possible there to continue with my former research interests – microwave masers and Josephson computer circuits – for lack of facilities and support. So in 1976 I took a summer school course in photon correlation techniques under E. Roy Pike and Herman Z. Cummins as a start for activities in this field with applications to hydrodynamics. But I must admit that then I was by no means an expert in these fields and it was not at all easy for me to define and direct meaningful research projects. But we were lucky. Just when Klaus arrived, I received a preprint by Joe Erdmann of Boeing Research in which he described a novel "rate correlation technique". It offers the possibility of measurement of velocity correlation functions in real time. I asked Klaus to investigate this new method and, sure enough, he developed an interesting and valuable variant of Erdmanns technique. This work is described in his Diplom thesis of 1978.

Even before that was completed officially, my group attended a Photon Correlation Conference in Stockholm in order to present, among other topics, the results of Klaus' Diplom work. There he impressed E. Roy Pike and others of the Royal Signals and Radar Establishment in Malvern (the birth place of the digital correlator) by his deep insights and fast understanding of new problems. I remember Roy saying: "This boy is quick!" It was agreed that Klaus would spend about one year at Malvern, from mid 79 to mid 80. It was fortunate that Klaus was a reserve officer so that he had NATO clearance. Otherwise he would not be permitted to walk around unaccompanied at Malvern. From there he brought home the subject of his doctoral thesis. He studied the deflection of a laser beam after passing hot air rising from a narrow gap. If the number of turbulence elements in the transit is small, then the central limit theorem does not apply and the beam deflection would show a non-Gaussian statistic. Klaus found the predicted K-distribution in which a numerical parameter (near one) gives the average number of turbulence elements. With this work he graduated summa cum laude in 1982.

From then on he started building up his own group; he managed to get the necessary funds and invented or developed a number of highly successful measuring tools for the study of colloidal systems. Since then, several of these systems are produced commercially and they belong to the standard equipment of many laboratories.

In Schätzels fast digital correlator, the time increment is doubled repeatedly. Thus a logarithmic time scale is generated, from 12 ns up to several seconds, for which the correlation function is generated in real time. Some noise in measured correlation functions is eliminated by using a symmetric normalization algorithm. This feature allows the identification of very slow processes which heretofore escaped detection. At present, the third generation of this correlator is marketed by ALV of Langen, Hessen.

When studying small particles with small Zeta-potential, the violent Brownian motion masks a simultaneous electrophoretic motion. Schätzels method of the "Amplitude Weighted Phase Structure Function" allows a clearcut separation of both effects with high resolution. For example, electrophoretic displacements as small as 1 Å may be resolved.

Considerable interest centers on concentrated colloidal systems because, among other facts, their interparticle potential may be manipulated chemically. In these systems the interaction between particles leads to a modification of the classical Brownian motion and to spatial structures. At the same time there is more multiple light scattering. But while single light scattering allows a simple and straight forward evaluation of measured curves, this is not so for multiple scattering. Here the two-color system by Schätzel and Drewel offers a solution. It is based on the fact that the single-scattering correlation function depends only on the scattering wave vector, hence it may be obtained with two different wavelengths as long as the scattering wave vector is the same. The contributions by multiple scattering, by contrast, are uncorrelated for different wave lengths and hence do not contribute to the cross-correlation function. The apparatus developed for this type of measurements is likewise produced and marketed commercially by ALV.

With polydispersity, that is particles of different size, the photon correlation technique does not give the size distribution with good resolution. This is for principal reasons in that the required mathematical inversion is "an ill-posed problem" in the sense of Hadamard. Here comes the help from Schätzels method of single particle tracking. Light of a laser diode is focused on a single particle and, as its Brownian motion proceeds, the focus follows that motion by means of a feedback system and piezo mirrors. The particle displacement is obtained from the piezo voltages, and from the mean-squared displacement the particle diameter is given by the Einstein formula. Experience has shown that in 1 s of tracking the diameter is obtained with 10% accuracy.

Experiments with this system showed further effects. Due to the light pressure, the particle is pushed forward in the direction of the beam, and at the same time it is drawn radially towards the center of the beam. This opens a new field of investigation with interesting possibilities yet to be explored.

In 1987 Klaus passed the Habilitation, in the German university scheme a prerequisite to become professor. In

Progr Colloid Polym Sci (1997) 104:1–3
© Steinkopff Verlag 1997

1988 he obtained a Heisenberg scholarship, a highly prestigious award open only to the best young university teachers which allowed him freedom of travel and to study fields of his choice. During this time he organized two conferences "Static and Dynamic Light Scattering", 1988 at Kiel-Schilksee and 1993 in Burg on Fehmarn Island. In 1992 he substituted for Professor Lohmann at Erlangen University while the latter was on leave of absence.

Here I should tell of many colleages around the world who came to work with Klaus or whom he came to work with; of ambitious projects like the correlator in a NASA space experiment; of his young family which went along with him on many trips and stayed abroad with him for longer periods; of the many gifted physics students who were fascinated by Klaus' many extraordinary talents and loved to work with him – all this others can tell better than me.

Klaus Schätzels death at the early age of 41 is a great loss to the international community of colloid scientist. His friends, coworkers, and colleages at Kiel and Mainz and many other places around the globe will not forget him. Perhaps I may express my feelings by a quotation from the German philosopher Johann Gottlieb Fichte (1762–1814):

Glücklich der Lehrer, dessen Schüler größer ward als er.

Progr Colloid Polym Sci (1997) 104:4–5
© Steinkopff Verlag 1997

W. Peters

Reviewing my time of personal cooperation with Klaus Schätzel

W. Peters (✉)
ALV-Laser Vertriebsgesellschaft m.b.H.
Robert-Bosch-Straße 46
63225 Langen, Germany

Meeting a person of extraordinary abilities and being capable to think ahead analytical, clear and at the first glance, to predict future developments is indeed of rare occasion in ones life.

Today I have great difficulties to determine the exact date in 1979 when I was introduced to Klaus Schätzel in Kiel by Professor Schulz-DuBois, however, in contrast I clearly remember how much I was impressed by Klaus Schätzel after my first long conversation with him on various matters of digital correlation and optics. Already at this meeting we both tried to get to the bottom of the problem how to improve digital correlation for the benefit of DLS measurements, yet here I was able to contribute with my personal view of the "future expectation by the users of such equipment" in this field based on my technical and commercial expertise obtained in distributing the Malvern Instruments Ltd "single-bit" correlator in the years before. To my surprise he supported my argumentation with no restriction and we finally left each other by claiming with a good portion of humor in mind the day may come and we both will do the "improvement". Much later in time I realized this was actually the beginning of the most efficient cooperation between Klaus Schätzel as the scientists and myself, as the engineer and manager financing the development and commercializing our both new ideas.

During 1980 I was able to win Klaus Schätzel as the independent scientific consultant for the development of a new digital correlator, a project of great financial risk for the newly in 1980 formed company ALV-GmbH. The times that followed were very hard for both of us and required our utmost of involvement until we were finally in a position to deliver the first ALV-3000 Digital Structurator/Correlator in 1984 to the customer. During this time we also established great respect to each other, even we covered very different fields clearly related to our best personal abilities, and as a result our motivation was growing to contribute additional new developments to the field of light scattering in the future.

Since 1987 Rainer Peters assisted me in all matters of new development projects and soon took over cooperating much closer with Klaus Schätzel, thus technically commercializing all "his correlators" and others ideas, and e.g. supporting these with the required software expansions etc., or in the case of the ZENO space project finalize and successfully complete this challenging opportunities with the delivery and acceptance of the ALV-5000/VME unit by the University of Maryland for NASA.

Progr Colloid Polym Sci (1997) 104:4–5
© Steinkopff Verlag 1997

The opportunity to share a part of our life time with and to participate in Klaus Schätzel's extraordinary abilities as scientists, theoretician and technical experimenter until October 14, 1994 indebted us to continue in his spirit and has left its great and permanent influence to all of us in ALV Company.

Progr Colloid Polym Sci (1997) 104:6–7
© Steinkopff Verlag 1997

W. Peters
R. Peters

Reviewing almost two decades of digital correlator development

Klaus Schätzels contribution on the past development of correlation techniques in general and digital correlators in special are well known and highly recognized by the scientific community. Starting from his first digital correlator development, a software correlator powered by a NOVA computer in the late 70's and the first digital 64 channel hardware structurator/correlator in the early 80's, all his major developments concerning applied correlator techniques were conducted together with ALV-GmbH, being responsible for commercialization, from 1982 on. From this time on, numerous milestone developments resulted from this partnership, ranging from the introduction of general concepts like "decoupled data input and processor stage concept", "4×4 bit computation", "random preset scaling for multi-bit correlators" and, most important, the "Multiple Tau Correlation Technique" and "Symmetric Normalization" to complete units, such as the ALV-3000 Digital Structuator/Correlator, an absolutely revolutionary design at the time (1983/4), which already incorporated all above key concepts and with this was and still is one of the most flexible correlators ever developed, to the first commercially available PC-correlator, the ALV-4000 (1987), to the probably most successful single board correlator family of the last decade, the ALV-5000-E-WIN correlator family (ALV-5000 in 1989, ALV-5000/E in 1992 and presently the ALV-5000/E/ WIN with software operation under MS-WINDOWS®).

In 1990, ALV and Klaus Schätzel started the development of a special "Space Flight Correlator" for the University of Maryland to be used in their μ-gravity experiment (ZENO) in 1990, which, as a result of this, was sent to orbit on March 4th 1994 on the Space Shuttle Flight STS-62 and was sent to orbit again in March 1996 on-board of mission STS-75.

In both cases, the ALV-5000/ VME correlator took over 500 correlation functions of light scattered by a Xenon sample near the critical temperature. It seems for certain, that the enormous ease and simplicity of using this correlator, a direct result of using again such key concepts as "Multiple Sampling Times" and "Symmetric Normalization", had been of great help in conducting such an experiment from a ground base, with the actual experiment being located several 100 km above the earth in orbit and no direct interaction possible.

Future development projects of ALV Company in the field of digital correlation will be used on these concepts of Klaus Schätzel, thus keeping his spirit alive.

W. Peters (✉) · R. Peters
ALV-Laser Vertriebsgesellschaft m.b.H.
Robert-Bosch-Straße 46
63225 Langen, Germany

Progr Colloid Polym Sci (1997) 104:6–7
© Steinkopff Verlag 1997

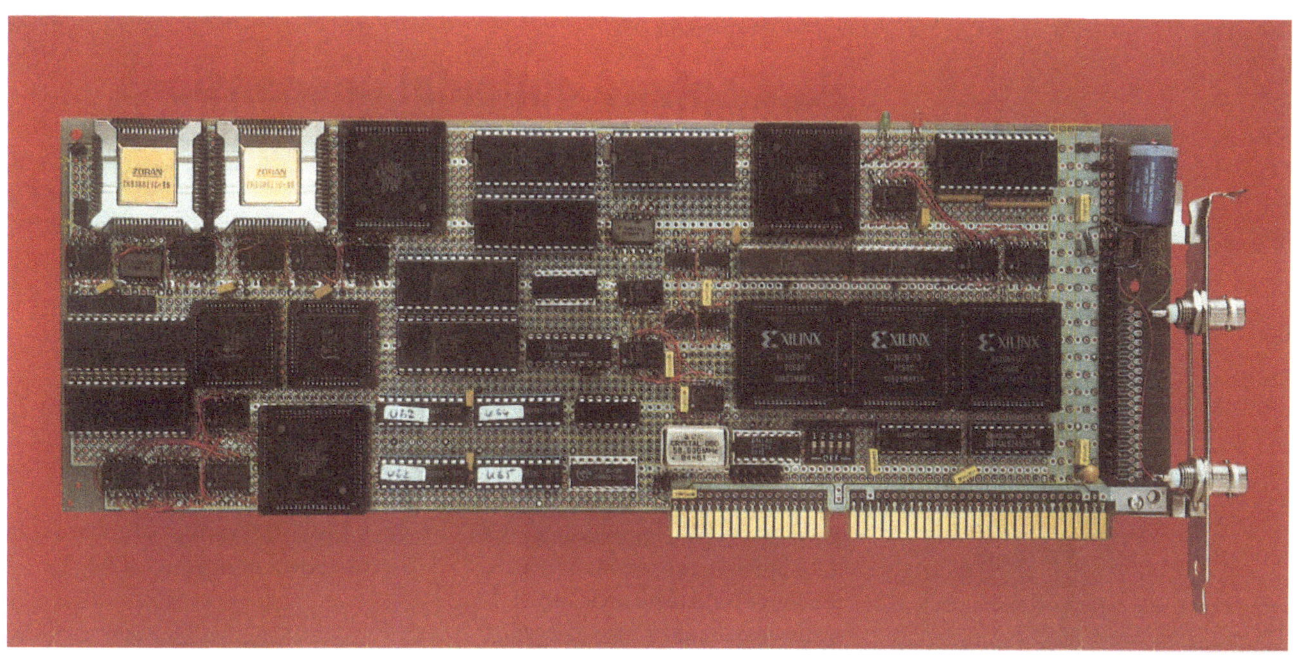

Progr Colloid Polym Sci (1997) 104:8–11
© Steinkopff Verlag 1997

P.N. Pusey
P.N. Segrè
O.P. Behrend
S.P. Meeker
W.C.K. Poon

Hard-sphere colloidal suspensions studied by two-colour dynamic light scattering

Dr. P.N. Pusey (✉) · P.N. Segrè
O.P. Behrend · S.P. Meeker · W.C.K. Poon
Department of Physics and Astronomy
The University of Edinburgh
Mayfield Road
Edinburgh EH9 3JZ, United Kingdom

Abstract Recent measurements of the diffusion properties of hard-sphere colloidal particles in suspension, made by two-colour dynamic light scattering, are described briefly.

Key words Dynamic light scattering – hard-sphere colloids – diffusion

Introduction

Over the past 25 years, dynamic light scattering (DLS) has proved to be a powerful technique for investigating the Brownian dynamics of a wide range of systems, including colloidal suspensions. Only if the sample is relatively transparent, so that single scattering dominates, it is possible to relate the quantity measured by DLS to simple properties of the scattering medium (Eq. (1)). The light scattered by optically turbid media contains contributions from both single and multiple scattering. The relationships between the properties of doubly, triply, etc., scattered light and those of the medium are complicated. Thus, the application of DLS, in its usual form, is restricted to transparent media.

One of Klaus Schätzel's many innovative contributions to light scattering instrumentation was to develop two-colour dynamic light scattering (TCDLS) into a versatile, almost routine, technique. TCDLS effectively suppresses multiple scattering and selects just the single scattering (see Section 3), allowing the study of optically turbid samples. Here we describe recent TCDLS studies of concentrated suspensions in a liquid of colloidal particles which interact like hard spheres.

These experiments have yielded two unexpected findings whose interpretation challenges theory. First, the rate of structural relaxation (defined below) of the suspensions shows the same dependence on suspension concentration as the inverse of their zero-shear-rate viscosity. Second, the intermediate scattering functions, measured by TCDLS, show an interesting scaling property which suggests that structural relaxation is controlled by self-diffusion of the particles.

Detailed descriptions of this work have been published recently. Thus, this paper will give only a very brief summary which directs the reader to the literature. A more detailed, but still concise, summary is given in Ref. [1].

Background

Two colour dynamic light scattering (and ordinary DLS for transparent samples) measures the normalised autocorrelation function of the amplitude of the singly scattered light field. This quantity is equal to the normalised intermediate scattering function $f(Q, \tau)$ of the suspension,

$$f(Q, \tau) \equiv \frac{F(Q, \tau)}{F(Q, 0)}, \tag{1}$$

where the intermediate scattering function $F(Q, \tau)$ is given by

$$F(Q, \tau) = \frac{1}{N} \left\langle \sum_{j=1}^{N} \sum_{k=1}^{N} \exp\{i\mathbf{Q} \cdot [\mathbf{r}_j(0) - \mathbf{r}_k(\tau)]\} \right\rangle, \tag{2}$$

and

$$F(Q, 0) = S(Q) \, , \tag{3}$$

where $S(Q)$ is the static structure factor. Here N is the number of particles, $\mathbf{Q}$ is the scattering vector and $\mathbf{r}_j(t)$ the position of particle j at time t. As can be seen from its definition, in general the intermediate scattering function measures a collective motion of the particles and can be recognised as the autocorrelation function of spatial Fourier components of the sample's density fluctuations of wavelength $2\pi/Q$.

In concentrated suspensions, where the fraction ϕ of the suspension's volume which is occupied by the particles may be as large as 0.5, the static structure factor resembles that of simple atomic liquids, showing a pronounced diffraction peak at $2\pi/Q \approx 2R$, where R is the particles' radius. The dominant structure in the suspension, which gives rise to this peak, is the short-ranged ordering, or cage, of particles surrounding a given particle.

In a dilute suspension, where interactions between the particles can be neglected, the intermediate scattering function takes the simple form $f(Q, \tau) = \exp(-D_0 Q^2 \tau)$, where D_0 is the free-particle (Stokes–Einstein) diffusion coefficient. In a concentrated suspension, due to both direct and hydrodynamic interactions between the particles, $f(Q, \tau)$ has a more complicated dependence on $Q^2\tau$, the slowest decay being found at the peak of $S(Q)$. Furthermore $f(Q, \tau)$ decays via a two-stage process: an initial exponential decay,

$$f(Q, \tau) = \exp[-D_S(Q)Q^2\tau], \quad \tau \ll \tau_R \, , \tag{4}$$

where τ_R is the "structural relaxation time", essentially the lifetime of a particle's cage of neighbours; and a second, slower, approximately exponential decay at long times,

$$f(Q, \tau) \propto \exp[-D_L(Q)Q^2\tau], \quad \tau \gg \tau_R \, , \tag{5}$$

with $D_L(Q) < D_S(Q)$.

Experimental

The basic idea of multiple scattering suppression in DLS, due to Phillies [2], is to use two illuminating laser beams and two detectors whose outputs are cross-correlated. The experiment is arranged so that, although the beam-detectors pairs have different geometries, their associated scattering vectors are identical. Thus, for single scattering, each detector "sees" the same spatial Fourier component of the sample's density fluctuations. However, for multiple scattering it can be shown that this degeneracy is broken. As a consequence, the time-dependent part of the measured intensity cross-correlation function (proportional to the square of the intermediate scattering function, Eq. (1)) reflects only single scattering, and multiple scattering contributes merely to the time-independent "baseline".

Phillies' original experiment in 1981 used counter-propagating laser beams of the same colour with detectors set at 90° on either side of the beams. While this experiment demonstrated clearly the suppression of multiple scattering, the arrangement could not be readily adapted to other scattering angles. In 1990, Schätzel and co-workers [3] proposed and demonstrated a more versatile equipment, based on the same principle, which by using laser beams of two different colours (the blue, 488 nm, and green, 514.5 nm, lines of the argon ion laser) can be operated over a range of angles, $\sim 20-140°$, similar to that of conventional single-beam DLS equipment. This TCDLS equipment was subsequently developed in a collaboration between Schätzel and ALV, Langen, Germany into the commercial instrument used in the present work. A detailed description of the equipment and its operation is given in [4].

The suspensions consisted of sterically stabilised particles of poly-methylmethacrylate in *cis*-decalin (see e.g. [5]). Due to a slight difference between the refractive indices of the particles and the liquid, these samples were quite turbid resulting in strong single scattering and significant multiple scattering. The multiple scattering was suppressed by the TCDLS technique, and the strong single scattering dominated that from dust and the sample cell walls allowing the collection of accurate data.

Short-time diffusion

The short-time diffusion coefficients $D_S(Q)$ describe the average motions of the particles over distances small compared to their radius and reflect both direct and hydrodynamic interactions between the particles. Extensive measurements were made of $D_S(Q)$ as functions of both scattering vector Q and suspension concentration ϕ [5]. These were compared with the predictions of theory and computer simulation, good agreement being found with the latter [5].

Long-time diffusion

The long-time diffusion coefficients $D_L(Q)$ describe motions of the particles over distances comparable to, or larger than, their radius. There is no satisfactory theory to date of long-time diffusion. The new results outlined below may provide insights which will stimulate theoretical developments.

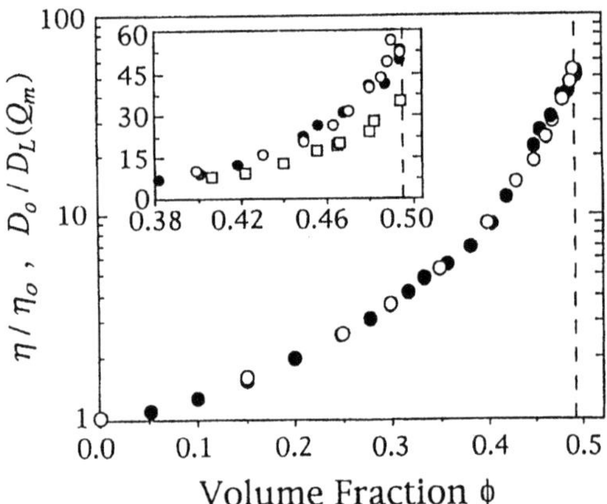

Fig. 1 Relative viscosity η/η_0 (open circles) and inverse rate of structural relaxation $D_0/D_L(Q_m)$ (filled circles) versus volume fraction ϕ of suspensions of PMMA spheres (from [4], q.v. for an explanation of the data points represented by squares in the inset)

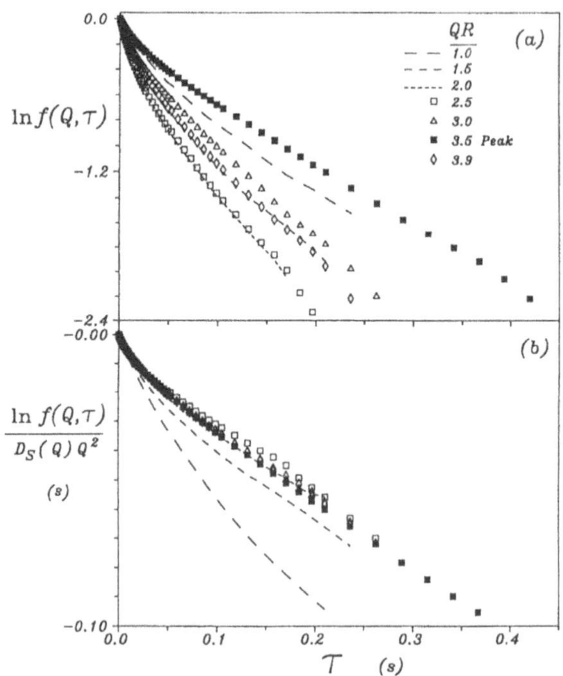

Fig. 2A Logarithm of normalised intermediate scattering functions $\ln f(Q,\tau)$ versus time τ for a PMMA suspension of volume fraction $\phi = 0.465$ for different values of QR as indicated. **B** Same data plotted as $\ln f(Q,\tau)/D_S(Q)Q^2$ versus τ, showing scaling for $QR > 2.5$ (from [5])

As noted above, in a dense fluid-like assembly of hard spheres the dominant structure, which gives rise to the main peak in $S(Q)$ at $Q = Q_m$, is the cage of particles surrounding a given particle. Thus, it can be argued that the long-time decay of $f(Q_m, \tau)$, the intermediate scattering function measured at $Q = Q_m$, reflects the dominant structural relaxation of the system so that $D_L(Q_m)$ is a measure of the rate of structural relaxation. By comparing measurements of $D_L(Q_m)$ with measurements of the zero-shear-rate viscosity η of the suspensions [6] we have found that the rate of structural relaxation shows the same dependence on suspension concentration as the inverse of the viscosity over the whole range $0 < \phi < 0.50$, i.e., that

$$\frac{D_L(Q_m)}{D_0} = \frac{\eta_0}{\eta}, \tag{6}$$

where η_0 is the viscosity of the liquid in which the particles are suspended (see Fig. 1). While one would certainly expect these two quantities to show similar dependences on concentration – the processes of simple shear flow and structural rearrangement both involve the relative motions of neighbouring particles – the apparent identity found experimentally is surprising and remains to be explained by theory.

Scaling of the intermediate scattering functions

During an attempt to understand better the mechanism of structural relaxation we made a second surprising discovery [7]. This was that for $QR > 2.7$, a range of scattering vector Q which encompasses most of the strong variation of structure factor $S(Q)$ including the main peak, plots of $\ln f(Q,\tau)/D_S(Q)Q^2$ against τ, measured at different values of Q, lay on a master curve. As can be seen from Eq. (4), this way of plotting the data ensures that they superimpose at short times (since $\ln f(Q,\tau)/D_S(Q)Q^2 = -\tau$, for $\tau \ll \tau_R$). What is surprising is the additional superimposition of data at intermediate and long times, $\tau \geq \tau_R$ (see Fig. 2). As noted in [7], this finding implies that, for $QR > 2.7$, the intermediate structure factor can be written

$$f(Q,\tau) \approx \exp\left(-\frac{D_S(Q)}{D_S(\infty)} Q^2 \langle \Delta r^2(\tau)\rangle/6\right), \tag{7}$$

where $\langle \Delta r^2(\tau)\rangle$ is the mean-square displacement of a single particle, suggesting that structural relaxation is controlled by self-diffusion. Although previous work [8] has suggested a connection between structural relaxation and self-diffusion, the detailed scaling implied by Eq. (7) awaits a full theoretical explanation.

References

1. Pusey PN, Segrè PN, Behrend OP, Meeker SP, Poon WCK (1996) Physica A, in press
2. Phillies GDJ (1981) J Chem Phys 74:260; (1981) Phys Rev A 24:1939
3. Drewel M, Ahrens J, Podschus U (1990) J Opt Soc Am 7:206; Schätzel K, Drewel M, Ahrens J (1990) J Phys: Condens Matter 2:SA393; Schätzel K (1991) J Mod Optics 38:1849
4. Segrè PN, van Megen W, Pusey PN, Schätzel K, Peters W (1995) J Mod Opt 42:1929
5. Segrè PN, Behrend OP, Pusey PN (1995) Phys Rev E52:5070
6. Segrè PN, Meeker SP, Pusey PN, Poon WCK (1995) Phys Rev Lett 75:958
7. Segrè PN, Pusey PN (1996) Phys Rev Lett 77:771
8. de Schepper IM, Cohen EGD, Pusey PN, Lekkerkerker HNW (1990) Physica A 164:12

Progr Colloid Polym Sci (1997) 104:12–16
© Steinkopff Verlag 1997

M. Heckmeier
G. Maret

Dark speckle imaging of colloidal suspensions in multiple light scattering media

M. Heckmeier (✉) · G. Maret
Institut Charles Sadron (CRM-EAHP)
6, rue Boussingault
67083 Strasbourg Cedex, France

Abstract Quasielastic multiple light scattering experiments have been performed on suspensions of particles in brownian motion embedded inside a solid turbid medium. The photon transport mean free path of the suspension and of its solid environment where adjusted to be identical. We show that by selecting a minimum intensity spot of the speckle pattern generated by the solid medium, it is possible to visualize objects which would be undetectable with common optical techniques. Our results are shown to be in agreement with a diffusion theory for the location of strongly absorbing objects inside multiple scattering media.

Key words Colloidal suspensions – multiple light scattering – imaging – heterogeneous media – speckle fluctuations

Recently, increasing interest has developed in dynamic multiple light scattering on macroscopically heterogeneous samples. The major aim of this work is to locate and to visualize objects which are embedded inside a turbid medium by analyzing the time dependent speckle fluctuations of the multiple scattered light. The underlying physical idea is that, on their random walk through the sample, photons which have crossed the object contribute by a different dynamic phase shift to the multiple scattering process than the photons which did not cross the object. By spatial resolved measurements of the dynamic autocorrelation function of the multiple scattered light, it is possible to construct a low resolution image of a colloidal inclusion inside a strongly scattering medium [1]. Flow of a colloidal suspension embedded in the very same suspension could be visualized, exploiting the different dynamics of the object and its environment without requiring any static scattering contrast [2]. It was shown theoretically [1] that the temporal depolarized electric field autocorrelation function at a position $\mathbf{r}$ inside a multiple scattering medium $G_1(\mathbf{r}, \tau) = \langle E(\mathbf{r}, t) E^*(\mathbf{r}, t + \tau) \rangle$ can be approximated by a steady state diffusion equation. Excellent agreement between this theoretical approach and multiple light

scattering data from dynamic heterogeneities has been found [3]. All the reported work [1–3] has one common feature, that is the embedded object was placed at a depth x inside the turbid medium which was comparable with the size d of the object itself.

Here we present results of multiple light scattering experiments on a white solid sample containing a cylindrical cavity. The cavity was filled with a concentrated suspension of colloidal spheres in brownian motion. The concentration of spheres was adjusted such that the photon transport mean free path was identical inside and outside the cavity. We introduce a new technique which allows to locate the filled cavity up to depths of about five times its diameter, thus substantially increasing the range of imaging. This method is compared with the techniques used so far [1–3]. Our data are in agreement with a theory developed for the location of absorbing and scattering objects inside a multiple scattering medium [4].

In our experimental setup, the beam of a vertically polarized mono-mode Ar^+ laser is incident on the light scattering sample as indicated in Fig. 1. About one coherence area of the backscattered light is collected onto

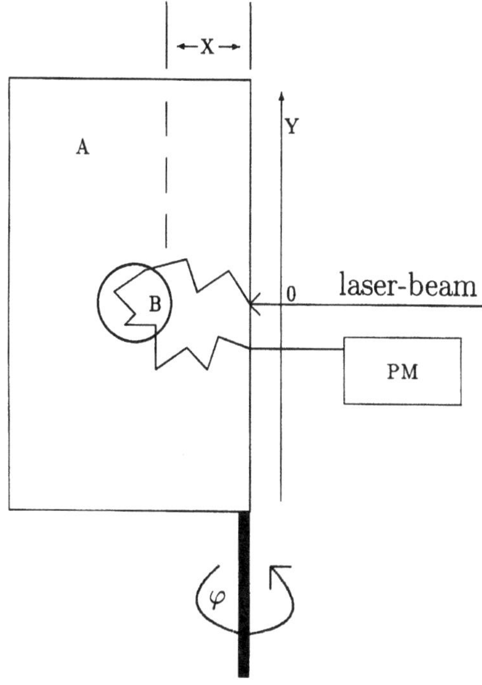

Fig. 1 Side-view of our light scattering cell. A block of teflon ($8\,\mathrm{cm} \times 7\,\mathrm{cm} \times 2\,\mathrm{cm}$) contains a cylindrical cavity B (diameter $d = 2\,\mathrm{mm}$) which is filled with a colloidal suspension. The height (y-position) of the block can be altered with respect to the incident laser beam. $y = 0$ denotes centered laser incidence with respect to the cylinder. The system can be rotated around the y-axis as indicated. $\varphi = 0$ corresponds to an orientation of the surface normal vector of the block parallel to the direction of detection. Different depths x were realized by using different blocks

a photomultiplier tube. In order to reduce the contributions of very short photon paths and single scattering events, we detect depolarized light in a so-called V–H-configuration. The intensity autocorrelation function of the multiple scattered light is determined with a computer-controlled correlator. Our sample consists of a solid teflon block A (photon transport mean free path $l^* \approx 280\,\mu\mathrm{m}$). A monodisperse suspension of polystyrene beads (particle diameter $1.14\,\mu\mathrm{m}$, volume fraction $\phi \approx 0.01$) in water with identical photon transport mean free path is filled in the cylindrical cavity. Since the motions of scatterers only occur in the spatially confined region B, photon paths that did not cross the cavity B create a stationary speckle pattern, while the paths that cross the suspension generate time dependent speckle fluctuations. This means that the time average and ensemble average of the detected light intensity become different and we are dealing with a non-ergodic sample. In dynamic light scattering experiments, non-ergodicity is usually taken into account by averaging the stationary speckle. This can be done by moving the scattering cell [5]. Along these lines, in the first part of this paper we rotate the cell, and obtain images of the cavity

B up to depths similar to the case of dynamic heterogeneities studied earlier [3]. In the second part, we demonstrate, that by measuring the autocorrelation function at a point where the static speckle pattern exhibits a local minimum, the relative contribution of the photons that crossed the suspension is significantly higher than in the case of averaging. This allows to obtain a measurable light scattering signal of the embedded heterogeneity up to positions much deeper inside the teflon block.

Figure 2 shows typical data of the normalized temporal intensity autocorrelation function $g_2(t)$, measured in the described backscattering geometry. There are two well-separated time constants in the decay of $g_2(t)$. The larger one ($t \approx 10^{-1}\,\mathrm{s}$) is due to the rotating block. To average the stationary speckle pattern, the block is rotated from $\varphi = -5°$ to $\varphi = +5°$ during the measurement time of 150 s. The smaller one ($t \approx 10^{-4}\,\mathrm{s}$) is due to the brownian motion of the colloidal spheres inside the cylindrical cavity. Its amplitude depends on the relative position of the cavity with respect to the incident laser beam. For non-central incidence of the beam ($\triangledown$) this amplitude Δg becomes smaller, since less photons scan the liquid region on their way through the sample. Δg therefore provides a parameter which is sensitive to the objects position and can be used to visualize the embedded suspension.

In Fig. 3, the variation of the amplitude parameter Δg with the y-position is shown. For $x = 1.5\,\mathrm{mm}$ ($\triangledown$) one obtains a clearly visible profile around the real cavity position ($|\,|\,|\,|\,|\,|\,|\,|$), illustrating the possibility of imaging a fluctuating object in a solid environment without any static scattering contrast. For the larger depth $x = 2.5\,\mathrm{mm}$ the values of Δg become smaller, but nevertheless provide a means to obtain the position of the embedded object and, as long as x is not much larger than d, to determine its approximate size. This is in agreement with the situation of a dynamic heterogeneity inside a multiple scattering liquid [3] where the object's size corresponds approximately to the depth x.

If the depth x becomes larger than the diameter of the cylindrical cavity, the contrast parameter Δg becomes unmeasurably small. As indicated in Fig. 4 ($\square$), there is no detectable short time decay in the angular averaged intensity autocorrelation function $g_2(t)$ and the described method fails to locate the colloidal suspension. However, by measuring the autocorrelation function at a fixed angular position where the scattering intensity exhibits a local minimum, the short time decay reappears ($\bullet$). To illustrate this idea, Fig. 5 shows five correlation functions which have been measured for a fixed spatial position of the teflon-block. The angle between the block's surface normal and the axis of detection was slightly varied to detect different static intensities. For smaller

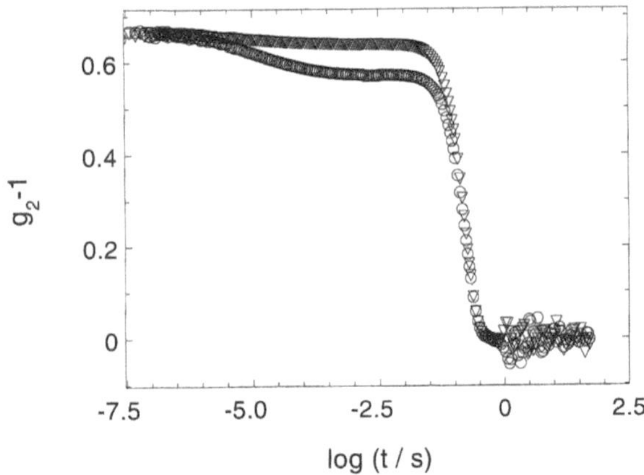

Fig. 2 Experimental intensity correlation functions for (○) centered laser incidence ($y = 0$) and (▽) $y = 2$ mm. The cavity is located at $x = 1.5$ mm

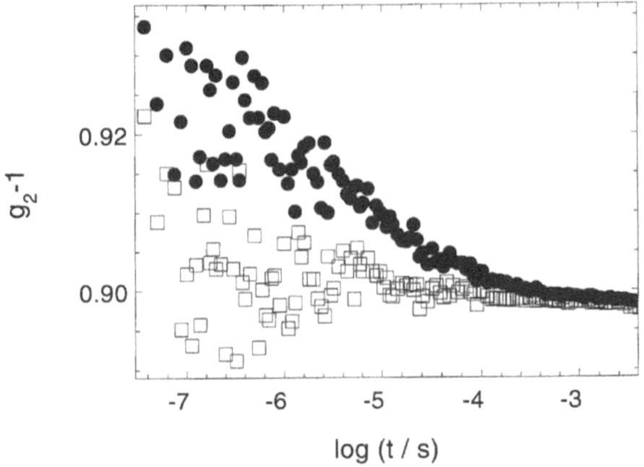

Fig. 4 For $x = 5.5$ mm a correlation function with an angular average over the static speckle pattern (□: mean count rate: ≈ 100 kHz) is compared to the case of a measurement at a fixed angular position at a local minimum of the static speckle pattern (●: mean count rate: ≈ 10 kHz). The curves are vertically shifted to coincide for large correlation times

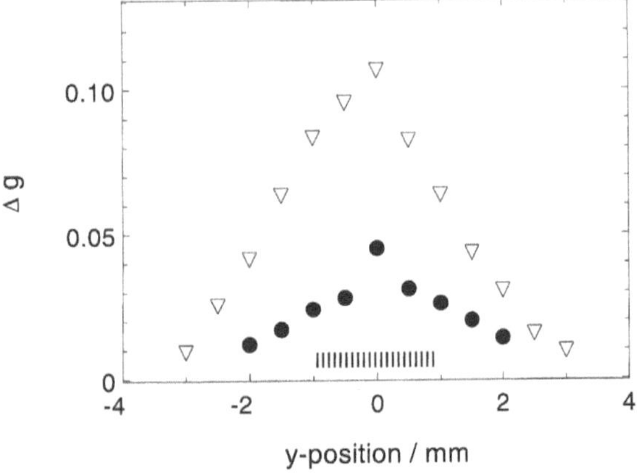

Fig. 3 For two different depths $x = 1.5$ mm (▽) and $x = 2.5$ mm (●), the amplitude of the fast process is shown as a function of the y-position of the cavity. ||||||| denotes the actual position of the cylindrical inclusion

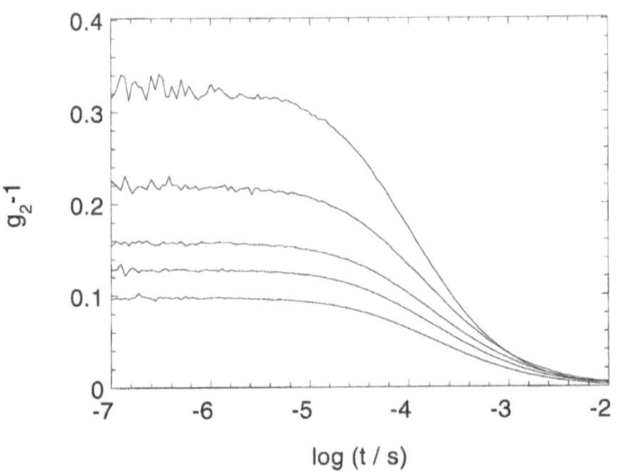

Fig. 5 For $x = 1.5$ mm and $y = 0$ (centered laser incidence), correlation functions are shown at angular positions where the static scattering intensity exhibits different values. (50, 100, 200, 300 and 400 kHz from the upper to the lower curve.)

intensities, i.e., dark spot of the static speckle pattern, a larger amplitude Δ of the short time decay is found. Thus the relative contribution of the light that crossed the suspension is increased by measuring at small intensity values of the speckle pattern. Nevertheless, as illustrated in Fig. 6 (▽), the amplitude Δ of the decay has a very noisy position dependence making it difficult to locate the embedded object. However, taking into account, that correlation functions measured for different y-positions of the sample are only comparable if the measured intensity in the minimum is identical for every run, it is possible to find a new parameter that allows imaging as follows: Since we have to measure at very low count rates (about 5–10 kHz at the speckle minimum) the required time of

a single run becomes relatively large (about one hour). During this time the static scattering intensity cannot be kept exactly constant because of slow rearrangements of the speckle pattern. We take this effect into account, by multiplying the dynamic amplitude parameter Δ with the mean square scattering intensity $\langle I \rangle^2$ during the measuring time. Thus, points with higher static scattering intensity are weighted more than others where the intensity is smaller. This is consistent with the measured amplitudes in Fig. 5. The product of Δ and the mean square static intensity of a run (Fig. 6, □) shows a clear variation with the cylinder's position and can thus

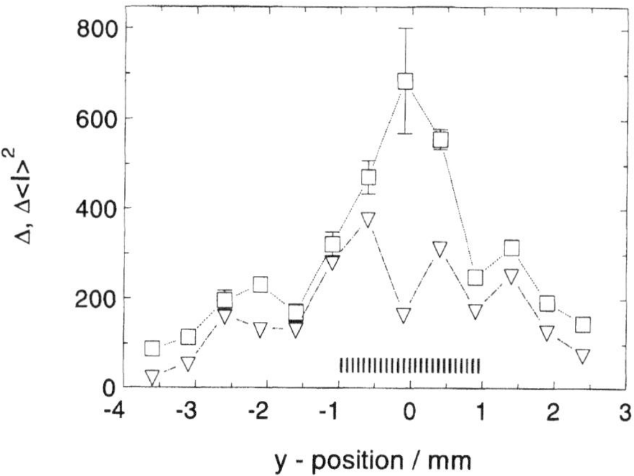

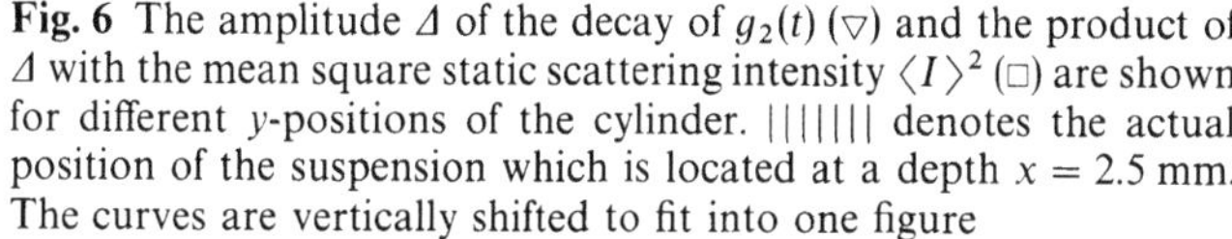

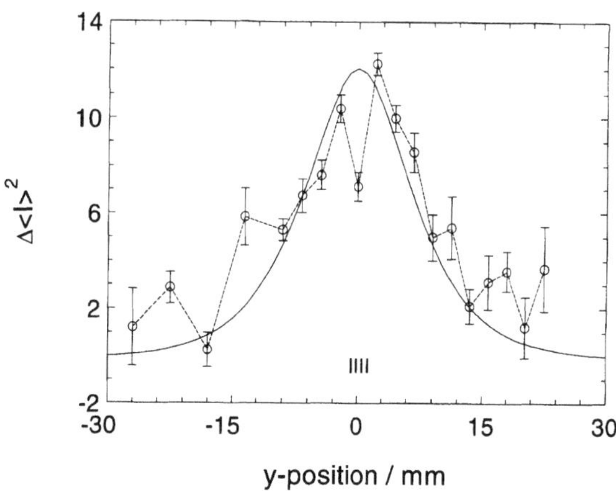

Fig. 6 The amplitude Δ of the decay of $g_2(t)$ ($\triangledown$) and the product of Δ with the mean square static scattering intensity $\langle I \rangle^2$ ($\square$) are shown for different y-positions of the cylinder. |||||||| denotes the actual position of the suspension which is located at a depth $x = 2.5$ mm. The curves are vertically shifted to fit into one figure

Fig. 7 Variation of the parameter $\Delta \langle I \rangle^2$ with the y-position of the cylinder for a depth $x = 10.5$ mm inside the teflon block. |||| shows the actual position of the embedded cavity. The solid line is a theoretical curve according to Ref. [4]

be used as a new parameter to visualize the dynamic heterogeneity.

The limits of this approach are illustrated in Fig. 7. For a depth $x = 10.5$ mm we still find a clear variation of the parameter $\Delta \langle I \rangle^2$ as a function of the y-position of the block of teflon [6]. Note that this depth x is larger than five diameters of the cylinder and corresponds to approximately 37 times the photon transport mean free path of the sample. Thus the possible range to visualize an embedded dynamic heterogeneity is significantly increased [3]. Since the object is placed much deeper inside the sample than its typical spatial extension, the variation of our parameter does not indicate any longer the object's size. It rather corresponds to the depth of the cylinder, reflecting the approximate linear diffusive spreading of the photons that crossed the object on their way back to the sample's surface.

We now compare this result with a theory made to localize absorbing or transmitting objects inside a multiple scattering medium [4] (solid line in Fig. 7). In our case, every photon that crossed the colloidal suspension contributes to the decay of the measured autocorrelation function, that means, *correlation* is absorbed inside the object. In the case of strong absorption inside the embedded object, photons that penetrate the object are absorbed and hence give rise to a decreasing backscattered intensity. Therefore one expects that the backscattered intensity has a similar position dependence than our parameter $\Delta \langle I \rangle^2$. By taking Eq. (17) of Ref. [4] for the backscattered intensity, neglecting the dipole term p for scattering objects and setting $\alpha = 0$ (no surface absorption at the embedded object), the charge term q can be

expanded for short absorption lengths ($\kappa \rightarrow \infty$). In Fig. 7 we show the relative change in the backscattered intensity as a function of position, considering only the 0th and first order in the multiple expansion of Eq. (17) [7]. The agreement between our data and this theory shows the similarity of the underlying physical mechanisms which are photon-absorption and picking up dynamic phase shifts, respectively. For smaller depths x the experimental results compare less satisfactorily with this approach. Higher orders in the multipole expansion reflecting the geometric details of the embedded object would have to be taken into account. This provides different expressions for our embedded cylinder and the spherical inclusion treated in Ref. [4].

In conclusion, a new method exploiting dark static speckle spots to localize and visualize a liquid colloidal suspension included inside a static environment is presented and compared with the conventional technique of working with an ensemble averaged correlation function. The visualization range of this minimum-speckle-method is about five times higher than in former experiments [3] and in agreement with a theory developed for the imaging of absorbing static objects inside a turbid medium [4]. We think that the improved image contrast illustrated by our experiments will stimulate further research in imaging and locating objects in turbid media with particular applications in medicine and biology.

Acknowledgment We are grateful to W. Leutz for fruitful discussions. M.H. acknowledges financial support by the Deutscher Akademischer Austauschdienst (DAAD) through the HSPII/AUFE-program.

References

1. Boas DA, Campbell LE, Yodh AG (1995) Phys Rev Lett 75:1855–1858
2. Heckmeier M, Maret G (1996) Europhys Lett 34(4):257–262
3. Heckmeier M, Skipetrov SE, Maret G, Maynard R (1997) J Opt Soc Am A 14:185–191
4. den Outer PN, Nieuwenhuizen ThM, Lagendijk A (1993) J Opt Soc Am A 10: 1209–1218
5. Xue J, Pine DJ, Milner ST, Wu X-l, Chaikin PM (1992) Phys Rev A 46: 6550–6563
6. It has to be noted that these measurements are rather involved. For a count rate of about 20 kHz in the speckle minimum, we had to measure for 5400 s at every position of the teflon block. This time was divided into 90 multi-runs of 60 s. After every run, there was a short interrupt in data acquisition to correct slightly the angular position of the block to get an approximate constant static scattering intensity during the whole run
7. The continous line is $1 - (I/I_0)$ multiplied with a constant factor (1905) (I and I_0 being the backscattered intensity with and without absorbing object) in order to compare with our data ($\Delta \langle I \rangle^2$) which are in arbitrary units

interference term. In fact, it can be shown [9] that the phase of the scattered field is shifted by $\pi/2$ with respect to the incident field. A simple way to obtain an interference signal is that of inserting a quarter-wave plate between the scattering cell and the analyzer in order to compensate the $\pi/2$ phase shift between the transmitted field and the depolarized scattered field [8]. It is interesting to note that the final sequence of optical elements is exactly the same as employed in an electric birefringence apparatus (see, for instance, Refs. [10, 11]). The forward scattering experiment can be seen as an electric birefringence experiment at zero electric field in which the spontaneous fluctuations of birefringence are investigated.

In this paper we describe in a simple and more intuitive way the heterodyne configuration proposed in Ref. [8], and we discuss some experiments performed by using dispersions of fluorinated polymer colloids [12] which are particles having a partially crystalline internal structure and represent the ideal system to test the new approach to forward dynamic light scattering.

Forward scattering

Consider an incident plane wave with real amplitude E_0, linearly polarized in the vertical direction. The wave goes through a scattering medium consisting of identical particles dispersed in a solvent. We consider a dilute dispersion of anisotropic particles which are characterized by a polarizability tensor $\boldsymbol{\alpha}$. We assume cylindrical symmetry for $\boldsymbol{\alpha}$ along the main optical axis, and call α_1, α_1 and α_3 the diagonal components of the polarizability tensor in the particle-fixed frame. The average polarizability of the particle is: $\alpha = (\alpha_3 + 2\alpha_1)/3$. We call β the anisotropy of the particle polarizability, $\beta = \alpha_3 - \alpha_1$.

We assume that the particles can be treated as Rayleigh scatterers, so that the approach outlined in [1, Ch. 7] can be used. The scattered field will contain a vertically polarized contribution E_{VV} and a horizontally polarized term E_{VH}. Under the assumption that the particles behave as independent scatterers and that multiple scattering effects are negligible, the first-order correlation functions of the two contributions are

$$G_{VV}^{(1)}(k, t) = I_0 N d \, \frac{\pi^2 n_s^2}{\lambda^2 A_s} \left(\alpha^2 + \frac{4}{45} \beta^2 e^{-6D_0^r} \right) e^{-k^2 D_0^t} , \qquad (1)$$

$$G_{VH}^{(1)}(k, t) = I_0 N d \, \frac{\pi^2 n_s^2}{15\lambda^2 A_s} \beta^2 e^{-6D_0^r} e^{-k^2 D_0^t} , \qquad (2)$$

where I_0 and λ are, respectively, intensity and wavelength of the incident laser beam, n_s is the index of refraction of the solvent, N is the number of particles per unit volume, d is the pathlength in the scattering cell, A_s is the cross-section of the laser beam, D_0^t and D_0^r are the translational and rotational diffusion coefficients of the single particle, and k is the modulus of the scattering vector.

We consider, for the sake of simplicity, ellipsoidal particles having a geometric symmetry axis coincident with the symmetry axis of the intrinsic optical anisotropy. We call a, a, b the lengths of the semiaxes of the ellipsoid. The components of α are expressed as [9]

$$\alpha_1 = V_p \, \frac{n_s^2(n_{p1}^2 - n_s^2)}{n_s^2 + L_1(n_{p1}^2 - n_s^2)} , \qquad (3)$$

$$\alpha_3 = V_p \, \frac{n_s^2(n_{p3}^2 - n_s^2)}{n_s^2 + L_3(n_{p3}^2 - n_s^2)} , \qquad (4)$$

where V_p is the particle volume, n_{p1} and n_{p3} are, respectively, the index of refraction of the particle for polarization perpendicular and parallel to the symmetry axis, L_1 and L_3 are known functions of the ratio b/a.

Normally, $\alpha \gg \beta$, so that the time-dependence of $G_{VV}^{(1)}(k, t)$ is essentially controlled by translational diffusion. If we are interested in the value of rotational diffusion, we have to measure the intensity correlation function $G_{VH}^{(2)}(k, t)$ at different scattering angles and extrapolate the value of the first cumulant at $k = 0$. An example of such a procedure is given in Ref. [13] where the dependence of rotational diffusion on the volume fraction Φ was studied for colloidal hard spheres. In principle, the simplest thing would be to measure the correlation function of E_{VH} at $k = 0$, as first proposed by Wada et al. [2]. This would require the blocking of the transmitted beam with a crossed polarizer (analyzer). In practice, considering that $I_{VH} \ll I_0$, even by using the best available polarizers, the portion of the transmitted beam which leaks through is never smaller than I_{VH}. Wada et al. assume that the presence of some direct laser light on the detector can be exploited to perform the measurement in the reference-beam (often called heterodyne) configuration. However, this is not generally true for the following reasons: (i) the field acting as local oscillator must present the same polarization as E_{VH}, (ii) even if the polarization is the same, there is no mixing if the two fields are in quadrature. Concerning point (i), a simple method to control the strength of the local oscillator is to set the analyzer at an angle $\pi/2 - \theta$ with the vertical axis. Concerning point (ii), it turns out that the phase of the field scattered by a collection of many particles is shifted by $\pi/2$ with respect to that of the incident beam [9]. Assuming that the analyzer is not completely crossed with the incident polarization, the total field $E_1(t)$ arriving on the detector can be written as the superposition of two contributions which have the same polarization, but present a relative phase-shift of $\pi/2$. If θ is small

$$E_1(t) = \theta E_0 + i|E_{VH}(k = 0, t)| . \qquad (5)$$

Progr Colloid Polym Sci (1997) 104:17–22
© Steinkopff Verlag 1997

Depolarized forward light scattering from anisotropic particles

V. Degiorgio
T. Bellini
R. Piazza
F. Mantegazza

Dr. V. Degiorgio (✉) · T. Bellini · R. Piazza
Istituto Nazionale
per la Fisica della Materia
Dipartimento di Elettronica
Università di Pavia
Via Ferrata 1
27100 Pavia, Italy

F. Mantegazza
Istituto di Scienze Farmacologiche
Università di Milano
20133 Milano, Italy

Abstract The rotational correlation function of anisotropic colloidal particles can be measured by observing the fluctuations of the depolarized intensity scattered in the forward direction. In this work we present a theoretical treatment of forward depolarized light scattering. Our calculations show the importance of the insertion of a quarter-wave plate between the scattering cell and the analyzer. The role of the quarter-wave plate is to generate a heterodyne signal between the transmitted beam and the depolarized forward scattered field. In order to illustrate the theoretical calculations, we present experimental results concerning the measurement of the rotational diffusion coefficient in dispersions of fluorinated polymer colloids.

Key words Depolarized light scattering – colloids – macro-molecules – rotational diffusion

Introduction

It is well known that the scattering of polarized light from anisotropic Brownian particles gives rise to a depolarized component in the scattered field [1]. The decay time of the autocorrelation function of the depolarized field scattered at a non-zero scattering angle θ is related to both the translational and the orientational Brownian motion of the particles. In the particular case of forward scattering ($\theta = 0$), the shape of the depolarized correlation function is determined only by the rotational diffusion of the particles. Of course, the design of the experiment has to take into account that the forward scattered beam is superposed to the much more intense transmitted beam. The first measurement of dynamic forward scattering was performed by Wada et al. [2] on solutions of tobacco mosaic virus (TMV). The experiment was carried out by blocking the transmitted beam with a crossed polarizer (analyzer) which ideally selects only that part of the forward scattered light which is polarized orthogonally to the incident beam. Subsequently, several other experiments of zero-angle de-polarized light-scattering were performed on solutions of biological macromolecules [3, 4], solutions of synthetic polymers [5, 6], and dispersions of polydisperse fluorinated polymer colloids [7], but the method was never analyzed in depth. In a real experiment, the analyzer always leaks a portion of the transmitted beam, and this raises the question whether such a contribution can act as a local oscillator. In some experiments [2, 4–6] the heterodyne correlation function was obtained, whereas others [3, 7] present homodyne correlation functions. In the case in which the analyzer is perfectly crossed with the polarization of the incident beam, only the non-ideality of the optical components could give rise to a horizontally polarized component of the reference beam which might act as a local oscillator for the heterodyne measurement. A very simple method of generating in a controlled way a component of the reference beam having the same polarization as the signal is that of slightly offsetting the analyzer with respect to the crossed position. However, as we have discussed in a recent article presenting a systematic treatment of the forward scattering method [8], the offset of the analyzer is not a sufficient condition for generating an

Note that the vertically polarized transmitted beam contains also a contribution coming from depolarized scattering. However, such a contribution represents a small random addition to a large deterministic term, and, consequently, is neglected in our treatment. The total intensity on the detector is

$$I_1(t) = |E(t)|^2 = \theta^2 I_0 + I_{VH}(t) \ . \tag{6}$$

We see from Eq. (6) that there is no interference term. The intensity correlation function is given by

$$\langle I_1(0)I_1(t)\rangle = \theta^4 I_0^2 + 2\theta^2 I_0 \langle I_{VH}\rangle + G_{VH}^{(2)}(t) \ . \tag{7}$$

Under the assumption that the probability distribution of the particle orientations is fully isotropic and that no correlation exists among the orientations of different particles, the average depolarized intensity is given by the expression [8]:

$$\langle I_{VH}\rangle = I_0 N d \frac{k^2 \beta^2}{60 A_s} \ . \tag{8}$$

We also recall that [8]

$$G_{VH}^{(2)}(t) = \langle I_{VH}\rangle^2 \left[1 + 2|g_{VH}^{(1)}(t)|^2\right] \ , \tag{9}$$

where $g_{VH}^{(1)}(t)$ is the normalized first-order correlation function. Note that Eq. (9) is different from the Siegert relation. A full discussion of the statistical properties of the forward scattered field will be presented in a separate paper [14]. Eq. (7) shows that the laser contribution gives to the measured correlation function an incoherent background which reduces the visibility of the signal correlation function.

As discussed in Ref. [8], a simple method to perform heterodyning is to insert before the analyser a quarter-wave plate, with fast axis along the vertical direction. The effect of the quarter-wave plate is to compensate the $\pi/2$ phase shift between the transmitted field and the forward scattered field. We call $E_2(t)$ the total field impinging on the detector in the presence of the quarter-wave plate. The total intensity on the detector becomes:

$$I_2(t) = |E_2(t)|^2 = \theta^2 I_0 + I_{VH}(t) + 2\theta E_0 |E_{VH}(t)| \ . \tag{10}$$

The intensity correlation function is now given by

$$\langle I_2(0)I_2(t)\rangle = \theta^4 I_0^2 + 2\theta^2 I_0 |G_{VH}^{(1)}(t)| + G_{VH}^{(2)}(t) \ . \tag{11}$$

Under the assumption that $\langle I_{VH}\rangle \ll I_0$, the third term on the right-hand side of Eq. (11) can be neglected, so that the only significant term is the one containing the first-order correlation function.

In a real experiment, partial mixing can be observed even in the absence of the quarter-wave plate because of the non-ideality of the used optical components, such as the presence of stress-induced birefringence in the cell windows. The discussion presented above makes clear that it is not sufficient to generate a horizontally polarized reference beam, but the non-ideality of the optics should also yield a phase difference of the reference field with respect to the scattered field which is different from $\pi/2$. In the case in which heterogeneity arises from the non-ideality of the optics, no control of the reference signal is possible, so that it may not be known whether the experiment is performed in homodyne or heterodyne or in an intermediate situation [8]. At this point it is interesting to comment on the previous measurements of the dynamics of depolarized forward scattering. The experiments of Refs. [2, 3, 5–7] use two crossed polarizers ($\theta = 0$) without quarter-wave plate. Wada et al. [2], Han and Yu [5], and Crosby et al. [6] have recorded a heterodyne signal which was attributed to the presence of a substantial amount of stray depolarized light. Schurr and Schmitz [3] have been able to perform homodyne measurements by using a very long pathlength to raise the scattered intensity of TMV solutions relative to the level of the transmitted direct beam. Russo et al. [7] have also performed a homodyne measurement by using polydisperse fluorinated polymer colloids. Thomas and Fletcher [4] have inserted a quarter-wave plate before the scattering cell with the aim of introducing some compensation for the birefringence of lenses and cell windows. They observe a heterodyne signal, again attributed to the presence of a depolarized leakage background much greater than the forward-scattered depolarized signal.

It should be noted that most of the difficulties encountered in standard heterodyne experiments are not present in the forward depolarized scattering configuration we have proposed and tested. In fact the signal-to-reference ratio is easily controlled by altering the direction θ of the analyzer. Furthermore, the coherence area for forward scattering coincides with the incident beam cross-section and does not depend on the length of the optical path inside the scattering cell: this implies that a good overlap of the wavefronts of signal and reference field is automatically ensured, and that one can use a long pathlength. Clearly, the mechanical stability of the optical components is much less critical than in a standard heterodyne experiment.

Experiments

Generally speaking, the study of the dynamics of fluctuations of the depolarized scattered intensity is made easier by maximizing the ratio $\rho = \langle I_{VH}\rangle/\langle I_{VV}\rangle$. For Rayleigh scatterers, ρ can be derived by recalling that $\langle I_{VV}\rangle = G_{VV}^{(1)}(t=0)$, $\langle I_{VH}\rangle = G_{VH}^{(1)}(t=0)$, and using

Eqs. (1) and (2). We obtain

$$\rho = \frac{3\beta^2}{45\alpha^2 + 4\beta^2} \,, \tag{12}$$

In the case of particles made of isotropic material, depolarized scattered light can arise if the shape is non-spherical. As an example, we take an ellipsoidal Rayleigh scatterer with a symmetry axis, having semiaxes a, b, b. We present in Fig. 1 a plot of ρ as a function of the semiaxes ratio b/a for the case of a strong optical mismatch against the solvent, $n_p - n_s = 0.17$. As expected, $\rho = 0$ for the sphere. Note that ρ remains quite small also for elongated ellipsoids, attaining the value 2.3×10^{-3} for an axial ratio equal to 10. At fixed axial ratio, ρ is a decreasing function of the mismatch $n_p - n_s$. For instance, if we take $n_p - n_s = 0.07$, the value of ρ corresponding to $b/a = 10$ decreases to 4×10^{-4}, as it is shown by Fig. 1. On the contrary, if the particle is made of anisotropic material, much larger values can be obtained also for spheres, and ρ takes its maximum value, 0.75, at index matching. Note that, rigorously speaking, the index-matching condition for ellipsoids possessing an intrinsic anisotropy depends on the axial ratio, but the effect is rather small.

The method discussed in the previous section was tested by using aqueous dispersions of colloidal particles of tetrafluoroethylene copolymerized with perfluoro-methylvinylether (MFA), prepared and kindly donated to us by Ausimont, Milano, Italy. The latex is obtained by a dispersion polymerization process in the presence of an anionic surfactant [12]. By a careful control of the nuclea-

tion steps, the process yields fairly monodisperse spherical particles (standard deviation in volume below 5%). MFA particles are partially crystalline. Their internal structure is probably a conglomerate of some tens of microcrystallites dispersed in an amorphous matrix [12]. Each crystallite is a folded ribbon of polymer chains packed in a regular crystalline structure. The crystallinity is about 30%, with a chain folding length of the order of 50 nm. The latex particles bear a negative surface charge which is due in part to adsorbed surfactant and in part to the end groups of the polymer chains (fluorinated carboxyl ions) generated by the decomposition of the initiator.

The used particles have a radius of 110 nm, an average index of refraction $n_p = 1.352$, and an intrinsic anisotropy $\Delta n_p \approx 0.5 \times 10^{-2}$. They are dispersed in an index-matched solvent (18% by weight urea–water mixture) at a volume fraction of 2.5%. 100 mM NaCl was added to the dispersion in order to screen the electrostatic interparticle interactions.

Note that the particles are too large to be considered Rayleigh scatterers. However, they satisfy the Rayleigh–Debye (also called Rayleigh–Gans) approximation. Indeed, for particles made of isotropic material the condition of validity of the Rayleigh–Debye approximation is: $(4\pi/\lambda)R(n_p - n_s) \ll 1$, where λ is the wavelength of light and R the size of the particle. In the case of anisotropic particles, there is an additional condition for the validity of the approximation: $(4\pi/\lambda)R\Delta n_p \ll 1$. One should therefore expect that ρ at index-matching takes the value 0.75, were it not for the fact that the particles are polycrystalline and contain amorphous regions occupying a volume fraction which might fluctuate from particle to particle. As a consequence, the MFA particles are optically polydisperse. The effect of optical polydispersity is that ρ at index-matching takes for the MFA particles a value of 0.50–0.55 instead of 0.75 [12].

In order to illustrate the theoretical considerations developed in the previous section, we show an experimental comparison between the standard dynamic light scattering technique and the heterodyne forward scattering technique. The description of the apparatus used for the standard depolarized light-scattering measurement can be found in Ref. [13] which presents a detailed study of the Brownian dynamics of these particles by intensity correlation measurements at non-zero scattering angles in a wide range of volume fractions. The heterodyne forward scattering measurement was performed by using the optical set-up schematized in Fig. 2. The experiment is described in some detail in Ref. [15]. The components are: a low power He–Ne laser, two Glan–Thompson polarizers having an extinction ratio better than 10^{-7}, a mica quarter-wave plate, and a cylindrical scattering cell with a 10 mm path-length and very low residual stress-induced

Fig. 1 The ratio ρ between I_{VH} and I_{VV} plotted as a function of the axial ratio b/a for an ellipsoidal Rayleigh scatterer which presents only form anisotropy ($n_{p3}-n_{p1} = 0$. The upper line refers to the case $n_p - n_s = 0.17$, and the lower line to the case $n_p - n_s = 0.07$)

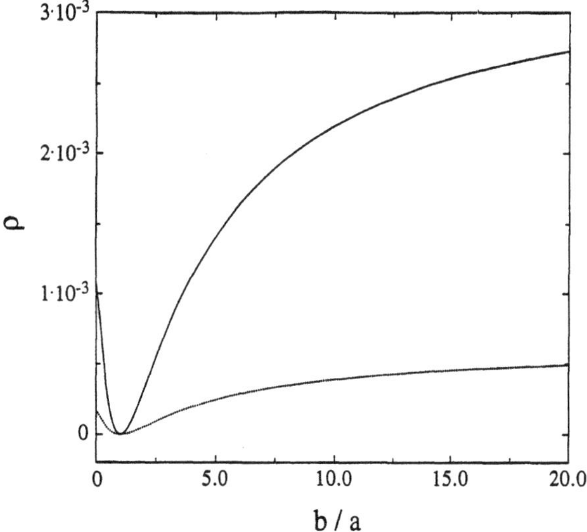

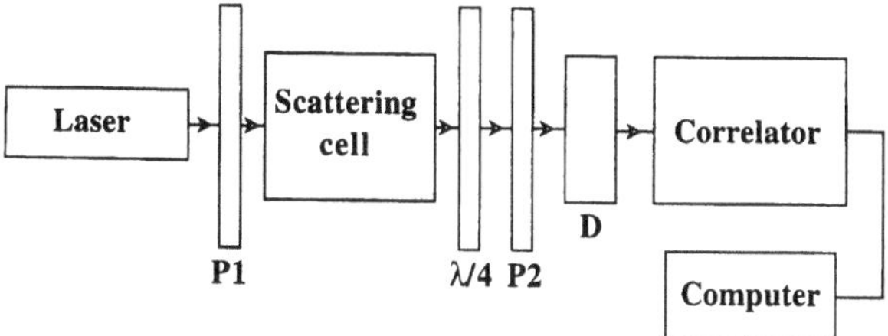

Fig. 2 Scheme of the experimental set-up for the forward-scattering measurement in the heterodyne configuration. P1 and P2 are polarizers, $\lambda/4$ is a quarter-wave plate and D is a photodetector

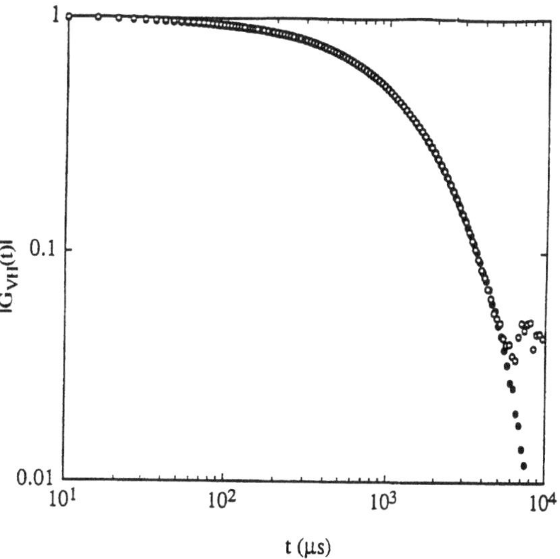

Fig. 3 Normalized time-dependent part of the heterodyne correlation function $|G_{VH}^{(1)}|$ measured in the forward-scattering configuration (full dots) compared to the normalized field-correlation function extracted from a homodyne intensity correlation measurement at a scattering angle of $15°$ (open dots). Both measurements refer to MFA suspensions at a particle volume fraction $\Phi = 0.1$

birefringence. The analyzer was offset from extinction by a small angle which was trimmed to give a signal-to-reference ratio between 10^{-3} and 10^{-2}. Despite the fact that only a very small fraction of the incident beam could reach the detector, the power of the laser beam had still to be attenuated before entering the scattering cell down to about $100\ \mu W$ in order to avoid excessive count rates. Taking into account that count rates are large, detection was made by using a fast H5783P Hamamatsu metal package photomultiplier and a high-speed discriminator. The accumulation time for the forward heterodyne measurement was about a factor of three shorter than for the homodyne measurement at $15°$. For both homodyne and heterodyne experiments the correlation function was measured by a Brookhaven B19000 multi-tau correlator.

The depolarized field correlation functions obtained with the two different techniques by using an MFA sample at a volume fraction $\Phi = 0.10$ are shown in Fig. 3. The homodyne measurement is performed at a scattering angle of $15°$. At such a small scattering angle, the contribution due to translational diffusion is so slow that the decay of the correlation function is controlled only by rotational diffusion. We see that the two curves are perfectly superposed. The logarithmic time-scale chosen for the plot puts evidence on the long-time behavior: it seems that the heterodyne measurement is less noisy than the homodyne at long times. This might be connected with the fact the correlation function is calculated from the homodyne data through a square-root operation.

The measurements of the rotational correlation function were performed at various volume fractions up to the region in which the hard-sphere colloidal crystal is formed [13, 15]. As discussed in Ref. [13], the shape of the correlation function departs considerably from the exponential behavior when $\Phi > 0.1$. We have derived from the first cumulant the short-time rotational diffusion coefficient D_S^r as a function of Φ. We present in Fig. 4a a comparison between the results obtained with the two techniques, and also a comparison between the experimental results and the theoretical predictions of Refs. [13, 16] which are expressed in terms of a series expansion of D_S^r in powers of

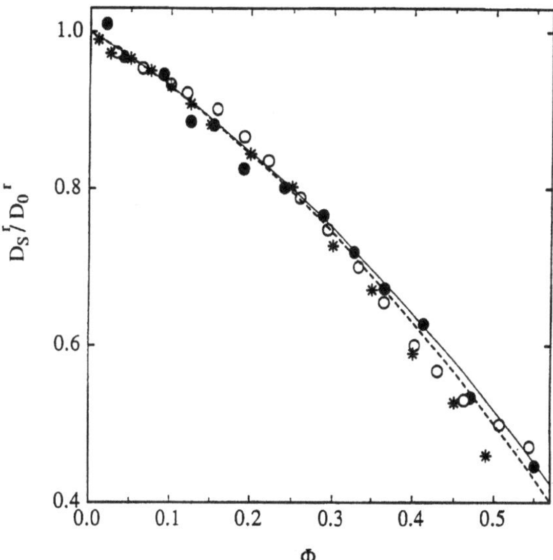

Fig. 4 Φ-dependence of the short-time rotational self-diffusion coefficient D_S^r divided by the rotational diffusion coefficient of the independent particle D_0^r. Full dots and open dots refer, respectively, to the heterodyne forward scattering measurements and to the homodyne non-zero scattering angle experiment. The full and the broken lines are the theoretical results of Refs. [13, 16]. Stars show numerical simulation results of Ref. [17]

Φ truncated at the quadratic term. We also show in the same figure numerical simulation results obtained by a Stokesian-dynamics method [17]. Fig. 4 shows that the two techniques yield results of comparable quality. The forward scattering measurement is much quicker and requires a simpler set-up. The comparison between theory and experiment is also very satisfactory, and is discussed in more detail in Refs. [13, 15]. It should be mentioned that a very recent paper by Watzlawek and Nägele [18] presents a calculation of the short-time rotational self-diffusion coefficient as a function of Φ for charged spherical particles with various amounts of added electrolyte. The results reported in [18] fully justify the assumption used in the interpretation of the experimental results [13, 15] that charged particles at the ionic strength of 100 mM behave as hard spheres.

Conclusions

By observing the fluctuations of the depolarized intensity scattered in the forward direction by a dispersion of anisotropic particles the measurement of the rotational correlation function can be performed directly without invoking any decoupling approximation which enables to separate the orientational correlation function from the translational one. We have presented a treatment of dynamic light-scattering experiments performed in the forward configuration. Our calculations show that it is important to insert a quarter-wave plate between the scattering cell and the analyzer. The role of the quarter-wave plate is to generate a heterodyne signal between the transmitted beam and the depolarized forward scattered field. In practical cases, the stress-induced birefringence of the cell windows can give heterodyne effects similar to those produced by the quarter-wave plate, but it might be difficult to decide whether the measured correlation function is fully heterodyne or rather it contains both homodyne and heterodyne contributions. Some features of the problem are better described by using an approach based on Gaussian beams, such as the one developed by Rička [19]. Such an approach will be presented in a forthcoming article. In order to illustrate the theoretical calculations, we have also reported some experimental results obtained with dispersions of fluorinated polymer colloids in a wide range of volume fractions. The observed rotational dynamics is in good agreement with the theoretical calculations of the effect of hydrodynamic interactions on the rotational behavior of colloidal hard spheres.

References

1. Berne BJ, Pecora R (1976) Dynamic Light Scattering Wiley, New York
2. Wada A, Suda N, Tsuda T, Soda K (1969) J Chem Phys 50:31
3. Schurr JM, Schmitz KS (1973) Biopolymers 12:1021
4. Thomas JC, Fletcher GC (1979) Biopolymers 18:1333
5. Han CC, Yu H (1974) J Chem Phys 61:2650
6. Crosby CR III, Ford Jr NC, Karasz FE, Langley KH (1981) J Chem Phys 75:4298
7. Russo PS, Saunders MJ, DeLong LM, Kuehl S, Langley KH, Detenbeck RW (1986) Anal Chim Acta 189:69
8. Degiorgio V, Bellini T, Piazza R, Mantegazza F (1996) Physica A 235:279
9. van de Hulst HC (1976) Light Scattering by Small Particles. Dover, New York
10. Piazza R, Degiorgio V, Bellini T (1986) Opt Commun 58:400
11. Fredericq E, Houssier C (1973) Electric Dichroism and Electric Birefringence. Oxford University Press, London
12. Degiorgio V, Piazza R, Bellini T, Visca M (1994) Adv Colloid Interface Sci 48:61
13. Degiorgio V, Piazza R, Jones RB (1995) Phys Rev E 52:2707
14. Degiorgio V, Bellini T, Piazza R, Mantegazza F, Ricka J, manuscript in preparation
15. Piazza R, Degiorgio V (1996) J Phys Condens Matter 8:9497
16. Clercx HJH, Schram PPJM (1992) J Chem Phys 96:3137
17. Phillips RJ, Brady JF, Bossis G (1988) Phys Fluids 31:3462
18. Watzlawek M, Nägele G (1996) Physica A 235:56
19. Rička J (1993) Appl Opt 32:2860

Progr Colloid Polym Sci (1997) 104:23–30
© Steinkopff Verlag 1997

K.D. Hörner
M. Budwitz
E.J. Röhm
M. Töpper
M. Ballauff

Assessment of the depletion forces in mixtures of a latex and a non-adsorbing polymer by turbidimetry

K.D. Hörner · M. Budwitz · E.J. Röhm
M. Töpper · Prof. Dr. M. Ballauff (✉)
Polymer Institut
Universität-TH-Karlsruhe
Kaiserstraße 12
76128 Karlsruhe, Germany

Abstract A turbidimetric analysis of the depletion interaction in mixtures of latexes and non-adsorbing polymers is presented. It is demonstrated that turbidimetry explores the small-angle part of the structure factor $S(q)$ ($q = (4\pi/\lambda)\sin(\theta/2)$, θ is the scattering angle, and λ the wavelength of the radiation in the medium) and allows to extrapolate to $S(0)$ in very good accuracy. Turbidimetry is insensitive towards multiple scattering and the analysis can therefore be extended to high latex concentrations. The measured turbidity τ is fully determined by the scattering intensity of the latex spheres, whereas the dissolved polymer does not contribute appreciably to τ. Therefore, $S(0)$ thus obtained can be used to assess the attractive depletion forces between the latex particles in these mixtures. It is shown that the volume-exclusion potential as modified by Lekkerkerker et al. [17] allows to account quantitatively for the dependence of $S(0)$ on latex concentration as well as on polymer concentration.

Key words Latex – light scattering – structure factor – turbidimetry – depletion forces

Introduction

Since the early days of colloid science [1] it has been known that non-adsorbing polymers may promote flocculation of colloid particles. The correct explanation was given later by Asakura and Oosawa [2] and independently by Vrij [3]: If the colloid particles get closer than the coil diameter of the dissolved polymer the osmotic pressure in the depleted region between the particles becomes lower than in the remainder of the bulk solution. As a consequence, there is an attractive depletion force between the colloid particles which may lead to reversible phase separation.

This relatively simple approach which is based on purely hard-sphere interaction of the colloid particles and the polymer coils has been successful in predicting the onset of demixing in many experimental systems [4–11].

Also, first successful attempts to measure directly the depletion forces have been reported [12–14]; a review over theoretical and experimental studies of the depletion forces has been given by Lekkerkerker and Stroobants [11] and by Poon and Pusey [15].

A statistical-mechanical theory of the phase equilibrium due to depletion forces has been worked out by Lekkerkerker, Pusey, Poon and coworkers [16–18]. The theory in agreement with experiment demonstrates that if the size of the polymer is small compared to the size of the colloid spheres, depletion interaction will lead to a solid/gas phase equilibrium; for larger size ratios a liquid phase becomes possible.

Most theories discussed so far treat the polymer coils as ideal and do not include the effect of coil–coil interactions. As shown by Sharma and Waltz [19], and later by Mao et al. [20, 21] and by Poon et al. [22], the mutual exclusion of the small particles representing the polymer coils leads to depletion repulsion as the two large particles first approach; at closer distance the forces switch sign and a significant depletion attraction results.

Only a few attempts to assess the depletion forces by scattering methods have been reported so far [23–27]. A principal difficulty of static light scattering arises from the very strong light scattering [28, 29] of particles of typical colloidal dimensions (ca. 10^2 nm) which leads to the problem of multiple scattering. Up to now, this effect has been circumvented by working with very small particles and/or small concentrations [23, 24, 27].

It has been shown [30–32] that strongly scattering latex systems can be studied by turbidimetry which is practically insensitive towards multiple scattering. Following Vrij and coworkers [33–35] the full analysis of the measured turbidity as a function of wavelength and latex concentration has been worked out to yield the structure factor in the region of small angles.

Here we give a survey over recent turbidimetric studies [36, 37] of particle interaction in a polystyrene latex when a non-adsorbing polymer is added. In particular, we focus on the experimental results using hydroxyethylcellulose (HEC) as soluble polymer [37]. The theoretical background will be delineated which forms the base of the discussion of these experimental results. It will become evident that turbidimetry is a highly sensitive technique for the study of the depletion forces.

Theory

Consider a binary mixture of monodisperse latex particles and monodisperse non-adsorbing polymers. The scattering intensity $I(q)$ ($q = (4\pi/\lambda)\sin(\theta/2)$ with θ the scattering angle and λ the wavelength of the radiation in the medium) of such a mixture may be calculated within the Rayleigh–Debye approximation by (cf. Refs. [28, 29])

$$I(q) = K\left[\left(\frac{dn}{dc_1}\right)^2 M_1 c_1 P_1(q) S_{11}(q) + 2\left(\frac{dn}{dc_1}\right)\left(\frac{dn}{dc_2}\right)\right.$$
$$\times (M_1 M_2)^{1/2} (c_1 c_2)^{1/2} (P_1(q) P_2(q))^{1/2} S_{12}(q)$$
$$\left. + \left(\frac{dn}{dc_2}\right)^2 M_2 c_2 P_2(q) S_{22}(q)\right], \tag{1}$$

where the (dn/dc_i) are the refractive increments of the compounds, M_1 and M_2 denote the molar masses of the latex particles and the polymer, respectively, c_1 and c_2 are the mass concentrations of the compounds, the $P_i(q)$ are their respective form factors, and the $S_{ij}(q)$ are the partial structure factors. The constant K follows in this approximation as $K = (2\pi^2 n_0^2)/(N_A \lambda_0^4)$ with n_0 being the refrac-

tive index of the medium and N_A is Avogadro's number [29]. For the system under consideration here, it can be shown [37] that the front factor of $S_{11}(q)$ pertaining to the colloid particles is greater than the respective front factor of $S_{12}(q)$ by nearly three orders of magnitude; the respective front factor of $S_{22}(q)$ is smaller by five orders of magnitude. For the present analysis which aims at $S(0)$ it suffices therefore to consider only the first term of Eq. (1).

In order to relate the partial structure factor $S_{11}(q)$ to the attractive depletion forces, we consider the model [37] of colloid–colloid interaction in the presence of dissolved polymers as shown in Fig. 1. Here the colloid spheres are characterized by their diameter σ_L and their effective diameter d of interaction. For latexes the description of particle–particle interaction at high ionic strength in the low-q region of $S(q)$ by a single parameter d, i.e. in terms of hard sphere interaction is fully justified by the experimental data [30–32]. Therefore, the partial structure factor $S_{11}(q)$ of the latex spheres may be approximated by the respective expression furnished by the Percus–Yevick theory [38].

Depletion interaction is treated in terms of a zone which the centers of gravity of the polymer cannot enter ([2, 3]; cf. Fig. 1). Therefore the minimum distance of the centers of gravity of the particles and the coils is given by the parameter $\bar{\sigma}$ defined by

$$\bar{\sigma} = \tfrac{1}{2}(\sigma_L + \sigma_p), \tag{2}$$

where $\sigma_p/2$ is the thickness of the depleted zone as defined in Fig. 1. In what follows, σ_p is the only adjustable parameter characterizing the spatial extensions of the polymer coils.

Fig. 1 Definition of the parameters characterizing the volume-exclusion potential equation (3)[37]. The larger spheres of diameter σ_L refer to the latex particles, whereas the smaller spheres symbolize the dissolved polymer. The centers of gravity of both components cannot approach closer than a distance $\bar{\sigma} = (\sigma_L + \sigma_p)/2$. Therefore, σ_p provides a measure for the depletion zone around the latex spheres

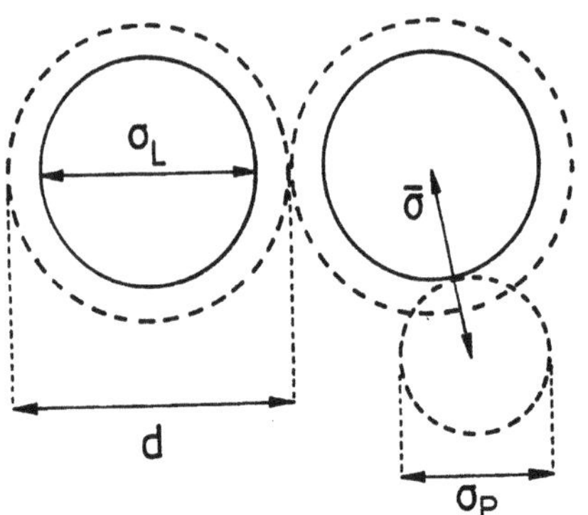

The depletion-potential U_{dep} follows as [2, 3, 17, 18]

$$U_{\text{dep}}(r) = \begin{cases} \infty, & r < d\,, \\ -\Pi^* V_{\text{overlap}}, & d \leq r \leq 2\bar{\sigma}\,, \\ 0, & r > 2\bar{\sigma}\,, \end{cases} \tag{3}$$

where

$$V_{\text{overlap}} = (1 - \tfrac{3}{4}\rho + \tfrac{1}{16}\rho^3)\tfrac{4}{3}\pi\bar{\sigma}^3 \tag{4}$$

with $\rho = r/\bar{\sigma}$.

According to Lekkerkerker, Poon, Pusey and coworkers [16–18] the osmotic pressure Π^* of the dissolved polymer is calculated from the concentration of the polymer molecules c_p^* in the free volume V_{free} not occupied by the colloid particles or their depletion zones. The effect of mutual interaction between the polymer molecules may be taken into account by the second virial coefficient A_2. With $c_p^* = c/\alpha_{\text{free}}$,

$$\Pi^* = \frac{RT}{M_n}c_p^* + 2A_2 c_p^{*2} + \cdots\,, \tag{5}$$

where M_n is the number-average molecular weight of the dissolved polymer. Lekkerkerker [16] has shown that the ratio α_{free} of V_{free} to the volume of the system V can be obtained by using the scaled particle theory:

$$\alpha_{\text{free}} = \frac{V_{\text{free}}}{V} = (1 - \phi_{\text{L}})\exp[-a\gamma - b\gamma^2 - c\gamma^3]\,, \tag{6}$$

where $\gamma = \phi_{\text{L}}/(1 - \phi_{\text{L}})$ and $a = 3\xi + 3\xi^2 + \xi^3$, $b = 9\xi^2/2 + 3\xi^3$, $c = 3\xi^3$ with $\xi = \sigma_p/\sigma_{\text{L}}$.

The depletion potential (Eq. (3)) can be introduced into $S(q)$ as a perturbation using the random-phase approximation (RPA) [39]. The RPA consists of equating the direct correlation function $c(r)$ of the latex spheres to

$$c(r) = c_0(r) - \frac{U_{\text{dep}}(r)}{kT}\,. \tag{7}$$

Here $c_0(r)$ is the direct correlation function of the hard-sphere reference fluid which can be obtained from the Percus–Yevick theory. The structure factor $S(q)$ of a system with the number density of particles N/V is given by [38]

$$S(q)^{-1} = 1 - \frac{N}{V}c(q)\,. \tag{8}$$

Following Grimson [39] the WCA-separation [38] of the depletion potential has been used. Therefore, $U_{\text{dep}}(r) = U_{\text{dep}}(d)$ for $r < d$ and $U_{\text{dep}}(r) = U_{\text{dep}}(r)$ for $r \geq d$. Fourier transformation of $c(r)$ given by Eq. (7) allows in turn to obtain directly $S_{11}(q)$ of the latex spheres in the presence of dissolved polymer. The resulting expressions are rather lengthy and we only give $S(q)$ in the limit of vanishing q [36, 37]:

$$S_{11}(0)^{-1} - S_{11,0}(0)^{-1} =$$

$$\quad - 8A_{\text{dep}}\phi_{\text{eff}}\frac{\bar{\sigma}^3}{d^3}\left[1 - \frac{3}{16}\frac{d^4}{\bar{\sigma}^4} + \frac{1}{32}\frac{d^6}{\bar{\sigma}^6}\right], \tag{9}$$

where A_{dep} is

$$A_{\text{dep}} = \frac{\tfrac{4}{3}\pi\bar{\sigma}^3\Pi^*}{kT}\,. \tag{10}$$

Here $S_{11,0}(0)$ is the value of $S_{11}(0)$ is the absence of dissolved polymer.

It is evident that the above calculation of $S(q)$ is not exact because of the approximate nature of the RPA. To check the validity of this procedure, it is expedient to use the invariant Q defined by

$$Q = \int\limits_0^\infty q^2\,dq\,I(q)\,. \tag{11}$$

Porod has shown (cf. Ref. [40]) that in case of a two-phase system, Q must be proportional to $\phi(1 - \phi)$ where ϕ is the volume fraction of one phase. Thus, in a system consisting of interacting particles, Q must scale with $\phi(1 - \phi)$ for any potential. Numerical calculation of Q for a system of homogeneous spheres using $S(q)$ as given by Eq. (8) shows that this criterion is fulfilled indeed for the set of parameters used herein.

The turbidity τ is related to $I(q)$ by [29]

$$\tau = 2\pi \int\limits_0^\infty I(q)\sin\theta\,d\theta\,. \tag{12}$$

For a system of monodisperse particles the dependence of τ on wavelength and latex concentration c_{L} may be factorized into the integrated form factor $Q(\lambda^2)$ and the integrated structure factor $Z(\lambda^2, c_{\text{L}})$ by [30–32]

$$\tau = K^* c_{\text{L}}\left(\frac{n_0\pi\sigma_{\text{L}}}{\lambda_0}\right)^3\frac{16\pi}{3}Q(\lambda^2)\,Z(\lambda^2, c_{\text{L}})\,, \tag{13}$$

where K^* defines the optical constant by

$$K^* = \frac{3n_0}{4\lambda_0\rho_{\text{L}}}\left(\frac{m^2 - 1}{m^2 + 2}\right)^2 \tag{14}$$

with ρ_{L} being the density of the particles, n_{L} the refractive index of the latex particles and $m = n_{\text{L}}/n_0$ their relative refractive index.

For polydisperse systems both $Q(\lambda^2)$ and $Z(\lambda^2, c)$ are replaced by the respective "measured" quantities $Q_{\text{M}}(\lambda^2)$ and $Z_{\text{M}}(\lambda^2, c)$ [31]:

$$\left(\frac{\tau}{c_{\text{L}}}\right) = K^*\frac{16\pi}{3}\left(\frac{n_0\pi\sigma_\tau}{\lambda_0}\right)^3 Q_{\text{M}}(\lambda^2)Z_{\text{M}}(\lambda^2, c_{\text{L}}) \tag{13'}$$

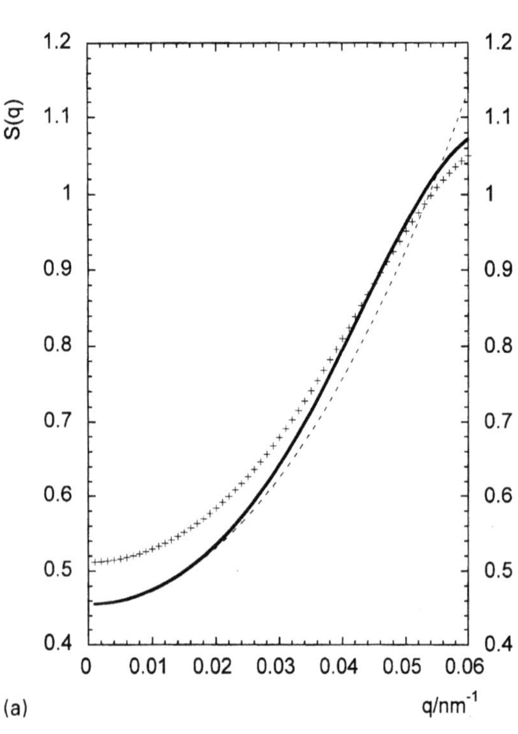

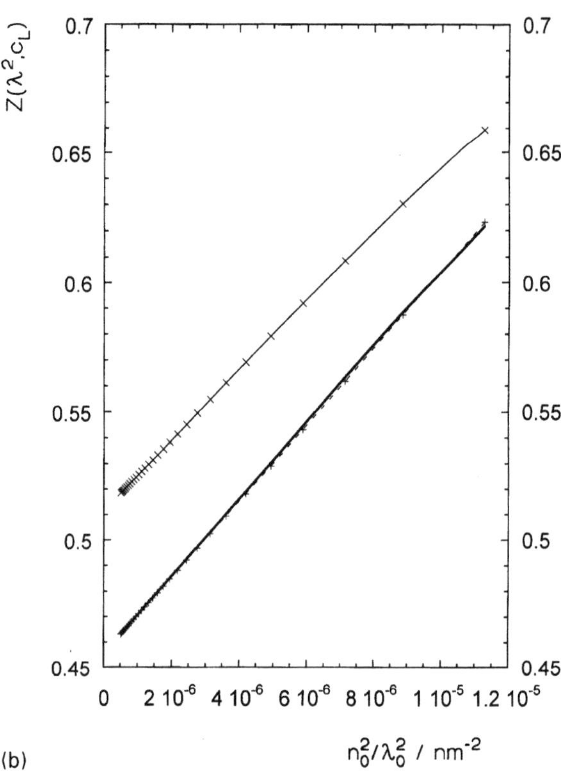

Fig. 2 Comparison of the structure factor $S(q)$ (a) and the integrated structure factor $Z(\lambda^2, c)$ (b) in absence of attractive interaction (Percus–Yevick theory [40], solid lines) and in presence of depletion interaction (crosses). In both calculations a system of monodisperse spheres with a diameter $\sigma_L = 75.5$ nm and an effective diameter d of interaction of 89.3 nm has been assumed. The effective volume fraction ϕ_{eff} characterizing the hard-sphere interaction is given by 0.1. The dashed line shows the result from the series expansion $S(q) = S(0) + \alpha q^2$ calculated for a system of hard spheres. The magnitude of the depletion zone is characterized by $\sigma_p = 40$ nm. The osmotic pressure has been calculated by neglecting the viral correction (Eq. (5)) assuming a number-average molecular weight of 66.000 g/mol and a polymer concentration c_p of 0.3 g/l

where σ_τ is the turbidity-average diameter defined by [22]

$$\sigma_\tau^3 = \frac{\sum_i c_i \sigma_i^3}{\sum_i c_i} . \tag{15}$$

Model calculations showed, however, that for the narrow size distributions (Gaussian size distribution with standard deviation of the order of 10%) of the latexes studied herein, the effect of polydispersity is of minor importance as compared to the alterations effected by depletion interaction [30]. Therefore, the discussion of the influence of attractive forces between the latex particles may be presented in terms of monodisperse systems.

The function $Z(\lambda^2, c_L)$ can be obtained from the experimental data by

$$Z(\lambda^2, c_L) = \frac{(\tau/c_L)}{(\tau/c_L)_0} , \tag{16}$$

where $(\tau/c_L)_0$ is the specific turbidity extrapolated to vanishing latex concentration c_L. The function $Z(\lambda^2, c_L)$ may be expanded into powers of the size parameter $n_0 \pi \sigma / \lambda_0$

leading to [30–32]

$$Z(\lambda^2, c_L) = S_{11}(0) + 8\alpha \left(\frac{n_0\pi}{\lambda_0}\right)^2 + B\left(\frac{n_0\pi}{\lambda_0}\right)^4 + O(\lambda^{-6}) . \tag{17}$$

Here α is the first coefficient of the expansion of $S(q)$ in powers of q^2, whereas the coefficient B contains factors pertaining to the interaction of the particles as well as to their optical radius of gyration [31]. If the size parameter $n_0 \pi \sigma / \lambda_0$ is not too big, the third and fourth terms in Eq. (17) give only minor contributions. In this case $S_{11}(0)$ can be extrapolated from experimental data with very good accuracy (see below; [31, 32]). For strongly scattering systems and high concentrations turbidimetry thus furnishes a quantity difficult to measure by other methods.

In order to discuss the changes effected by attractive interactions on $I(q)$ and the turbidity, it is expedient to compare $S(q)$ and $Z(\lambda^2, c_L)$ for a system of hard spheres and for a system of spheres with depletion interaction (Fig. 2a). The parameters have been chosen to match the system studied in detail recently (cf. Ref. [37] and below): The system is assumed to consist of monodisperse latex particles with a diameter $\sigma_L = 75.5$ nm and an effective diameter of interaction d of 89.3 nm. The effective volume fraction ϕ_{eff} characterizing the hard sphere interaction is given by 0.1. The solid line in Fig. 2a is $S(q)$ calculated for a system of monodisperse spheres with the above ϕ_{eff} and d using the Percus–Yevick theory [38, 41]. The dashed line shows the result from the truncated series expansion $S(q) = S(0) + \alpha q^2$.

The dissolved polymer is characterized by an effective diameter $\sigma_p = 40$ nm which determines the magnitude of the depleted zone according to Eq. (2). Its osmotic pressure derives from the number-average molecular weight ($M_n = 66.000$ g/mol; [37]) and the polymer concentration c_p chosen to be 0.3 g/l.

First of all, the structure factor of the pure hard spheres (solid line in Fig. 2a) can be well approximated by the first two coefficients of its expansion into powers of q^2 up to $q \cong 0.03$ nm^{-1}. This region therefore may be called "Guinier region" of the structure factor. Furthermore, the comparison of $S(q)$ calculated for the pure latex and the mixture of latex and polymer shows that the effect of depletion attraction is small and restricted to the region of lowest angles. The main alteration is effected on $S(0)$ which is raised considerably.

While these small changes are difficult to measure by a small-angle technique which would have to start at a minimum q of 0.01 nm^{-1} at least, the integrated structure factor $Z(\lambda^2, c_L)$ shown in Fig. 2b is markedly shifted. This is due to the fact that $Z(\lambda^2, c_L)$ is governed by $S(q)$ in the immediate vicinity of $q = 0$: For the longest wavelength of 1100 nm which can be used in experimental studies [30–32, 36, 37], the maximum q-vector, i.e. the value of q resulting for this wavelength at a scattering angle of 180°, is given by 0.0152 nm^{-1}; for 600 nm it is given by 0.028 nm^{-1}. Therefore, the turbidimetric measurement explores the "Guinier-region" of $S(q)$ for the particular case under consideration here. As a consequence, the integration equation (12) leads only to a trivial factor as first shown by Vrij and coworkers [34]. The expansion of the integrated structure factor is thus mainly determined by the first two coefficients (cf. dashed line in Fig. 2b) and $S(0)$ together with α can be obtained with good accuracy. If higher coefficients in Eq. (17) must be taken into account, it has been shown [30–32] that $S(0)$ and α are accessible for typical latex systems as well.

Experimental

Up to now, two systems have been studied by the above turbidimetric technique: (i) a mixture of a polystyrene latex

with the anionic polyelectrolyte sodium poly(styrene sulfonate) [36], and (ii) a mixture of a polystyrene latex and the non-adsorbing uncharged hydroxyethyl cellulose (HEC) [37]. The former system has clearly the advantage that the added polymer has no tendency to adsorb on the surface of the polystyrene particles but the spatial extensions of the polyelectrolyte will depend strongly on the ionic strength. In the second type of mixtures studied intensively by previous workers [5–7, 9] the surface of the particles had to be covered by the nonionic surfactant Triton X-405 to avoid bridging flocculation. On the other hand, recent data on the adsorption of Triton X-405 on a polystyrene latex obtained by small-angle scattering [42] allow to determine the amount of the surfactant needed to cover the surface of the particles but avoid the formation of free micelles. Thus, the second system seems to be well-suited for a quantitative investigation. It can be shown that for this system the contribution of the cross term to $I(q)$ (see Eq. (1)) can be dismissed indeed and the measured turbidity is fully determined by the contribution of the latex spheres.

The experimental details of the studies of systems (i) and (ii) were given elsewhere [36, 37]. Therefore, it suffices to delineate the main problems and difficulties encountered during these measurements. The measured turbidity is extremely sensitive to small traces of coagulum or dust particles and filtration of the mixtures turns out to be a central step in sample preparation. Insufficient removal of these traces reveals itself immediately by a strong upturn of the integrated form factor $Q(\lambda^2)$ at long wavelength [32]. Hence, the turbidimetric technique allows to scrutinize the system for even weak bridging of particles and to relegate contaminated samples.

Meaningful measurements of the turbidity τ as a function of wavelength require fulfillment of the Lambert–Beer law better than 1%, i.e. the extinction must be strictly proportional to the length of the optical path. This can be checked by using an optical path length varying between 1 and 50 mm. Another problem of turbidity measurements is the possible forward scattering into the detector of the UV/VIS-spectrometer. A sensitive check of possible disturbances due to forward scattering is provided by mounting slits of various width in the optical path, thus narrowing the angle of uptake of the detector. These experiments gave no indication that the measurements of τ are seriously disturbed by forward scattering.

In practice, measurements of the turbidity in aqueous latex systems can be performed using wavelengths between 600 and 1100 nm. For longer wavelengths the absorption of water becomes very strong and renders a meaningful determination of τ impossible. In the region between 500 and 1100 nm most common polymers have no absorption and τ is fully determined by light scattering.

The possibility of changing the wavelength into the near infrared alleviates many problems encountered in static light scattering at a fixed wavelength in the visible range: Due to the λ^{-4} dependence of the front terms in Eq. (13) the use of long wavelength decreases τ considerably. The extinction of the suspension is thus shifted into a range convenient to measure by a conventional spectrometer. Forward scattering is much less pronounced for longer wavelengths and can be dismissed in agreement with the experiment. As already discussed above, the use of long wavelengths gives access to the small-angle part of $S(q)$ governed by $S(0)$ and the first coefficient α. Therefore, the extension of the measurements to maximum wavelength greatly facilitates the extrapolation of $S(0)$ from $Z(\lambda^2, c_L)$ [31, 32].

Results and discussion

The evaluation of the measured structure factor $Z(\lambda^2, c_L)$ from latex suspensions has been discussed at great length previously [30–32, 36, 37]. Therefore, is suffices here to discuss $Z(\lambda^2, c_L)$ obtained for mixtures of a polystyrene latex ($\sigma_L = 75.5$ nm; $d = 89$ nm; [37]) and HEC ($M_n = 66.000$ g/mol; [37]). Figure 3 displays $Z(\lambda^2, c_L)$ obtained for a latex concentration of 5 wt% and three different HEC concentrations c_p. The marked change upon addition of HEC is immediately obvious and the data can be used to extrapolate $S_{11}(0)$ of the latex spheres.

The $S_{11}(0)$ thus obtained now allow a critical comparison of Eq. (9) with the experiment. Figure 4a displays the variation of the left-hand side of Eq. (9) with the effective volume fraction ϕ_{eff} of the latex for given polymer concentrations c_p, whereas Fig. 4b shows a plot of $S_{11}^{-1}(0)$ vs. polymer concentration at a given latex concentration.

The solid lines in Fig. 4a give the resulting fit according to Eq. (9). The only fit parameter indicated in the graph is σ_p which defines the apparent magnitude of the depleted zone around the polystyrene particles (see Fig. 1). Despite the error incurred through small differences of large numbers a good agreement of theory and experiment is seen. The magnitude of σ_p is in quantitative agreement with the results of Leal Calderon et al. [9]. In this context it must be noted, however, that the comparison of Eq. (9) with the experimental data allows only to discuss the product of $\bar{\sigma}^6$ and Π^*, whereas a direct determination of the magnitude of the depleted zone would require the respective comparison of theory and experiment at finite q, i.e. at finite wavelength.

The principal result of Fig. 4a is the strong downward curvature which is due to the factor α_{free} (Eq. (6)): Despite of the constant polymer concentration c_p the osmotic pressure Π^* increases with increase of ϕ_{eff} because the free volume V_{free} in which the centers of gravity of the polymers can be located will decrease. This point first raised by Lekkerkerker, Poon, Pusey and co-workers [16, 17] presents a most important extension of the original theory of the depletion forces [2, 3]. The present experimental results confirm these ideas in a quantitative fashion.

The plot shown in Fig. 4b presents the variation of $S_{11}(0)^{-1}$ with polymer concentration c_p. The solid lines have been calculated taking into account the small decrease of σ_p with increase of c_p [37]. Since no additional parameter has been added for this comparison, Fig. 4b provides a consistency check and the good agreement confirms the above conclusions. It is thus evident that Eq. (9) provides a valid description of the experimental data measured to latex concentrations up to 10 wt% and

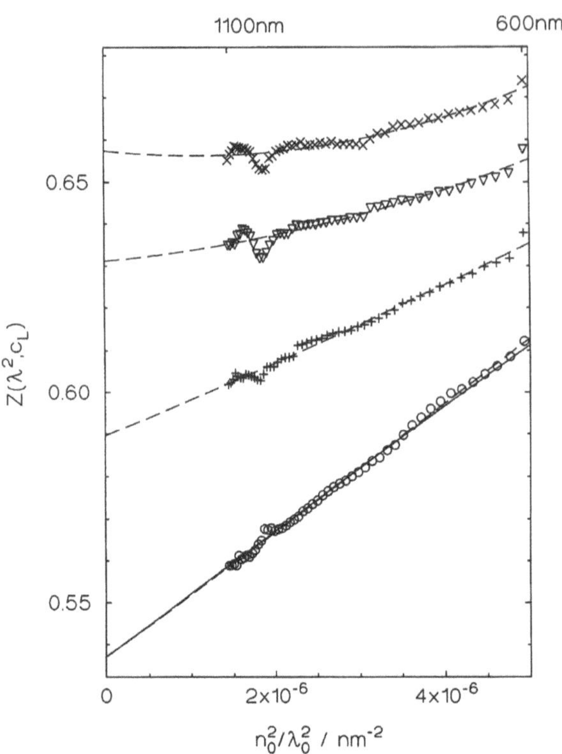

Fig. 3 Integrated structure factor $Z(\lambda^2, c)$ of the latex spheres calculated from experimental results according to Eq. (14) [37]. The system consists of a polystyrene latex ($\sigma_L = 75.5$ nm, d = 89 nm; latex concentration $c_L = 50$ g/l) without added HEC (O) and in presence of HEC (+ : 0.3 g/l HEC; V: 0.5 g/l HEC; x: 0.7 g/l HEC). The dashed line presents a second-order polynomial fit in powers of $(n_0/\lambda_0)^2$ serving for the extrapolation to infinite wavelength. The solid lines present a fit of the integrated structure factor for a system of monodisperse hard spheres [30–32]. The intercept of the dashed lines is given by $S_{11,0}(0)$ or $S_{11}(0)$, respectively, i.e. the osmotic compressibility of the latex spheres without HEC and in presence of HEC

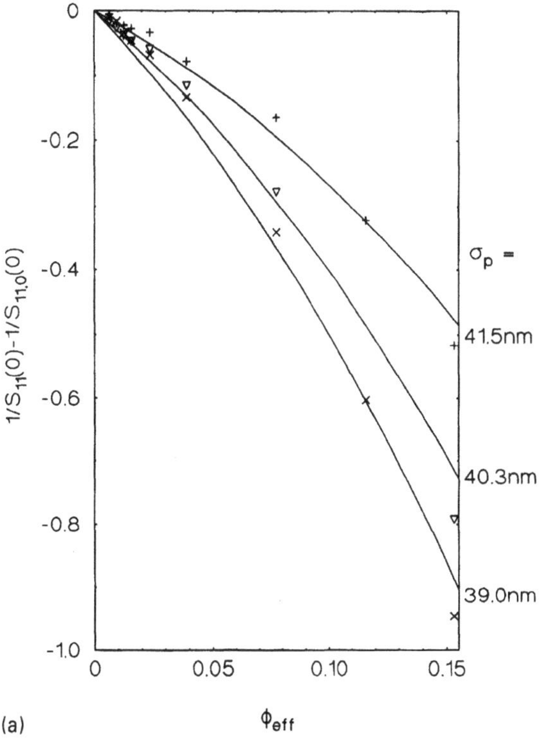

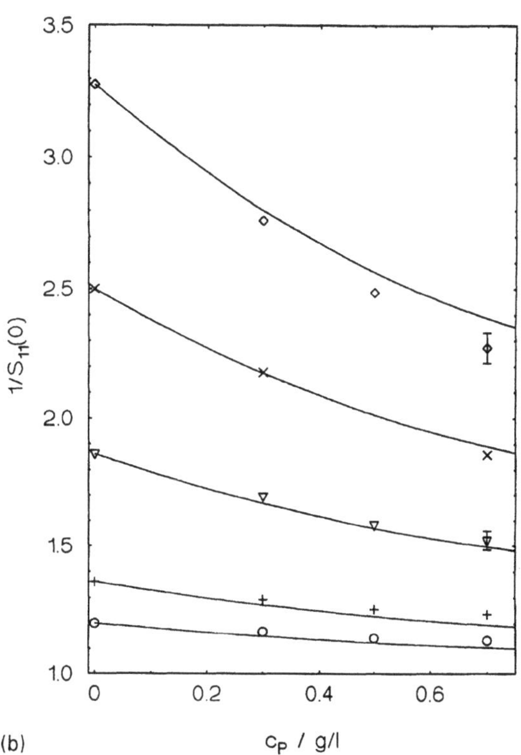

Fig. 4 Test of the validity of Eq. (9). (a) Dependence of $1/S_{11}(0) - 1/S_{11}(0)$ on the effective volume fraction ϕ_{eff} of latex spheres. The parameter σ_p deriving from this fit is indicated in the graph. O: without added HEC; +: 0.3 g/l HEC; ∇: 0.5 g/l HEC; ×: 0.7 g/l HEC. (b) Plot of $1/S_{11}(0)$ versus polymer concentration c_p at five different latex concentrations c_L:O: 15 g/l; +: 25 g/l; ∇: 50 g/l; ×: 75 g/l; $\diamond$: 100 g/l. See text for further explanation

a polymer concentration reaching into the semidilute regime [37]. On the other hand, the present results are not a direct confirmation of the potential equation (3), of course, but mainly validate the dependence of Π^* on c_p and ϕ_{eff} according to Eqs. (5) and (6). As mentioned above, a more detailed investigation of the spatial extension of the depletion forces can be done by comparing the integrated structure factor $Z(\lambda^2, c_L)$ to theoretical results. A study of this problem is under way.

Acknowledgment Financial support by the Deutsche Forschungsgemeinschaft, by the Bayer AG, Geschäftsbereich Kautschuk, by the Bundesministerium für Bildung und Forschung, and by the AIF (project 9749) is gratefully acknowledged.

References

1. Traube J (1925) Gummi Ztg 39:434; Bondy C (1939) Trans Faraday Soc 35:1093
2. Asakura S, Oosawa F (1954) J Chem Phys 22:1255; Asakura S, Oosawa F (1958) J Polym Sci 33:183
3. Vrij A (1976) Pure Appl Chem 48:471
4. de Hek H, Vrij A (1981) J Colloid Interf Sci 84:409
5. Sperry PR, Hopfenberg HB, Thomas NL (1981) J Colloid Interf Sci 82:62
6. Sperry PR (1982) J Colloid Interf Sci 87:375
7. Gast AP, Hall CK, Russel WB (1983) J Colloid Interf Sci 96:251
8. Russel WB, Saville DA, Schowalter WR (1989) Colloidal Dispersions. Cambridge University Press, Cambridge
9. Leal Calderon F, Bibette J, Biais J (1993) Europhys Lett 23:653
10. Ilett SM, Orrock A, Poon WCK, Pusey PN (1995) Phys Rev E 51:1344
11. Lekkerkerker HNW, Stroobants A (1994) Il Nuovo Cim 16D:949
12. Richetti P, Kékicheff P (1992) Phys Rev Lett 68:1951
13. Milling A, Biggs S (1995) J Colloid Interf Sci 170:604
14. Mondain-Monval O, Leal Calderon F, Phillip J, Bibette J (1995) Phys Rev Lett 75:3364
15. Poon WCK, Pusey PN (1995) In: Baus M, Rull LR, Ryckaert JP (eds) Observation, Prediction and Simulation of Phase Transitions in Complex Fluids, NATO Advanced Study Institute, Series C: Mathematical and Physical Sciences, Vol 460 Kluwer Academic, Dordrecht
16. Lekkerkerker HNW (1990) Colloids Surf 51:419
17. Lekkerkerker HNW, Poon WCK, Pusey PN, Stroobants A, Warren PB (1992) Europhys Lett 20:559
18. Poon WCK, Ilett SM, Pusey PN (1994) Il Nuovo Cim 16D:1127
19. Walz JY, Sharma A (1994) J Colloid Interf Sci 168:485

20. Mao Y, Cates ME, Lekkerkerker HNW (1995) Physica A 222:10
21. Mao Y (1995) J Phys II (France) 5:1761
22. Warren PB, Ilett SM, Poon WCK (1995) Phys Rev E 52:5205
23. de Hek H, Vrij A (1982) J Colloid Interf Sci 88:258
24. Tong P, Witten TA, Huang JS, Fetters LJ (1990) J Phys (France) 51:2813
25. Bibette J, Roux D, Pouligny B (1992) J Phys II (France) 2:401
26. Snowden MJ, Williams PA, Garvey MJ, Robb ID (1994) J Colloid Interf Sci 166:160
27. Ye X, Narayanan T, Tong P (1996) Phys Rev Lett 76:4630
28. Klein R, D'Aguanno B (1996) In: Brown W (ed) Dynamic Light Scattering. Clarendon Press, Oxford, pp 30–102
29. Kerker M (1969) The Scattering of Light and other Electromagnetic Radiation. Academic Press, San Diego
30. Apfel U, Grunder R, Ballauff M (1994) Colloid Polym Sci 272:820
31. Apfel U, Hörner KD, Ballauff M (1995) Langmuir 11:3401
32. Weiss A, Pötschke D, Ballauff M (1996) Acta Polym 47:333
33. Jansen JW, de Kruif CG, Vrij A (1986) J Colloid Interf Sci 114:492
34. Rouw W, Vrij A, de Kruif CG (1988) Colloids Surf 31:299
35. Penders MGHM, Vrij A (1990) J Chem Phys 93:3704
36. Röhm EJ, Hörner KD, Ballauff M (1996) Colloid Polym Sci 274:732
37. Hörner KD, Töpper M, Ballauff M (1996) Langmuir, in press
38. Hansen JP, McDonald IR (1986) Theory of Simple Liquids. Academic Press, London
39. Grimson MJ (1983) J Chem Soc Faraday 2 79:817
40. Glatter O, Kratky O (1982) Small Angle X-Ray Scattering. Academic Press, London
41. Ashcroft NW, Lekner J (1966) Phys Rev 145:83
42. Bolze J, Hörner KD, Ballauff M (1996) Langmuir 12:2906

Progr Colloid Polym Sci (1997) 104:31–39
© Steinkopff Verlag 1997

G. Nägele
M. Watzlawek
R. Klein

Hard spheres versus Yukawa particles: Differences and similarities

Dr. Nägele (✉) · M. Watzlawek · R. Klein
Fakultät für Physik
Universität Konstanz
Postfach 55 60
78434 Konstanz, Germany

Abstract Whereas structural properties of suspensions of hard spheres of diameter σ are well approximated by analytic expressions, it is necessary to use numerical solutions of integral equations to calculate these properties when the pair potentials have a soft part. The finite range of repulsive pair potentials gives rise to a correlation hole, meaning that the pair correlation function $g(r)$ is essentially equal to zero up to a well-defined nearest-neighbour separation larger than σ. The aim of this work is to show that because of the correlation hole various dynamic properties of charge-stabilized suspensions are qualitatively different from those of hard spheres. It will be argued that the observed non-linear volume fraction dependencies of the short-time self-diffusion coefficients and of the sedimentation velocity can be understood in terms of a model of effective hard spheres with diameter σ_{EHS} which depends on the volume fraction. Moreover, the long-ranged electrostatic repulsion gives rise to an unexpected enhancement of the long-time self-diffusion coefficient due to hydrodynamic interactions, in contrast to what is known for hard spheres. This enhancement is also understood in terms of an effective hard sphere model.

Key words Self-diffusion – collective diffusion – hydrodynamic interactions – charge stabilized suspensions – effective hard spheres

Introduction

When colloidal particles are suspended in a solvent they attract each other because of the van der Waals interaction. To prevent particles from aggregating, it is necessary to stabilize the suspension, which can be accomplished either by steric or by charge stabilization [1].

In the first case, polymer molecules are grafted to the surface, forming a polymer brush on each colloidal particle. If the dispersion medium is a good solvent for the polymer, brushes on different particles repel each other rather strongly. Since the thickness of the brushes is usu-ally small compared to the diameter σ of the particles, the colloidal pair potential $u(r)$ can be well approximated by a hard sphere model

$$\beta u(r) = \begin{cases} \infty; & r < \sigma\,, \\ 0; & r > \sigma\,. \end{cases} \tag{1}$$

On the other side, charge-stabilization is achieved by ionizable surface groups which dissociate in a polar solvent. Therefore, small counter-ions surround the particles which may carry a charge $Q = eZ$ as large as several hundreds of the elementary charge e. In such systems, the direct interaction between two colloidal particles can be described by an effective pair potential, consisting of a hard sphere potential with diameter σ and a screened

Coulomb (or Yukawa) potential for $r > \sigma$ [2–5]

$$\beta u(r) = A \frac{e^{-\kappa(r-\sigma)}}{r} . \tag{2}$$

Here $\beta = (k_B T)^{-1}$ and $A = L_B (Z/(1 + \kappa\sigma/2))^2$ with $L_B = \beta e^2/\varepsilon$ the Bjerrum length, and $Z = Q/e$ the effective macroparticle charge in units of the elementary charge e. The screening parameter κ is given by $\kappa^2 = 4\pi L_B[n|Z| + 2n_s]$, where n and n_s are the number densities of macroparticles and added 1–1-electrolyte ions, respectively. Here, we assume that the counterions are monovalent. Notice that in this model, known as the one-component macrofluid model, the counter-ions appear only through the screening parameter κ and the solvent through its dielectric constant ε.

Properties of hard-sphere-like suspensions are in many respects easier to describe quantitatively than those of charge-stabilized particles. This is particularly obvious for their structural and thermodynamic properties, since there are many results available for liquids of hard spheres [4]. The microstructure, for example, which is determined by static scattering experiments through a measurement of the static structure factor $S(k)$, is given rather precisely in terms of an analytical expression by the Percus–Yevick solution of the Ornstein–Zernike equation up to volume fractions $\phi = (\pi/6)n\sigma^3 \approx 0.4$. For charge-stabilized colloids with their longer-ranged interactions no such simple results exist; the Ornstein–Zernike equation has to be combined with a closure relation different from the Percus–Yevick scheme in order to achieve good agreement with experiment and computer simulation. The appropriate different schemes can only be handled numerically [5, 6], leading to results for $S(k)$ and the pair correlation function $g(r)$, which show large qualitative differences in the fluid structures of hard sphere suspensions and dispersions of charged particles. Nevertheless, recent work on the dynamical properties of charge-stabilized particles has shown that it is possible in a limited and qualitative sense to map the charge-stabilized particle together with its counterion cloud to an effective hard sphere (EHS) with diameter $\sigma_{EHS} > \sigma$ [5, 7–10].

After summarizing the mentioned differences in the static structures of hard spheres and charged colloidal suspensions, this article reviews some important results on self-diffusion, collective diffusion, and the sedimentation velocity of both type of suspensions. It will be shown that the volume fraction dependence of the translational and rotational short-time self-diffusion coefficients, D_s^t and D_s^r, and the sedimentation velocity are substantially different for charged and uncharged suspensions. Whereas it is known that the short-time dynamical properties of hard sphere suspensions, normalized to their values at infinite dilution, are well represented by virial expansions of the

form $1 + a\phi + \mathcal{O}(\phi^2)$ for small volume fractions, ϕ, the corresponding results for charge-stabilized systems can be fitted to $1 + b\phi^p$, where p is some fractional exponent. These results, obtained from proper theoretical treatments appropriate for charged particles include the important effect of hydrodynamic interactions (HI) and can be understood in terms of an EHS model [5, 8, 10, 11].

Interesting qualitative differences between charge-stabilized suspensions and hard spheres exist, however, also with respect to long-time collective diffusion and long-time self-diffusion. In particular, it has been shown very recently [12], that the long-time diffusion in charge-stabilized colloids is significantly influenced by HI for volume fractions even as low as $\phi \leq 10^{-3}$. In addition it is found that the long-ranged electrostatic repulsion leads to a hydrodynamically induced enhancement of the long-time self-diffusion coefficient D_l^t as compared to its value without HI [12]. This rather unexpected enhancement of D_l^t should be compared with hard spheres, where HI gives rise to an additional hindering of long-time diffusion.

Static properties

The basic quantity describing the microstructure of a colloidal suspension is the pair correlation function $g(r)$. From this function the thermodynamic properties and the angle-dependent scattered intensity

$$I(k) = N P(k) S(k) , \tag{3}$$

as obtained in a static light scattering experiment, can be calculated [5, 6]. Here N denotes the number of particles in the scattering volume, $P(k)$ is the form factor of a sphere and

$$S(k) = 1 + n \int d\mathbf{r}\, e^{i\mathbf{k}\cdot\mathbf{r}} (g(r) - 1) \tag{4}$$

is the static structure factor. The modulus k of the scattering vector $\mathbf{k}$ is related to the scattering angle ϑ by $k = (4\pi/\lambda)\sin(\vartheta/2)$, where λ is the wavelength of light in the medium.

The pair correlation function $g(r)$, or the total correlation function $h(r) = g(r) - 1$, satisfies the Ornstein–Zernike equation

$$h(r) = c(r) + n \int d\mathbf{r}'\, c(|\mathbf{r} - \mathbf{r}'|)h(r') , \tag{5}$$

where $c(r)$ is the so-called direct correlation function [13]. To make use of Eq. (5) a further relation between the correlation functions and the pair potential $u(r)$ is needed. This relation, known as closure relation, introduces approximations, and several such closures are available [5, 6, 13]. The quality of the closure relations depends on the type of pair potential. From comparison with Monte Carlo simulations it is known that the Percus–Yevick (PY) closure works well for short-ranged potentials, whereas the hypernetted chain (HNC) closure or the very accurate Rogers–Young (RY) closure have to be used for long-ranged potentials. For the hard sphere potential, Eq. (1), an analytical expression for the direct correlation function $c(r)$ exists in PY approximation, from which an analytical expression for $S(k)$ can be obtained by combining Eqs. (4) and (5), $S(k) = (1 - nc(k))^{-1}$, where $c(k)$ is the Fourier transform of $c(r)$. For charge-stabilized systems, for which

the numerically involved RY scheme performs best, it is found that nearly identical results for $S(k)$ are obtained using the numerically much simpler rescaled mean-spherical approximation (RMSA), provided that the effective valency Z is replaced by another effective parameter Z^{RMSA} somewhat larger than Z [5, 14].

Figures 1 and 2 show results for $g(r)$, demonstrating the qualitative differences in the microstructures of hard spheres and of charge-stabilized particles. In case of hard

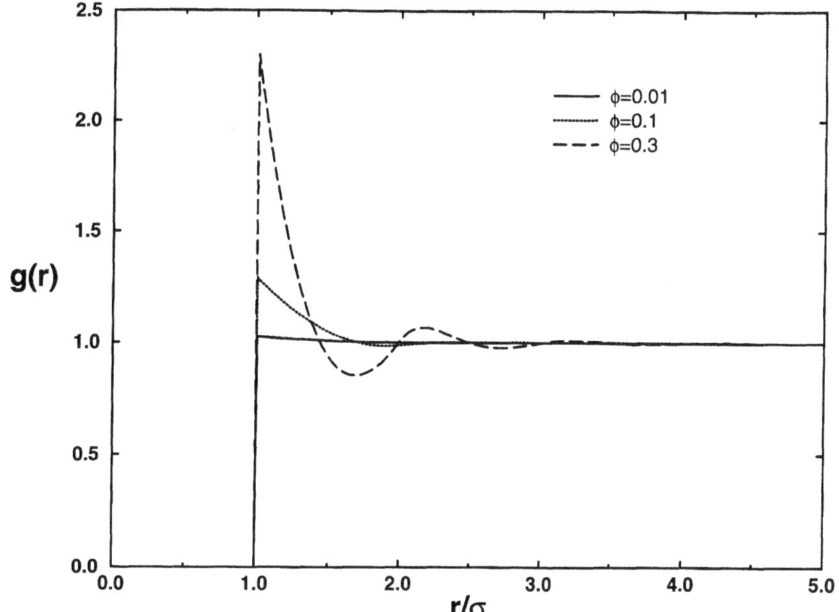

Fig. 1 Pair correlation functions $g(r)$ for hard sphere suspensions obtained within the Percus–Yevick (PY) scheme. The volume fraction is defined as $\phi = (\pi/6)n\sigma^3$

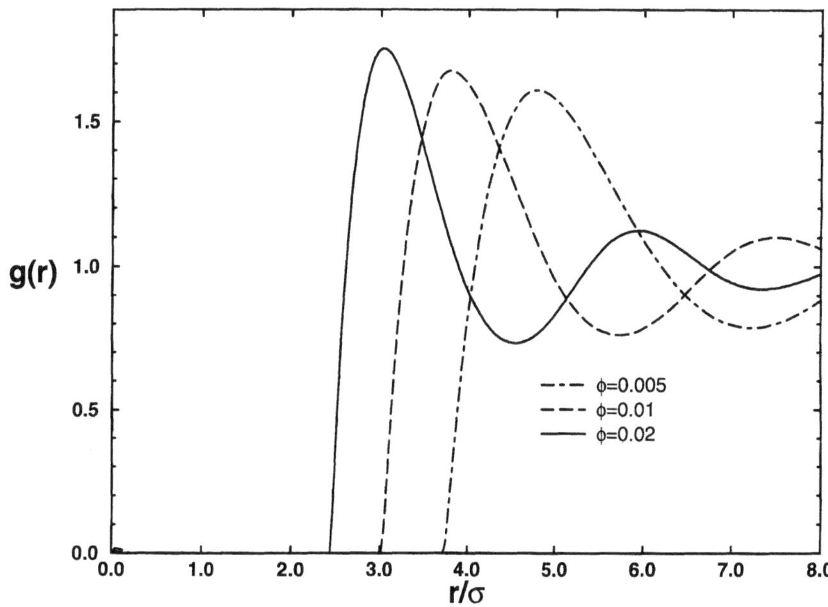

Fig. 2 Pair correlation functions $g(r)$ for a charged suspension with $Z = 200$, $\sigma = 90\,\mathrm{nm}$, $\varepsilon = 87.0$, $T = 294\,\mathrm{K}$ and $n_s = 0$. For the numerical calculation the RMSA scheme [14] was used

spheres, $g(r)$ has its maximum at contact $(r = \sigma)$, given by $g(\sigma^+) = (1 + \phi/2)/(1 - \phi)^2$ in PY approximation (cf. Fig. 1). As seen from Fig. 2, the longer-ranged repulsive pair potential for charged particles keeps such particles apart and creates the so-called correlation hole, i.e., $g(r)$ is zero for some range of distances r larger than σ, after which a well-defined shell of nearest neighbours exists, even for systems as dilute as $\phi = 0.005$. Increasing ϕ reduces the extent of the correlation hole, sharpens the nearest neighbours shell and increases the main maximum of $g(r)$. The position of this maximum, r_m, scales with volume fractions as $r_m \sim \phi^{-1/3}$ in case of deionized suspensions [5, 10].

We mention that the main peak of $g(r)$ increases and r_m stays practically constant, when Z is increased and ϕ remains fixed. Furthermore, addition of electrolyte increases the screening of the longer-ranged Yukawa potential, so that the pair potential becomes short-ranged and $g(r)$ approaches the result for hard spheres for $n_s \to \infty$ or $\kappa \to \infty$, respectively (cf. e.g. Refs. [5, 10]).

Noticing the rather different microstructures of uncharged and charged spheres, it can be expected that the dynamical properties of the two systems also will behave differently.

Dynamic properties

We will now consider several dynamic properties and will compare the corresponding theoretical results for hard spheres and Yukawa particles with each other and, when possible, with experimental data. These experimental data were obtained by quasi elastic light scattering measurements which resolve times larger than the velocity relaxation time $\tau_B = m/\zeta_0$ of the macroparticles; here m is the particle mass and $\zeta_0 = 3\pi\eta\sigma$ the friction coefficient of a sphere of diameter σ in a solvent of shear viscosity η. It is well established that the dynamics of the interacting suspended particles at times $t \gg \tau_B$ is described by the many-body Smoluchowski equation for the distribution function $P(\mathbf{r}^N, \mathbf{u}^N; t)$ of the center-of-mass positions $\mathbf{r}^N \equiv (\mathbf{r}_1, \ldots, \mathbf{r}_N)$ of the particles and the orientations $\mathbf{u}^N \equiv (\mathbf{u}_1, \ldots, \mathbf{u}_N)$ of axes fixed to the particles [10, 15, 16],

$$\partial P/\partial t = \Omega P . \qquad (6)$$

The Smoluchowski operator Ω depends on the total potential energy of interactions, which is assumed to be a sum of pair potentials $u(|\mathbf{r}_i - \mathbf{r}_j|)$, and on hydrodynamic diffusion tensors $\mathbf{D}_{ij}^{tt}(\mathbf{r}^N)$, $\mathbf{D}_{ij}^{tr}(\mathbf{r}^N)$, $\mathbf{D}_{ij}^{rt}(\mathbf{r}^N)$ and $\mathbf{D}_{ij}^{rr}(\mathbf{r}^N)$ [15, 17, 18]. Here the subscripts refer to particles, $i, j = 1, \ldots, N$, and the superscripts t and r refer to translation and rotation, respectively. Due to the many-body character of the hydrodynamic interactions, all diffusion tensors

depend on the particle configuration $\mathbf{r}^N$ of all spherical particles.

Dynamic light scattering experiments on monodisperse systems of optically homogeneous spherical particles determine the dynamic structure factor $S(k, t)$, when it is assumed that there is only single light scattering in the suspension. $S(k, t)$ satisfies the equation of motion [1]

$$\frac{\partial S(k, t)}{\partial t} = -k^2 D_{\text{eff}}(k) S(k, t) + \int_0^t d\tau\, M(k, t - \tau) \frac{S(k, \tau)}{S(k)} , \qquad (7)$$

which is obtained from the Smoluchowski equation.

The first term in Eq. (7) is sufficient to determine the short-time behaviour [5, 19] of the measured autocorrelation function, which is usually characterized by the first cumulant $\Gamma(k) = k^2 D_{\text{eff}}(k)$, where the effective wavenumber dependent diffusion coefficient is given by [1]

$$D_{\text{eff}}(k) = D_0 H(k)/S(k) . \qquad (8)$$

Here $D_0 = (k_B T)/\zeta_0$ is the translational diffusion coefficient at infinite dilution, and the hydrodynamic function

$$H(k) = \frac{1}{ND_0} \sum_{i, j = 1}^{N} \langle \hat{\mathbf{k}} \cdot \mathbf{D}_{ij}^{tt}(\mathbf{r}^N) \cdot \hat{\mathbf{k}}\, e^{i\mathbf{k} \cdot (\mathbf{r}_i - \mathbf{r}_j)} \rangle \qquad (9)$$

with $\hat{\mathbf{k}} = \mathbf{k}/k$, determines the effects of the hydrodynamic interactions (HI). Without HI, $H(k) = 1$ for all k. At $k \to 0$, $H(0)$ is related to the normalized sedimentation velocity, $H(0) = U/U_0$, if the suspension is sufficiently diluted, so that three-body effects of HI can be neglected (i.e., approx. $\phi < 0.08$). Here U is the sedimentation velocity of the interacting suspension, and U_0 is the sedimentation velocity of a single dispersed sphere [5, 7, 9]. For wavenumbers appreciably larger than k_m, which is the position of the main maximum of $S(k)$ of the suspension, $H(k \gg k_m) \to D_s^t/D_0$, where D_s^t is the short-time self-diffusion coefficient [8]. This coefficient is different, i.e., smaller than D_0 only when HI are of importance.

The combination of static and dynamic light scattering allows for a measurement of $H(k) = S(k)\Gamma(k)/(k^2 D_0)$ over a wide range of k [20, 21]. Fig. 3 shows experimental data (open circles) for charged silica spheres in an organic solvent which are compared with theoretical results obtained from different treatments of Eq. (9) [5, 7, 22, 23]. Several features should be noted: (a) The data are rather well described by using the pairwise additive approximation (PA) for the hydrodynamic tensors in far-field expansions up to terms of order r^{-11} [5, 7, 23]; the Beenakker–Mazur theory $(\delta\gamma)$ [22, 24], which includes in an approximative way many-body contributions of HI, does not change the results significantly. (b) The maximum of $H(k)$ has a position roughly coinciding with k_m, and a value larger than one for charged particles. (c) Treating the suspension as a system of (uncharged) hard spheres

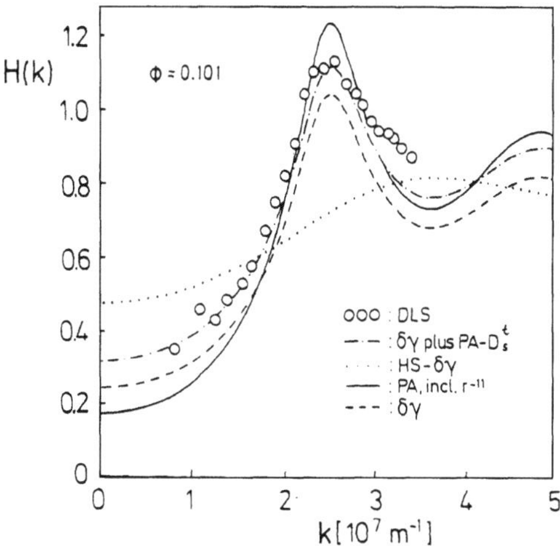

Fig. 3 Hydrodynamic function $H(k)$ of a suspension of charged silica spheres with system parameters as given in Refs. [7, 21]. Symbols as indicated in the figure. Reproduced from [7]

(HS-$\delta\gamma$) leads to results for $H(k)$ which are quite different: At long wavelength the effects of HI are less pronounced than for the charged particles, the maximum of $H(k)$ is less than one, and at large k, where essentially short-time self-diffusion is probed, the effects of HI are stronger for hard spheres than for charged spheres.

These different results can be understood qualitatively from the leading contributions of the hydrodynamic tensors $\mathbf{D}_{ij}^{tt}$, as needed in Eq. (9), and the different forms of the pair distribution functions $g(r)$ for hard spheres and charged spheres. At $k \to 0$, where the system is probed on long length scales, the most important contribution of HI is the well known Oseen tensor, which is rather long-ranged, i.e., proportional to a/r [17]. At $k \gg k_m$, where self-diffusion is probed, the leading contribution of HI is far more short ranged. Since only the tensor $\mathbf{D}_{11}^{tt}$ is needed for the calculation of $D_s^t/D_0 = H(k \gg k_m)$, the leading term is proportional to $(a/r)^4$ [7]. To obtain $H(k)$ according to Eq. (9), these different powers of a/r are multiplied by $g(r)$ and the results are integrated over r [7, 8]. For charged systems $g(r)$ has a correlation hole, which extends over several particle diameters (cf. Fig. 2). This leads to the fact that the effect of HI on the self-diffusion coefficient D_s^t of charged particles will be less important than for hard spheres, because the correlation hole extends over the range, where the hydrodynamic function $\propto (a/r)^4$ is perceptibly different from zero [8, 11, 25]. For $k \to 0$, on the other hand, the situation is reversed, since the Oseen tensor contribution is still significant where $g(r)$ for charged systems has its main maximum, leading to a strong influence of HI in $H(k)$ for small wavenumbers k. For a further intuitive explanation of the qualitatively

different behaviour of $H(k)$ for charged and uncharged particles we refer to Ref. [12].

This qualitatively different influence of HI on the short-time dynamics of charged and uncharged particles can best be demonstrated by studying various quantities at low volume fractions. $H(0)$, which is the normalized sedimentation velocity, has been calculated in dependence of volume fraction ϕ, both for hard spheres [26, 27] and charge-stabilized suspensions [5, 7–9]. The results for charged systems without added electrolyte ($n_s = 0$) can be fitted to

$$H(0) = 1 - a_c \phi^{b_c} \tag{10}$$

with $a_c = 1.8$ and $b_c = 0.33$. In contrast, the corresponding result for hard spheres is $a_c = 6.55$ and $b_c = 1$. Expressing the short-time self-diffusion coefficient as

$$\frac{D_s^t}{D_0} = 1 - a_t \phi^{b_t}, \tag{11}$$

it was recently found that $a_t = 2.59$ and $b_t = 1.3$ for charged systems [8, 11, 25], whereas $a_t = 1.83$ and $b_t = 1$ for hard spheres [27–29]. We notice that all these results only hold for dilute suspensions, because only two-body contributions to HI are taken into account. (For a consideration of three-body contributions to D_s^t see Ref. [24] for hard spheres and Ref. [11] for charged particles.) It is well known for hard sphere suspensions that an approximative evaluation of U/U_0 and D_s^t can be performed in terms of a virial expansion by separating the hydrodynamic effects of n-particle clusters ($n = 1, \ldots, N$) [15, 24, 26, 29]. The above results show, however, that the same procedure in case of charge-stabilized particles leads to results which are not expressable as ordinary virial expansions.

This fact applies also to the short-time rotational self-diffusion coefficient D_s^r, which can be measured by depolarized dynamic light scattering from spherical particles with anisotropic polarizability [10, 30, 31]. Again, the theoretical results can be expressed as

$$\frac{D_s^r}{D_0^r} = 1 - a_r \phi^{b_r}, \tag{12}$$

and it is found that $a_r \simeq 1.2$ and $b_r \simeq 2$ for charged spheres [10, 25], whereas $a_r = 0.63$, $b_r = 1$ for hard spheres. [16, 31]. Here D_0^r is the Stokesian rotational diffusion coefficient given by $D_0^r = k_B T/(\pi\eta\sigma^3)$.

Considering the large qualitative differences in the static structure functions of charged and uncharged particles, it seems to be difficult to understand the above results for charge-stabilized suspensions in terms of effective hard sphere models, since even the concentration dependence of the transport coefficients is qualitatively different for hard and soft spheres. The resolution of this problem lies in the recognition that the extent of the correlation hole of $g(r)$ is concentration dependent: the position r_m of the main maximum of $g(r)$ for charged particles is approximately equal to the geometrical mean interparticle distance $\bar{r} = n^{-1/3}$, which scales with volume fraction as $\phi^{-1/3}$ [5]. Using this property, we can obtain a rather rough but very useful approximation for the actual $g(r)$ of the observed charge-stabilized suspension: we crudely approximate $g(r)$ by a unit step function $g_{EHS}(r) = \Theta(r - \sigma_{EHS})$. The EHS diameter $\sigma_{EHS} = r_m \simeq \bar{r} > \sigma$ accounts for the extension of the correlation hole. When this approximation is used for the calculation of $H(0)$, and if only the leading contributions of the hydrodynamic tensors are included, one obtains the result [5, 32, 33]

$$H(0) = 1 + \phi\left[-6\left(\frac{\sigma_{EHS}}{\sigma}\right)^2 + 1 + \mathcal{O}\left(\frac{\sigma}{\sigma_{EHS}}\right)\right]. \tag{13}$$

Since σ_{EHS} is appreciably larger than σ for low volume fractions, it is sufficient to consider only the first term in the bracket, leading to $H(0) = 1 - a_c\phi^{1/3}$, when the scaling property $\sigma_{EHS} \propto \bar{r} \propto \phi^{-1/3}$ is used. So, this simple argument explains very nicely the exponent $1/3$, observed in the numerical calculations [5, 7, 9]. The prefactor a_c from Eq. (10), however, turns out to be different, i.e., $a_c = 3.90$, than the observed value of 1.8 if σ_{EHS} is approximated by $\bar{r}$ [5, 9]. This is not surprising, since the approximation $g_{EHS}(r)$ does not include the fact that the actual $g(r)$ has pronounced undulations and that it is significantly larger than one at $r = \sigma_{EHS}$, describing the enhanced density of particles in the shell of nearest neighbours around each particle. The same kind of reasoning can also be applied to a rough estimate of D_s^t/D_0. It turns out that the exponent in Eq. (11) is $b_t = 4/3$, which is again close to the value found by fitting the numerical theoretical results [8, 11, 25]. Finally, the exponent b_r for the rotational diffusion coefficient is $b_r = 2$ from this model of effective hard spheres, again in agreement with the numerical calculations [10, 25].

A prefactor a_c closer to 1.8 is obtained, when instead of the step function $\Theta(r - \sigma_{EHS})$ the Percus–Yevick hard-sphere expression $g_{PY}(r; \phi_{EHS})$ is used as an approximation for the actual $g(r)$, with $\phi_{EHS} = \phi(\sigma_{EHS}/\sigma)^3$. To show this we note that to leading order, i.e., on the Oseen-level, $H(0)$ is given by

$$H(0) = 1 + 12\phi\frac{1}{\sigma^2}\int_0^\infty dr\, rh(r)$$

$$= 1 + 12\phi\frac{1}{\sigma^2}\tilde{H}(s = 0), \tag{14}$$

where

$$\tilde{H}(s) = \int_0^\infty dr\, rh(r)e^{-sr} \tag{15}$$

denotes the Laplace transform of $rh(r)$, with $h(r) = g(r) - 1$. Equation (14) is a good approximation for the normalized sedimentation velocity of dilute suspensions of strongly charged particles. We approximate now $h(r)$ by the Percus–Yevick form $h_{PY}(r; \phi_{EHS})$ evaluated for effective hard spheres of diameter σ_{EHS}, and we take advantage of the fact that an analytic expression is known for the Laplace transform $\tilde{H}_{PY}(s)$ of $rh_{PY}(r)$ [34]. By performing the zero-s limit, we find after a lengthy but straightforward calculation [35]

$$\tilde{H}_{PY}(s = 0) = -\frac{10 - 2\phi_{EHS} + \phi_{EHS}^2}{20(1 + 2\phi_{EHS})}\sigma_{EHS}^2. \tag{16}$$

Next, we employ the scaling property $\sigma_{EHS}/\sigma = (\phi_{EHS}/\phi)^{1/3}$. This readily leads, together with Eqs. (14) and (15), to Eq. (10), with $b_c = 1/3$ and a_c determined as

$$a_c = \frac{3}{5}\phi_{EHS}^{2/3}\frac{10 - 2\phi_{EHS} + \phi_{EHS}^2}{1 + 2\phi_{EHS}}. \tag{17}$$

By approximating σ_{EHS} by $\bar{r}$, we obtain a value of $\phi_{EHS} = \pi/6$ independent of Z. Substitution of ϕ_{EHS} into Eq. (17) leads to $a_c = 1.76$, i.e., a prefactor close to the one determined from the numerical calculations. We mention that along the same lines, and again to leading order in the hydrodynamic far-field contributions, similar expressions for a_r and a_t can be given in terms of one-dimensional integrals over $s^4\tilde{G}_{PY}(s)$ and $s^2\tilde{G}_{PY}(s)$, respectively. Here, $\tilde{G}_{PY}(s)$ is the Laplace-transform of $rg_{PY}(r; \phi_{EHS})$, which is known analytically. Moreover, we point out that the analytic expression for $\tilde{G}_{PY}(s)$ is useful also for determining σ_{EHS} according to the Gibbs–Bogoliubov inequality, by mapping the one-component macrofluid model onto a reference hard-sphere suspension. Such a mapping has been implemented [5], and it is useful in conjunction with Eq. (17) to describe sedimentation in dilute suspensions also with added electrolyte.

Concerning the comparison of the above results for the short-time dynamics of colloidal suspensions with experiment, it can be stated that the theoretical predictions for hard sphere suspensions are in good agreement with measurements of the sedimentation velocity [36, 37] and the diffusion coefficients [30, 31, 38–41], respectively. In the case of strongly correlated charged particles, the predictions for $H(0) = U/U_0$ are in good agreement with measurements of the sedimentation velocity for deionized charge-stabilized suspensions, at least with respect to the value of the exponent b_c [9, 42]. With regard to the short-time translational self-diffusion coefficient D_s^t, we are not aware of experimental results which are sufficiently precise at low values of ϕ to distinguish the $\phi^{4/3}$ behaviour from the linear dependence for hard spheres. In the case of short-time rotational self-diffusion, recent depolarized dynamic light scattering experiments are in good agreement with the quadratic behaviour predicted for charged particles [43].

Turning back to Eq. (7), it is obvious that it is necessary to take into account the second term on the right-hand side of Eq. (7), when the long-time dynamics of the colloidal dispersion is investigated. This term is referred to as the memory term. The physical origin of the memory term, which becomes important at times when a particle has diffused at least a perceptible fraction of its own diameter, is the so-called caging effect: a particle gets temporarily trapped in a dynamic cage of neighbouring particles, with whom it interacts both hydrodynamically and by direct forces. Due to the memory term, the temporal decay of $S(k, t)$ is slowed down and becomes non-exponential. A global measure of the non-exponentiality of $S(k, t)$ is the quantity [44, 45]

$$\Delta(k) = 1 - \frac{\tau_s(k)}{\bar{\tau}(k)}, \tag{18}$$

referred to as non-exponentiality factor. Here $\bar{\tau}(k) = \int_0^\infty dt\, S(k, t)/S(k)$ is the mean relaxation time of $S(k, t)$, and $\tau_s(k) = (k^2 D_0 H(k)/S(k))^{-1}$ is the decay time of the short-time expression $S(k, t)$, given by $S(k, t) = S(k)\exp[-t/\tau_s(k)]$. This short-time form of $S(k, t)$ follows directly from Eq. (7), when the memory term is neglected (compare Eq. (8)). Notice that $0 \leq \Delta(k) \leq 1$, since memory effects always give rise to a slower decay of $S(k, t)$ as compared to its short-time behaviour.

In the past, memory effects in strongly correlated colloidal systems have been studied theoretically only disregarding the effects of HI. A quantitative theoretical analysis of the influence of HI on the long-time dynamics of Yukawa particles is given for the first time in Ref. [12]. In this work, the non-exponentiality factor $\Delta(k)$ and the long-time translational self-diffusion coefficient D_l^i are calculated using a novel mode-coupling scheme (MCA), which accounts for the dominant far-field contributions to HI. Figure 4 displays the MCA-results for $\Delta(k)$ with and without HI, and compared with experimental results for the non-exponentiality factor obtained by Müller and Schätzel [46]. Here k is scaled by the value k_m, where $S(k)$ has its maximum. We point out that the theoretical results shown for $\Delta(k)$ account for the small size polydispersity (relative standard deviation $s = 0.05$) which exists in the samples studied experimentally in Ref. [46]. Size polydispersity gives rise to an incoherent scattering contribution to the measurable dynamic structure factor. Therefore, in deionized suspensions surprisingly small amounts of size

polydispersity cause the measurable non-exponentiality factor $\Delta(k)$ to attain values close to one at $k = 0$. On the other hand, $\Delta(k) \to 0$ for $k \to 0$ in case of an ideally monodisperse suspension. Notice from Fig. 4 that the effect of HI in dilute charge-stabilized suspensions is to lower the value of $\Delta(0)$ by a rather small but noticeable amount.

In Fig. 5 we show MCA-results for the normalized long-time self-diffusion coefficient $D^* = D_l^i/D_0$ vs. volume fraction, taken from Ref. [12]. Very remarkably, in charge-stabilized suspensions with modest to small salinity, D^*, as calculated with regard of the effects of HI, is found to be larger than with HI being neglected. This result is in contrast to suspensions of hard spheres, where HI lead to reduction of D^* [1]. The surprising enhancement of D^* caused by HI is not an artifact of the MCA, but is corroborated by analytic results for D^*, again obtained for an effective hard sphere model of charged particles [12, 32].

An intuitive explanation for the hydrodynamic enhancement of D^* in dilute charge-stabilized suspensions is as follows: the Brownian motion of a charged tracer particle is predominantly affected by the far-field part of the HI, since the probability of two or more particles coming close to each other is very small. The far-field part, however, counteracts the caging of the tracer particle due to neighbouring host particles. Conversely, the probability of finding two hard spheres is largest at contact distance where $g(r)$ attains its maximum. For this reason, the caging caused by direct interparticle forces is enhanced by strong lubrication forces (i.e., near-field HI) which oppose the relative motion of closely spaced particles.

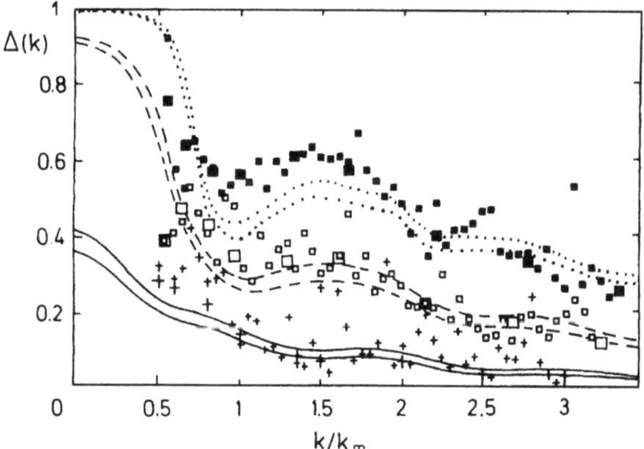

Fig. 4 Experimental $\Delta(k)$ of systems YZ (■), WX (□), and ST (+) studied by Müller and Schätzel [46], vs. the MCA-results of Ref. [12] calculated with (lower set of curves) and without HI (upper set). The MCA-$\Delta(k)$ for YZ (dotted lines), WX (dashed lines), and ST (solid lines) have been calculated for $s = 0.05$. Samples YZ, WX, and ST are aqueous suspensions of spherical particles with average diameter $\sigma = 100$ nm. YZ: $\phi = 1.3 \times 10^{-3}$, $Z = 364$, $\kappa\sigma = 0.28$; WX: $\phi = 1.28 \times 10^{-3}$, $Z = 475$, $\kappa\sigma = 0.7$; ST: $\phi = 1.15 \times 10^{-3}$, $Z = 329$, $\kappa\sigma = 0.9$. Reproduced from [12]

G. Nägele et al.
Hard spheres vs. Yukawa particles

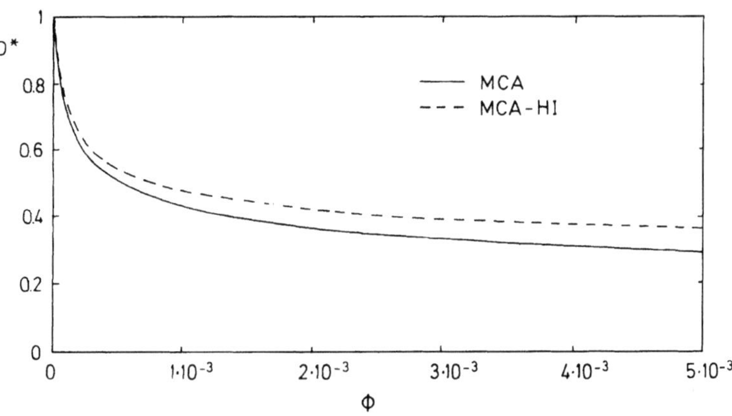

Fig. 5 Normalized long-time self-diffusion coefficient $D^* = D_s^l/D_0$ of system YZ (with $n_s = 0$) calculated in MCA with (dashed line) and without HI (solid line). Reproduced from [12]

Conclusions

As a conclusion it can be stated that the three short-time dynamical properties considered here, which behave rather differently for charge-stabilized suspensions and for suspensions of hard spheres, can at least qualitatively be understood by mapping the charged particles to effective hard spheres with a volume fraction dependent diameter given by the position of the main maximum of the pair distribution function of the charged particles. This position is nearly equal to the mean interparticle distance which scales with volume fraction as $\phi^{-1/3}$ in case of deionized (i.e., salt-free) suspensions. It is important to note that the differences in the concentration dependence of the short-time dynamic properties for charged and uncharged particles arise entirely from the presence of hydrodynamic interaction, since with neglect of HI $H(k) = 1$

and $D_s^r = D_0^r$ for both types of systems. Although the hydrodynamic diffusion tensors for charged and uncharged particles are identical, the hydrodynamic function $H(k)$ and the diffusion coefficients for charged and uncharged particles are different because of the fact that these quantities are determined by using static correlation functions besides these tensors. It is entirely the rather different static structure for the two types of suspensions which gives rise to the observed differences in the volume fraction dependencies of the short-time transport coefficients and the fractional exponents in case of charged particles. Finally, the qualitative differences between Yukawa particles and hard spheres carry over to long-time diffusion, when memory contributions are of importance. In particular, an unexpected enhancement of long-time self-diffusion is found in charge-stabilized suspensions arising from HI even at very small volume fractions.

References

1. Pusey PN (1991) Colloidal suspensions In: Hansen J-P, Levesque D, Zinn-Justin J (eds) Liquids, Freezing and Glass Transition: II, North Holland, Amsterdam
2. Medina-Noyola M, McQuarrie DA (1980) J Chem Phys 73:6279
3. Belloni L (1986) J Chem Phys 85:519
4. Löwen H, Hansen JP, Madden PA, (1993) J Chem Phys 98:3275
5. Nägele G (1996) Phys Rep 272:215
6. Klein R, D'Aguanno B (1996) Static scattering properties of colloidal suspensions, In: Brown W (ed) Light Scattering, Principles and Development. Clarendon Press, Oxford, pp 30–102
7. Nägele G, Steininger B, Genz U, Klein R (1994) Physica Scripta T 55:119
8. Nägele G, Mandl B, Klein R (1995) Progr Colloid Polym Sci 98:117
9. Thies-Weessie DME, Philipse AP, Nägele G, Mandl B, Klein R (1995) J Coll Int Sci 176:43
10. Watzlawek M, Nägele G (1997) Physica A 235:56
11. Watzlawek M, Nägele G, University of Konstanz, submitted
12. Nägele G, Baur P, University of Konstanz, Physica A, in press
13. Hansen J-P, McDonald IR (1986) Theory of simple liquids. Academic Press, London, 2nd ed
14. Nägele G, Medina-Noyola M, Klein R, Arauz-Lara JL (1988) Physica A 149:123
15. Jones RB, Pusey PN (1991) Annu Rev Phys Chem 42:137
16. Jones RB (1988) Physica A 150:339
17. Happel J, Brenner H (1973) Low Reynolds Number Hydrodynamics. Noordhoff, Leiden
18. Felderhof BU (1977) Physica A 89:373
19. Klein R, Nägele G (1994) Il nouvo cimento 16 D:963
20. Phalakornkul JK, Gast AP, Pecora R, Nägele G, Ferrante A, Mandl-Steininger B, Klein R (1996) Phys Rev E 54:661
21. Philipse AP, Vrij A (1988) J Chem Phys 88:6459
22. Genz U, Klein R (1991) Physica 171:26
23. Nägele G, Kellerbauer O, Krause R, Klein R (1993) Phys Rev E 47:2562

24. Beenakker CWJ, Mazur P (1983) Physica A 120:388
25. Watzlawek M, Nägele G, this volume
26. Batchelor GK (1972) J Fluid Mech 52:245
27. Cichocki B, Felderhof BU (1989) Physica A 154:213
28. Batchelor GK (1976) J Fluid Mech 74:1
29. Cichocki B, Felderhof BU (1988) J Chem Phys 89:1049
30. Degiorgio V, Piazza R, Bellini T (1994) Adv Coll Int Sci 48:61
31. Degiorgio V, Piazza R, Jones RB (1995) Phys Rev E 52:2707
32. Cichocki B, Felderhof BU (1991) J Chem Phys 94:556
33. Denkov ND, Petsev DN (1992) Physica A 183:462
34. Wertheim MS (1963) Phys Rev Lett 10:321
35. Nägele G, University of Konstanz, unpublished results
36. Buscall R, Goodwin JW, Ottewill RH, Tadros TF (1982) J Coll Int Sci 85:78
37. Kops-Werkhoven MM, Fijnaut HM (1981) J Chem Phys 74:1618
38. Pusey PN, van Megen W (1983) J Phys 44:285
39. Ottewill RH, Williams NSJ (1987) Nature 325:232
40. van Megen W, Underwood SM (1989) J Chem Phys 91:552
41. van Veluwen A, Lekkerkerker HNW, de Kruif CG, Vrij A (1987) J Chem Phys 87:4873
42. Ackerson BJ, Oklahoma State University, private communication
43. Bitzer F, Palberg T, Leiderer P, University of Konstanz, private communication
44. Nägele G, Baur P, Klein R (1996) Physica A 231:49
45. Baur P, Nägele G, Klein R (1996) Phys Rev E 53:6224
46. Müller J, Schätzel K, published in Müller J (1993) PhD thesis, Universität Kiel, Germany

Progr Colloid Polym Sci (1997) 104:40–48
© Steinkopff Verlag 1997

E. Bartsch
V. Frenz
S. Kirsch
W. Schärtl
H. Sillescu

Multi-speckle autocorrelation spectroscopy – a new strategy to monitor ultraslow dynamics in dense and nonergodic media

E. Bartsch (✉) · V. Frenz[1] · S. Kirsch
W. Schärtl · H. Sillescu
Institut für Physikalische Chemie
der Universität Mainz
Jakob-Welder-Weg 15
55099 Mainz, Germany

[1] *Present address:*
Cassella AG
Hanauer Landstraße 526
60343 Frankfurt/Main, Germany

Abstract We present a modification of the conventional dynamic light scattering set-up which allows to monitor the intensity fluctuations of many independent spatial Fourier components of the density fluctuations, i.e. "speckles", simultaneously by using a charge-coupled device (CCD) camera as area detector. By averaging over the intensity auto-correlation function the final 10–20% decay of the intermediate scattering function in very dense colloidal dispersions is obtained with much higher accuracy. At the same time this multi-speckle autocorrelation spectroscopy provides an alternative route for constructing ensemble-averaged intermediate scattering functions in nonergodic media by replacing the average over many independent sample volumes by an average over independent spatial Fourier components of the density fluctuations. We will survey the methods proposed so far to generate ensemble averages in nonergodic media and discuss their merits and limits. We then demonstrate the advantages of the new technique, taking as an example a colloidal dispersion where in the glassy state long-lived density fluctuations superimpose on the frozen ones. Finally, we make a direct comparison with another "speckle-averaging" technique, the "interleaved sampling" method, which has been proposed and applied to the same system recently [J. Müller, T. Palberg, Progr. Coll. Polym. Sci. (1996) 100:121–126].

Key words Photon correlation spectroscopy – nonergodic systems – ensemble averaging

Introduction

Photon correlation spectroscopy (PCS) [1, 2] has become a standard technique for determining the size of colloidal particles [3] and for obtaining detailed information about particle dynamics and interactions in colloidal dispersions [4]. Recently, much interest has been devoted to study very slow density fluctuations in highly concentrated dispersions. One of the aims has been to compare long-time diffusion coefficients measured at the peak of the static structure factor and long-time self-diffusion coefficients with the volume fraction dependence of the zero shear viscosity [5]. Another point of interest is the analysis of intermediate scattering functions $f(Q, \tau)$ measured in the neighbourhood of the colloid glass transition and the comparison with newly developed theoretical concepts [6–11]. Here, two specific problems have to be faced: (i) In order to extract reliable long-time diffusion coefficients or to analyze the so-called structural relaxation behaviour at the colloid glass transition one needs to know the last 20% decay of the intermediate scattering function with sufficient accuracy. In situations where the relevant dynamics occurs on time scales of 10^2–10^4 s this task requires extremely long measuring times with the concomitant problems of the long-time stability of the experimental setup. (ii) In a nonergodic system like the colloidal glass the Siegert relation [1], which connects the intermediate scattering function – an ensemble-averaged quantity – to the square root of the time-averaged normalized intensity autocorrelation function $g_T^{(2)}(Q, \tau)$ which is usually measured in a PCS experiment, is no longer applicable [12].

In the literature a number of strategies have been proposed and applied to overcome these problems. To obtain the final decay of the intermediate scattering function with better accuracy one can, in principle, use the heterodyne technique [2], which avoids the square root operation of the Siegert relation by directly measuring the autocorrelation function of the electric field, i.e. essentially $f(q, \tau)$ itself, instead of $g_T^{(2)}(Q, \tau)$. Since this method requires mixing of the scattered intensity with that of a local oscillator at the detector, it makes alignment of the instrument difficult and is much more sensitive to mechanical instabilities than the conventional set-up. Thus, the technique is rarely used in practice.

To obtain the true ensemble-averaged intermediate scattering functions for nonergodic systems one has basically three options: (i) the brute force method, i.e. constructing the ensemble average by averaging measurements of $g_T^{(2)}(Q, \tau)$ taken on a large number of independent sample volumes [13, 14], (ii) averaging over a large number of independent "speckles", i.e. Fourier components of the density fluctuations, by rotating the sample during the measurement [14, 15] and (iii) to calculate $f(Q, \tau)$ from one measurement of $g_T^{(2)}(Q, \tau)$ via an equation which takes the place of the Siegert relation in case of nonergodic media [12]. As will be discussed briefly in the next section, all these approaches have their specific merits and limitations and are, thus, applicable in certain conditions and time ranges.

Motivated by the problems encountered in obtaining reliable intermediate scattering functions in a certain type of colloidal glass, where pronounced long-time density fluctuations persist, even though the system shows all signatures of being nonergodic, we devised yet another strategy [10, 16]. It basically consists of replacing the photomultiplier tube used conventionally as detector by a CCD camera, thereby monitoring the intensity fluctuations of many independent speckles simultaneously and – after calculating $g_T^{(2)}(Q, \tau)$ for each speckle – constructing the "speckle average". This strategy – termed multi-speckle autocorrelation spectroscopy (MSCS) – increases the statistical accuracy at long decay times while dramatically reducing the overall measuring time by replacing time averaging partly by speckle averaging. Even though devised to deal with ultraslow fluctuations in nonergodic systems, it can also be used as an easily implemented

alternative to heterodyning for measuring the final decay of the time correlation functions in dense ergodic media.

In the following, we will first review the techniques used to obtain intermediate scattering functions in nonergodic media and discuss their applicability as well as their limitations. Then we will sketch the experimental set-up of the multi-speckle technique. After this we will demonstrate its advantages by applying it to a dispersion of micronetwork colloids where density fluctuations are partially frozen and make a comparison with results obtained with some of the other methods. In the conclusions we will point out possibilities for further development of the MSCS technique.

Constructing ensemble averages in nonergodic media

In this section we will briefly survey the methods used so far to construct ensemble-averaged intermediate scattering functions $f(Q, \tau)$ for nonergodic systems, i.e. systems where a significant amount of the density fluctuations is frozen during the light scattering experiment. We will compare the performance and practicability of the various methods with respect to a specific sample: a colloidal dispersion of polystyrene spheres where the polymer chains within the particle are chemically interconnected by a cross-linker (1 cross-link per 50 monomer units) to form a polymer network. Swollen in a good solvent they can be thought of as soft spheres (hydrodynamic radius $R_H = 100$ nm) which interact with a soft repulsive potential [17]. The dynamical behaviour around the colloid glass transition which occurs at a higher volume fraction $\varphi_g = 0.645$ than that of hard-sphere PMMA colloids ($\varphi_g \approx 0.58$ [7]) due to differences in interactions and/or size polydispersity ($\sigma_R = [\langle R^2 \rangle - \langle R \rangle^2]^{1/2}/\langle R \rangle = 0.16$) has been intensively studied by us [9–11]. One remarkable feature of this system is that the time decay of the intermediate scattering function does not stop at a finite plateau in the glassy state. Rather there appears a pronounced long-time decay, hinting at the existence of density fluctuations with relaxation times of the order of 10^3–10^4 s [11]. To check whether this feature is really due to sample physics and not merely an artifact of mechanical instabilities of our light scattering set-up, insufficient sampling times and/or problems to construct the true ensemble average was the main motivation of designing the multi-speckle technique.

To exemplify this point we show, in Fig. 1a, time-averaged intensity autocorrelation functions $g_T^{(2)}(Q, \tau)$ obtained by conventional photon correlation spectroscopy (PCS) experiments on independent sample volumes of a colloidal dispersion with $\varphi = 0.663 > \varphi_g$. The duration of one individual measurement of $g_T^{(2)}(Q, \tau)$ amounted to 2×10^5 s at a scattering vector Q close to the maximum of the static

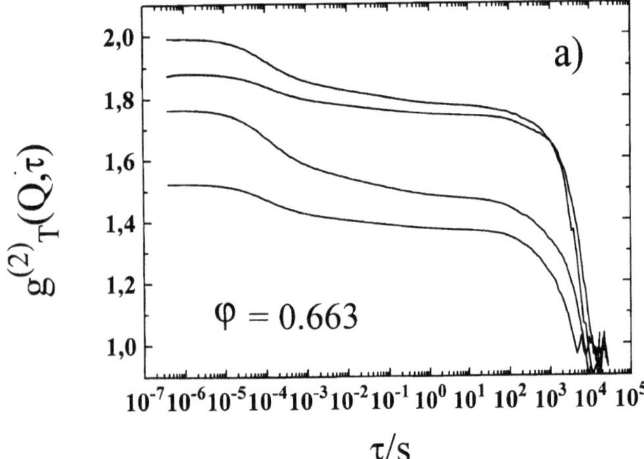

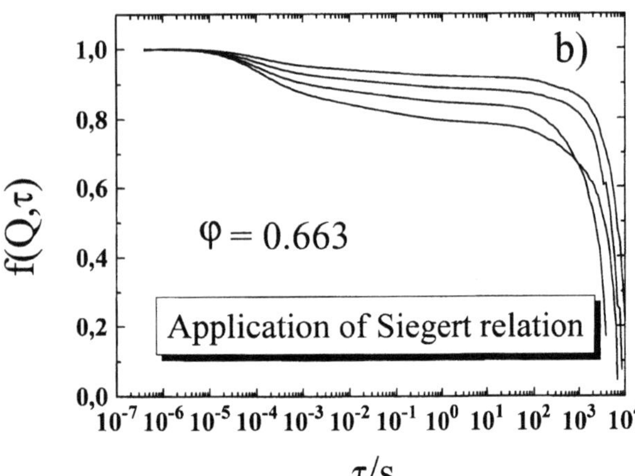

Fig. 1 a) Intensity autocorrelation functions of a dispersion of colloidal polystyrene micronetwork spheres (hydrodyanmic radius $R_H = 100$ nm) in ortho-dichlorobenzene above the colloid glass transition ($\varphi = 0.663 > \varphi_g = 0.645$) taken on different sample volumes at a scattering vector of $QR_H = 2.8$. The scatter of the intercept values $g^2_T(Q, \tau = 0)$ indicates the nonergodic character of the sample. b) Intermediate scattering functions $f(Q, \tau)$ determined via the Siegert relation (1) from the intensity autocorrelation functions shown in a). The failure of the Siegert relation to yield a unique $f(Q, \tau)$ curve indicates again the nonergodic character

structure factor ($QR_H = 2.8$) and we used the monomode fiber detection optics suggested by Rička [18] (see the next section for a complete description of our light scattering set-up). One clearly sees from Fig. 1a that the sample behaves nonergodically, different sample volumes giving rise to different $g^{(2)}_T(Q, \tau)$-curves with a considerable scatter in the intercept values $g^{(2)}_T(Q, 0)$. Application of the Siegert relation [1],

$$g^{(2)}_T(Q, \tau) = 1 + c|f(Q, \tau)|^2 , \tag{1}$$

to calculate $f(Q, \tau)$, thereby attributing the scatter of the intercept to the coherence factor c, which is just the inverse

number of coherence areas or speckles monitored by the detector, does not yield a unique, i.e. the true, ensemble-averaged intermediate scattering function. Instead, one obtains different results for each $g^{(2)}_T(Q, \tau)$, as demonstrated in Fig. 1b. How can one now proceed to construct reliable intermediate scattering functions for such a sample?

The brute force method

The conceptually and experimentally most simple and straightforward way to achieve the proper ensemble average is to perform many measurements of $g^{(2)}_T(Q, \tau)$ on different sample positions or orientations and to average them according to [7, 12]

$$g^{(2)}_{E(m)}(Q, \tau) = \frac{1}{m} \sum_{j=1}^{m} (\langle I(Q)\rangle^2_{T,j}/\langle I(Q)\rangle^2_E)g^{(2)}_{T,j}(Q, \tau) . \tag{2}$$

Here, T and E denote the time average and the ensemble average, respectively, and the ensemble-averaged static light scattering intensity $\langle I(Q)\rangle_E$ is obtained from the total number of photon counts measured while rotating and/or translating the sample [19]. Then, using the Siegert relation (1), with $g^{(2)}_T(Q, \tau)$ replaced by $g^{(2)}_{E(m)}(Q, \tau)$ yields an approximation of the ensemble-averaged intermediate scattering function $f(Q, \tau)$. Clearly as $m \to \infty$, $g^{(2)}_{E(m)}(Q, \tau) \to g^{(2)}_E(Q, \tau)$. It has been shown [7] that at the position Q_{max} of the peak of the structure factor averaging over more as 900 independent Fourier components may be necessary in order to achieve a reasonable estimate of the true ensemble average. Thus, even when allowing the detector to capture as many as 10 coherence areas (corresponding to $c = 0.1$, the lower limit as determined by the signal to noise ratio) or independent Fourier components during each individual measurement of $g^{(2)}_{T,j}(Q, \tau)$ by using an unfocused laser beam and a large detector aperture, m may have to be as large as 90. Given a system like ours, where density fluctuations of the order of 10^3–10^4 s are suspected to occur, measuring times in the order of 10^5 s (≈ 1 day!) that are necessary to capture a significant amount of the slowest fluctuations make the brute force method not only a tedious, but a practically impossible task.

The Chaikin method

A variant of the brute force method has been proposed by Chaikin et al. [14]. Here, the sample is translated or rotated at a constant speed or frequency while a measurement of the intensity autocorrelation function is taken. Since different sample volumes or sample orientations are scanned during one experiment, the resulting correlation function is equivalent to the outcome of the brute force method. However, the sample motion introduces an additional decorrelation of the intensity fluctuations, which acts like a cutoff, bringing the intermediate scattering function down to zero even in a nonergodic situation. This cutoff time is determined by the time it takes to move a speckle through the detector aperture and thereby related to the translation speed or the rotation frequency. Using convenient translation speeds or rotational frequencies this has limited the longest lag times τ to be sampled to about 1 s, making the method most profitable for nonergodic systems with fast relaxational processes like gels [14]. It would, in principle, be possible to monitor longer lag times with slower sample motions, however, at the expense of longer measuring times. Thus, the Chaikin method shares the problems of the brute force method with respect to nonergodic systems with ultraslow density fluctuations.

The interleaved sampling method

This technique, recently presented by Müller and Palberg [15], is a complement of Chaikin's method. It both cases the sample is rotated during the measurement. Contrary to the latter method, however, the observed intensity fluctuations are fed into a multichannel correlator, which can be conceptually thought to consist of a multichannel analyzer with a subsequent parallel processing software correlator. The sampling of the intensity fluctuations and the rotation of the sample cuvette are synchronized by an internal clock, such that intensity fluctuations belonging to one speckle are reproducibly funneled into the same channel after every rotational period. Thus, each channel monitors one independent spatial Fourier component of the density fluctuations. Depending on the scattering geometry 500 up to several thousand independent speckles can be monitored simultaneously and their calculated intensity correlation functions are then averaged essentially according to Eq. (2) in order to obtain the ensemble average, $g_E^{(2)}(Q, \tau)$ (see Ref. [15] for details). The rotation period sets a cutoff time here as well, but now it is a cutoff at short times, again typically at 1 s, while the largest lag time is essentially given by the total duration of the measurement. Thus, this technique is specifically suitable for probing slow density fluctuations. In combination with other techniques that focus on the fast dynamics like the Chaikin method, a time range of 10^{-7}–10^4 s can be covered [20]. Since the interleaved sampling method covers the same time range as the multi-speckle technique, we will defer a discussion of it to the results section, where a direct comparison of both methods is presented.

The Pusey–van Megen (PvM) method

A very elegant theoretical approach to the problem of constructing ensemble-averaged intermediate scattering functions in nonergodic media has been put forward by Pusey and van Megen [12]. Realizing that the frozen density fluctuations of a nonergodic system contribute a static component to the scattered light intensity in a similar manner as a local oscillator in a heterodyne experiment, they derived the expression

$$f(Q, \tau) = 1 + Y^{-1}\{[\langle g_T^{(2)}(Q, \tau)\rangle_T - \langle g_T^{(2)}(Q, 0)\rangle_T + 1]^{1/2} - 1\}, \tag{3}$$

$$Y = \langle I(Q)\rangle_E / \langle I(Q)\rangle_T.$$

Equation (3) allows to determine the ensemble-averaged intermediate scattering function from one single measurement of the time-averaged intensity autocorrelation function $g_T^{(2)}(Q, \tau)$ with the ensemble-averaged static intensity $\langle I(Q)\rangle_E$ being the only quantity needed in addition to a conventional measurement.

The validity of this procedure has been thoroughly tested against the brute force method in a number of cases [7, 19, 21] and it has been found to work quite well, reducing the total amount of measuring time considerably. However, its applicability depends on two premises that have to be fulfilled: the derivation of Eq. (3) assumes (i) an ideal coherence factor $c = 1$, corresponding to an intercept of $g_T^{(2)}(Q, 0) = 2$ in case of an ergodic sample and (ii) that density fluctuations giving rise to the static component of the scattered intensity remain frozen, or, phrased in other words, that particles remain trapped within their cages of nearest neighbours, throughout the experiment.

Condition (i), which means that only one single speckle is monitored by the detector, is usually guaranteed by using a very small detector aperture [7] or a monomode fiber detection [10]. However, there exist light scattering set-ups like the dual-color cross-correlation scheme developed by Schätzel et al. [22] where $c = 1$ cannot be realized [23]. Condition (ii) is a matter of the sample physics and we will show below that for our system of micronetwork colloids it is not fulfilled.

For this purpose we present in Fig. 2a the results obtained when applying the PvM-method to the $g_T^{(2)}(Q, \tau)$-curves of Fig. 1a. One clearly sees that the PvM-method provides a unique intermediate scattering function, collapsing the four individual $g_T^{(2)}(Q, \tau)$-curves onto one single $f(Q, \tau)$-curve which shows after a fast initial decay a crossover to a plateau-like region with a much weaker, though finite slope. However, this coincidence is only observed until a cutoff time of about 200 s, where two things happen simultaneously: the "plateau" is terminated by another crossover to a much faster decay of the correlation

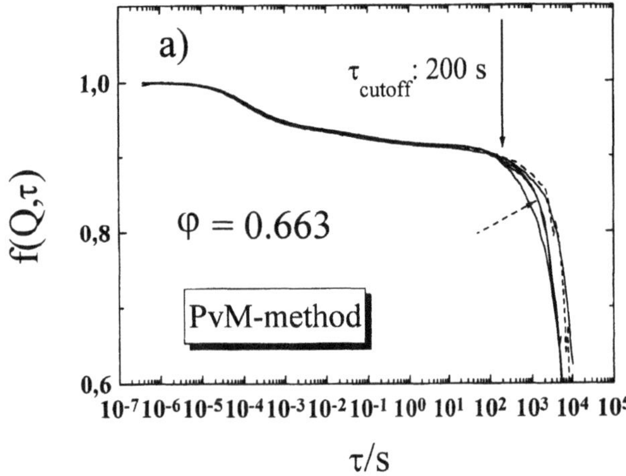

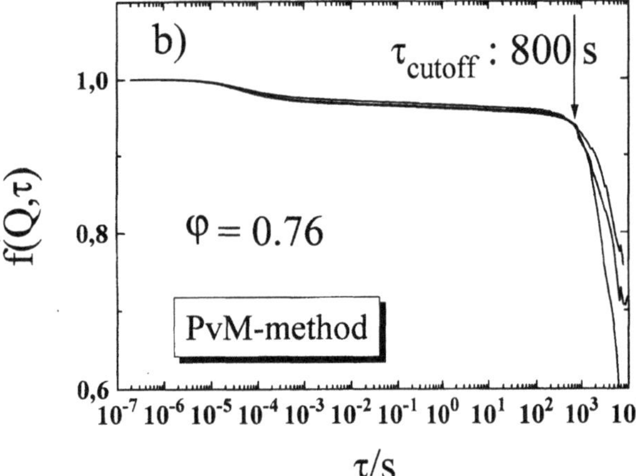

Fig. 2 a) Application of the Pusey–van Megen procedure, Eq. (3), to the $g_T^2(Q, \tau)$-curves in Fig. 1a provides intermediate scattering functions that are identical up to a lag time of 200 s (solid lines). The dashed line corresponds to the $f(Q, \tau)$ curve marked with the dashed arrow, but with the PvM-procedure applied to the intensity autocorrelation function after dividing through the $g_T^2(Q,\tau)$-curve measured on a piece of frosted glass (see text). b) same as a), but for a higher concentrated sample ($\varphi = 0.76$). Note that the cutoff time where the curves leave the plateau and start to split has now moved to 800 s

function and the splitting of the individual $f(q, \tau)$-curves for longer delay times. Two explanations of this behaviour are possible. First, it may be due to true sample physics, i.e. the particles escaping their cages on time scale τ_{cutoff}, thereby terminating the plateau and making the PvM-procedure invalid and leading to different long-time decays (possibly overlaid by the effect of insufficient sampling of the longest density fluctuations). Second, it may be just an artifact of mechanical instabilities and insufficient measuring time. A hint that the first might be the case comes from Fig. 2b, where the PvM-result for a dispersion with a higher volume fraction ($\varphi = 0.76$) is depicted. The same qualitative behaviour is observed, but now the cutoff time has

moved to 800 s. Since the total measuring time was the same for both dispersions, one would have expected the cutoff to occur at the same time if it were merely due to stability and an insufficient number of samples. To clarify this point the multi-speckle technique was introduced, whose set-up and performance are described in the next section and compared with the standard PCS set-up.

Experimental

Our set-up which is used both for standard photon correlation spectroscopy (PCS) and the new multi-speckle autocorrelation spectroscopy (MSCS) is sketched schematically in Fig. 3. The light scattering experiments were performed with a Nd:YAG solid state laser (DPSS-50, Coherent; $\lambda = 532$ nm, 50 mW). The sample cuvettes (Hellma 540 110-QS; 10 mm diameter) were mounted in the vertical axis of a X-ray goniometer (Siemens M 386 XA 3) equipped with an index match vat (Hellma) and a thermostatted sample holder. The PCS detection unit consisted of an ALV-PM15 photomultiplier, an ALV-PMPD discriminator/amplifier and an ALV5000/E correlator. In contrast to conventional light scattering experiments, the scattered light was not focused by the usual two pin-hole set-up onto the detector. Instead, it is coupled via an O/Z collimator (PMJ 33-3,5/125-488-3.0-1,2; Spindler&Hoyer), mounted on the 2ϑ circle of the goniometer, into a polarization preserving mono-mode optical fiber (HPUCO 488-$P - i = 200$, $f = 10$; Spindler&Hoyer) with a core diameter of $4/125$ μm which was connected to the photomultiplier. This replacement of the spatial filter by a mono-mode fiber, first suggested and tested by Rička [18], yields a high intercept value ($g_T^{(2)}(Q, 0) = 1.95 - 1.98$) in dilute samples while at the same time allowing count rates of more than 300 kHz. Thus, the achievement of high contrast, necessary for properly analysing light scattering experiments in nonergodic media [12], has not to be paid for by notoriously low count rates. This is usually the case when high intercepts are obtained by limiting the detection area to less than one coherence area or "speckle" through use of very narrow pinholes [19].

For MSCS experiments, the fiber-optical detection unit is replaced by a charge-coupled device (CCD) camera which is used as an area detector allowing to simultaneously monitor the intensity fluctuations of a large number of independent speckles. The CCD camera was a Hitachi KP 140. The CCD chip had an active area of 11.5×10 mm, divided into 580×500 pixels. The light sensitivity is 0.15 Lux. In order to optimally utilize the sensitivity range of our 8-bit CCD chip, the primary laser beam is attenuated using an adequately oriented polarizer. An

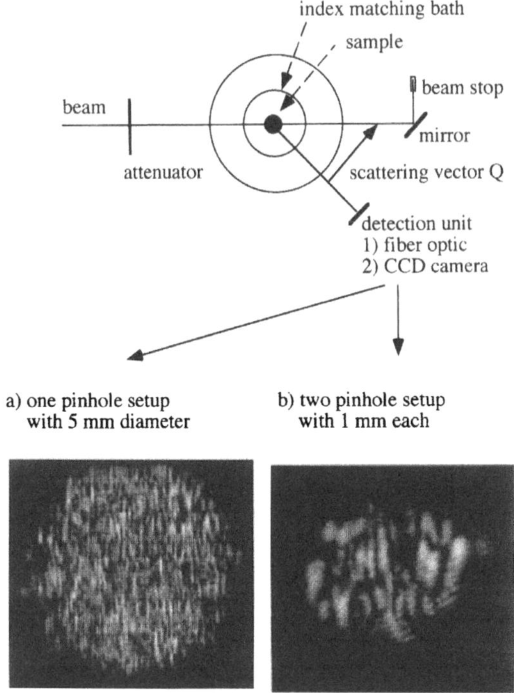

Fig. 3 Upper part: Schematic drawing of the experimental setup used for conventional dynamic light scattering experiments (using monomode fiber detection optics and a photomultiplier) and the multi-speckle autocorrelation spectroscopy, MSCS, (using pinholes and a CCD camera for detection). Lower part: Speckle patterns as monitored by the CCD camera for the indicated pinhole setups

additional outlet in the thermostatted sample cell surrounding the index match vat, opposite to the regular 2ϑ detection half-circle, allows to simultaneously perform PCS and MSCS experiments at $2\vartheta = 90°$. The scattered light is spatially filtered by a pinhole set-up in front of the CCD camera. We use two different pinhole set-ups (cf. Fig. 3, where examples of the respective speckle patterns observed by the camera are depicted). Note that two small pinholes (1 mm diameter each) lead to a better defined scattering angle ($\Delta 2\vartheta \approx \pm 0.1°$) while limiting the number of independent speckles. On the other hand, use of a single bigger pinhole (5 mm diameter) results in a large number of independent speckles, but also in a decrease of angular resolution ($\Delta 2\vartheta \approx \pm 1°$).

The pictures are digitized using a Macintosh personal computer supplied with a Datatranslation framegrabber board (DT2255). The computer collects images from the speckle pattern (512×256 pixel) from the camera at a maximum rate of 3 frames/s. The light intensity of each independent speckle (up to 50 speckles per picture) is determined via averaging the intensity of small square sections whose size (typically 2×2 pixel) and positions are user-selectable. These speckle intensities are stored in the RAM memory of the computer as a function of time. At present the data storage is limited to 5 MB, corresponding, for example, to 50 speckles and 100 000 pictures ($50 \times 100\,000 \times 8$ bit) or 20 speckles and 250 000 pictures. After data acquisition the time-averaged intensity autocorrelation function $g_T^{(2)}(Q, \tau)$ of each speckle is constructed by a software correlator according to

$$g_T^{(2)}(Q, \tau) = \frac{\langle I(t)I(t + k\Delta t)\rangle_T}{\langle I(t)\rangle_T^2}$$

$$= \frac{(N_{\text{pic}} - k) \sum_{n=0}^{N_{\text{pic}}-k} I(n\Delta t)I(n\Delta t + k\Delta t)}{\sum_{n=0}^{N_{\text{pic}}-k} I(n\Delta t) \sum_{n=0}^{N_{\text{pic}}-k} I(n\Delta t + k\Delta t)}, \quad (4)$$

with the symmetric normalization scheme of Schätzel [24] being implemented. N_{pic} is the total number of pictures whose speckle intensities have been stored as a function of time. Δt is the time spacing of two successive pictures and $k\Delta t$ represents the delay time τ. The ensemble-averaged intensity autocorrelation function $g_E^{(2)}(Q, \tau)$ is then obtained by averaging over the speckles according to Eq. (2) where m denotes now the number of independent speckles and $\langle I(Q)\rangle_E$ being found by averaging the time-averaged intensities of the individual speckles, $\langle I(Q)\rangle_{T,j}$ [25].

The available hardware restricts the smallest accessible lag-time to $\Delta t \geq 0.33$ s, while the largest lag-time is given by $10\,000\ \Delta t$ as limited by data size and statistical accuracy. Thus the typical time regime of MSCS covers lag-times from about 1 to 10^4 s, the total measuring time being determined by the number of correlation cycles, $N_{\text{pic}} + 10\,000$ times Δt. Due to the use of large pinholes each speckle corresponds to an average over several coherence areas. This is indicated by a much lower coherence factor, $c \approx g_T^{(2)}(Q, \tau = 0.33\ \text{s}) = 0.2$ (for two 1 mm pinholes), as compared to the standard set-up. Note, that no additional lens is inserted between the sample and the CCD camera. Thus, the size and the shape of the speckles, being related to the Fourier transform of the scattering volume, are only determined by the choice of pinholes. Use of smaller pinholes increases the speckle size (see Fig. 3), thereby increasing the contrast factor. Similarly, we found that decreasing/increasing the detection area (i.e. the number of pixels over which the intensities are summed) leads to a concomitant increase/decrease of the contrast. Thus, the contrast factor $c \approx 0.2$ reflects the number of coherence areas collected by one detection area. Even though, strictly spoken, each detection area of the MSCS-technique corresponds to roughly 5 coherence areas, we will keep the somewhat loose notation to call each detection area a "speckle". Since the true short-time limit of $g_T^{(2)}(Q, \tau)$ is not accessible to the MSCS method, the exact coherence factor is not known. Thus, a normalization to absolute scale is not possible and the MSCS data have always to be normalized to results obtained with the standard set-up under identical sample conditions. For this purpose a sufficient overlap between PCS and MSCS data is necessary (typically the time range between $\tau = 1$ s and $\tau = 100$ s).

To check the MSCS set-up and the data evaluation procedure we investigated a dilute dispersion of polystyrene latexes of 350 nm radius in glycerol which was cooled down to $10\,^\circ$C in order to shift the dynamics into the time window of the technique. We demonstrated [16] that the results agree well with those obtained by the standard PCS experiment on the same sample. Especially, it could be shown that the lower angular resolution caused by averaging over speckles which belong to slightly different scattering vectors does not distort the line shape of the decay curves. Even with the worst angular resolution of $\Delta 2\vartheta = \pm\,1^\circ$, tested by comparing speckle averages taken on opposite sites of a speckle pattern generated by use of one pinhole with 5 mm diameter (see Fig. 3), we found that the difference between the resulting intermediate scattering functions was well within the reproducibility limits of an conventional PCS experiment.

The question arises, if the correlation loss at long times seen in the intermediate scattering functions of our glassy samples obtained by the conventional DLS set-up (cf. Fig. 2) is simply due to mechanical instabilities and/or insufficient counting statistics. In order to roughly estimate these effects on the intermediate scattering functions of our nonergodic dispersions, we assumed that the intensity fluctuations caused by them are statistically independent of those due to particle dynamics. We then calculated a "background-corrected" intensity autocorrelation function $g^{(2)}_{T,\text{bc}}(Q,\tau) = g^{(2)}_T(Q,\tau)/g^{(2)}_{T,\text{back}}(Q,\tau)$, with $g^{(2)}_{T,\text{back}}(Q,\tau)$ being the intensity autocorrelation function of a piece of frosted glass, and obtained $f(Q,\tau)$ via the PvM method, Eq. (3). The resulting $f(q,\tau)$ curve, corresponding to the uncorrected $f(Q,\tau)$ curve marked by the dashed arrow in Fig. 2a, is included in the figure (dashed curve). The effect is large enough to be the origin of the splitting of the intermediate scattering functions measured on independent sample volumes at long lag times. However, it cannot account for the long time decay.

Comparison of MSCS with other techniques

We now proceed to compare MSCS results obtained on nonergodic samples with those provided by some of the other methods. Since the "brute force" method as well as the Chaikin method are inapplicable to systems with extremely long-lived density fluctuations, this only leaves the Pusey–van Megen (PvM) procedure and the interleaved sampling (IS) method.

In Fig. 4 we compare intermediate scattering functions obtained with the multi-speckle autocorrelation spectroscopy (MSCS) on two nonergodic samples of our mi-

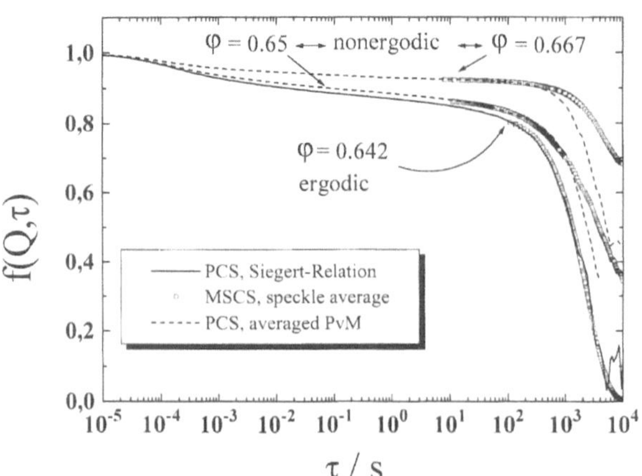

Fig. 4 Comparison of intermediate scattering functions obtained at $QR_{\text{H}} = 2.8$ for nonergodic samples of micronetwork colloids with volume fractions $\varphi = 0.65$ and 0.667 via standard photon correlation spectroscopy (PCS) and the Pusey–van Megen procedure, Eq. (3) (dashed lines), with those provided by the MSCS-technique (squares). Also included are corresponding data for the ergodic sample closest to the colloid glass transition ($\varphi = 0.642 < \varphi_{\text{g}} = 0.645$), where the Siegert relation (1), has been applied to determine $f(Q,\tau)$ from PCS data (solid line). For the PCS data on the nonergodic samples four individual measurements were averaged after applying the PvM procedure, while the PCS data for the ergodic sample represents an average over two measurements. Each individual measurement had a duration of 2×10^5 s. The MSCS data are averages over 250 independent spatial Fourier components of the density fluctuations (50 "speckles" times 5 coherence areas), where Eqs. (1) and (2) have been applied. The measuring time was 5×10^4 s

cronetwork colloids with corresponding results from the PvM-procedure. Also included are results of the ergodic sample closest to the colloid glass transition $\varphi_{\text{g}} = 0.645$ which have been evaluated using the Siegert relation (1) in case of the PCS experiment. The measuring time of the PCS experiments was $T = 2 \times 10^5$ s and for the nonergodic samples four measurements have been averaged after extracting $f(Q,\tau)$ via Eq. (3). The MSCS data represent averages over 50 speckles measured for $T = 5 \times 10^4$ s. Since each "speckle" collects intensity fluctuations from five coherence areas, 250 independent spatial Fourier components of the density fluctuations have been averaged altogether. Note that the MSCS data were normalized to absolute scale by matching them to the normalized PCS data. The figure shows that for the ergodic sample MSCS and PCS yield the same decay curves within experimental accuracy, and that the MSCS method provides more reliable results for the final 10–20% decay ($\tau > 2000$ s).

In case of the nonergodic samples we find that whereas the data sets agree perfectly up to about 200 and 400 s for $\varphi = 0.65$ and $\varphi = 0.667$, respectively, there are significant deviations at longer times. The PCS curves decay much faster than their MSCS counterparts, the discrepancy increasing at the higher volume fraction. This discrepancy cannot be compensated by the "background correction" discussed above. Obviously, the PCS experiments did not capture enough samples of the long-time fluctuations of the scattered intensity as compared to the MSCS method which collects a much larger number of samples. Nevertheless, even the MSCS curves indicate a pronounced, though slower decay of the density fluctuations at very long times. Here, the question arises, how one can be sure that the MSCS method itself does capture enough independent spatial Fourier components to achieve a good approximation to the true ensemble average. It is at least conceivable that the remnant long-time decay for the non-ergodic samples would disappear if more speckles were averaged, since at the peak of the structure factor maximum more than 900 independent Fourier components might be required [7].

It would therefore be instructive to make a comparison with the interleaved sampling technique, which is conceptually analogous to the MSCS method. In terms of the light scattering experiment the same spatial Fourier components are probed by rotating the sample with respect to a fixed detector position and by rotating the detector around the sample. Thus, the major difference between both approaches is that by keeping the CCD camera at a fixed position MSCS measures only an angular fraction of the Fourier components accessible in an IS experiment. Thus, the IS method, in principle, yields a closer approximation to the true ensemble average than MSCS in the present state of the art.

Since the interleaved sampling method was tested by using one of our nonergodic samples [15], we could make a direct comparison by remeasuring it with the MSCS technique. The result is shown in Fig. 5. The symbols are the data obtained in Ref. [15], where the squares correspond to a "brute force" ensemble average, achieved by averaging 100 measurements of 300 s duration. The circles represent the IS data obtained through averaging of 2000 speckles by rotating the sample at a frequency of 1 Hz. The measuring time was 10^4 s. The lines are our results. The dashed line gives the $f(q, \tau)$ curve provided by the PvM-procedure as stated above, while the solid line represents the MSCS data from averaging 250 Fourier components with a measuring time of 10^5 s. The agreement between the different measuring techniques cannot be called other than perfect, given that it extends up to such long lag times as 10^4 s. The only exception is the PvM-method which at lag times larger than 200 s yields a faster decay, similar to the behaviour shown in Fig. 4. The coincidence of the data provided by both methods makes now clear that the number of independent Fourier components collected in the MSCS experiment is sufficient to construct the true ensemble average. Furthermore, the long-time decay observed for our glassy samples is indeed due to particle

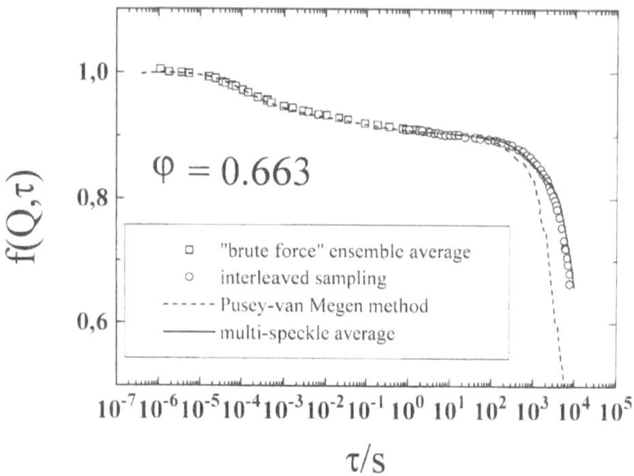

Fig. 5 Comparison between multi-speckle autocorrelation spectroscopy (MSCS) and the "interleaved sampling" (IS)-technique for a nonergodic sample of glassy micronetwork colloids ($\varphi = 0.663$) with long-lived density fluctuations. The MSCS data (solid line) represent averages over 250 spatial Fourier components of the density fluctuations (50 "speckles" times 5 coherence areas). The measurement took 10^5 s and $f(Q, \tau)$ was constructed according to Eqs. (1) and (2). The IS data (circles; taken from Ref. [15] and converted to $f(Q, \tau)$ via Eq. (1)) correspond to an average over 2000 speckles obtained by rotating the sample with a frequency of 1 Hz. The duration of the experiment was 10^4 s. Also included are results from the "brute force"-method (squares; taken from Ref. [15] and converted to $f(Q, \tau)$ via Eq. (1)) and the Pusey–van Megen (PvM) procedure (dashed line). For the "brute force"-method 100 measurements of 300 s duration have been averaged. The PvM-data were obtained as stated in Fig. 4. All data were taken at $QR_{\mathrm{H}} = 2.8$

dynamics and not merely an artifact of the measuring procedure.

So both the MSCS and the IS technique are viable alternatives to the PvM and the "brute force" method for constructing ensemble-averaged intermediate scattering functions in nonergodic media. Moreover, their use becomes imperative for systems where one wants to analyse very long-lived density fluctuations that are superimposed on essentially frozen ones. The decision between the two speckle averaging methods will depend on the specific problems posed by the sample or the instrumental requirements. Both methods are easily implemented and integrated into conventional light scattering set-ups, the MSCS method needing just a standard CCD camera and a suitable frame grabber board and being less prone to effects of mechanical long-time instabilities, since the relative positions of monitored speckle and detector do not change during the experiment. However, with the more complicated dual colour cross-correlation scheme, the interleaved sampling technique is the only choice. The major advantage of the IS method lies in the huge number of speckles available for averaging. This has to be compensated by a longer duration of the measurements in the MSCS technique. Thus, in situations where a sample undergoes time-dependent changes which one would like to resolve, e.g. aging processes in a glass, interleaved sampling may be preferable. On the other hand, the present limitation of the MSCS method to 50 speckles can be overcome by using a larger computer memory. The use of the MSCS method will be most advantageous where the rotation of the sample might induce unwanted side effects. Highly concentrated colloidal dispersions usually have a significant tendency to shear-thin. In this respect, the micronetwork colloids used to compare both techniques were rather unusual, since they require rather large forces to achieve a significant amount of shear thinning. It would be of interest to compare the performance of both methods on a sample which shear-thins more easily, say a dispersion of hard sphere colloids. Since both techniques need to be combined with results from other methods which cover the time range of the fast density fluctuations, the short-time limit imposed onto IS by the maximum rotational frequency compatible with sample requirements might become a problem if a significant overlap with short-time techniques like Chaikin's method is necessary. Even though MSCS is at present restricted to minimum lag times of 1 s as is the IS method, it can, in principle, be extended to shorter times by use of faster hardware like high-speed cameras.

Conclusions

We have shown that the multi-speckle autocorrelation spectroscopy (MSCS) is a very simple possibility to measure the final decay of the intermediate scattering function in dense, ergodic dispersions with high accuracy as well as to construct correct ensemble averages in nonergodic samples, where ultraslow density fluctuations on the time scale of 10^2–10^4 s superimpose on frozen ones. In such systems the "brute force" approach of measuring many sample volumes becomes impossible and the Pusey–van Megen procedure results in a too fast decay at long times due to an insufficient number of samples of the slow fluctuations being collected. We demonstrated that the MSCS method leads to identical results as yet another speckle averaging technique – the interleaved sampling method – when applied to the same sample. Thereby we could verify that both techniques yield the true ensemble average and that the long-time decay seen in our glassy micronetwork colloid dispersion originates from particle dynamics. At present the MSCS method is restricted to a minimal lag time of 1 s and to the averaging over 50 speckles. These hardware-posed limits can be overcome. Implementation of a larger computer memory will allow to monitor several hundreds of speckles simultaneously. Alternatively, the use of a high-speed camera together with faster digitizing hardware will make shorter lag times accessible, thereby achieving a larger overlap with ensemble-averaging methods restricted to the short-time ($\tau \leq 1$ s) domain.

Acknowledgments Financial support by the Deutsche Forschungsgemeinschaft (SFB 262) is gratefully acknowledged.

References

1. Berne BJ, Pecora R (1976) Dynamic Light Scattering. Wiley, New York
2. Chu B (1991) Laser Light Scattering. Basic Principles and Practice. Academic Press, New York
3. Schmitz KS (1990) An Introduction to Dynamic Light Scattering by Macromolecules. Academic Press, New York
4. Pusey PN (1991) In: Hansen JP, Levesque D, Zinn-Justin J (eds) Liquids, Freezing and the Glass Transition. North-Holland, Amsterdam, p 765
5. Segrè PN, Meeker SP, Pusey PN, Poon WCK (1995) Phys Rev Lett 75:958
6. van Megen W, Underwood SM (1993) Phys Rev E 47:248
7. van Megen W, Underwood SM (1994) Phys Rev E 49:4206
8. van Megen W (1995) In: Yip S (ed) Relaxation Kinetics in Supercooled Liquids – Mode Coupling Theory and its Experimental Tests, Transp Theory Stat Phys 24:1017
9. Bartsch E, Frenz V, Sillescu H (1994) J Non-Crystal Solids 172–174:88
10. Bartsch E, Frenz V, Baschnagel J, Schärtl W, Sillescu H (1997) J Chem Phys 106:3743
11. Bartsch E (1995) In: Yip S (ed) Relaxation Kinetics in Supercooled Liquids – Mode Coupling Theory and its Experimental Tests, Transp Theory Stat Phys 24:1125
12. Pusey PN, van Megen W (1989) Physica A 157:705
13. Pusey PN, van Megen W (1987) Phys Rev Lett 59:2083
14. Xue J-Z, Pine DJ, Milner ST, Wu X-L, Chaikin PM (1992) Phys Rev A 46:6550–6563
15. Müller J, Palberg T (1996) Progr Colloid Polym Sci 100:121–126
16. Kirsch S, Frenz V, Schärtl W, Bartsch E, Sillescu H (1996) J Chem Phys 104:1758
17. Renth F, Bartsch E, Kasper A, Kirsch S, Stölken S, Sillescu H, Köhler W, Schäfer R (1996) Prog Colloid Polym Sci 100:127–131
18. Ricka J (1993) Appl Opt 32:2860
19. Joosten JGH, Gelade ETF, Pusey PN (1990) Phys Rev A 42:2161
20. van Megen W, private communication
21. Joosten JGH, McCarthy JL, Pusey PN (1991) Macromol 24:6690
22. Schätzel K, Drewel M, Ahrens J (1990) J Phys: Condens Matter 2:SA393
23. Segrè PN, van Megen W, Pusey PN, Schätzel K, Peters W (1995) J Mod Opt 42:1929
24. Schätzel K (1987) Appl Phys B 32:193
25. Note that the weighing factor $\langle I(Q) \rangle_{T,j}/\langle I(Q) \rangle_E$, even though applied, was erroneously omitted in Eq. (3b) of Ref. [16])

Progr Colloid Polym Sci (1997) 104:49–58
© Steinkopff Verlag 1997

J. Rička
I. Flammer
W. Leutz

Single-mode DLS: colloids in opaque porous media

Dr. J. Rička (✉) · I. Flammer · W. Leutz
Institute of Applied Physics
University of Bern
Sidlerstraße 5
3012 Bern, Switzerland

Abstract Single-mode fiber optical receivers have become the instrumentation standard for Dynamic Light Scattering (DLS). In a regular homodyne experiment one values their superb signal-to-noise ratio as well as the simplicity of the optical setup. Moreover, mode-selective DLS enables the researcher to tackle seemingly hopeless experimental problems, such as colloidal motions inside an opaque porous medium consisting of a water filled packing of small glass grains. The particles to be measured are completely masked by strong diffuse scattering in the porous matrix. Nevertheless, mode-selective DLS makes it possible not only to detect the motions of the colloids within the pores but also to determine their diffusion coefficient and, simultaneously, their average convective speed. We outline the theoretical background of these measurements and present data on diffusion and convection of latex particles in dense packings of glass-beads in a chromatographic column. Our technique allows an accurate determination of the tortuosity of the interstitial flow.

Key words Single-mode dynamic light scattering – opaque porous media – diffusion – convection – tortuosity

Introduction

The porous medium we have in mind is a dense but disordered packing of tiny glass beads, the voids between the beads are filled with a dilute aqueous suspension of colloidal particles (Figs. 1 and 2). The colloids move within the pore system of the packing (Fig. 2). They exhibit Brownian motion, but they are also carried in the flow field that percolates the pores when a pressure gradient is applied. In view of the abundance of similar porous systems in our natural and technological environment, there is a substantial interest in measuring the colloidal dynamics within the pores. In the context of colloid dynamics, one inevitably thinks of the powerful technique of Dynamic Light Scattering (DLS). However, in our case the conventional DLS technique is useless, because of the refractive index mismatch between glass and water: the porous medium under investigation is completely opaque. In principle one could employ index matching in order to render the sample transparent [1]. But this would mean to use a very special mixture of organic solvents instead of water which is, in view of the relevance to biological and environmental problems, the solvent of greatest interest.

We developed a new technique that does not depend on the refractive index matching, but rather on its mismatch. We take advantage of the multiple scattering regime of light propagation in the packing: the light propagation is completely disordered and, in the statistical sense, isotropic. This information on the light field turns out to be sufficient to obtain quantitative information on motion of colloidal particles by dynamic light scattering; an

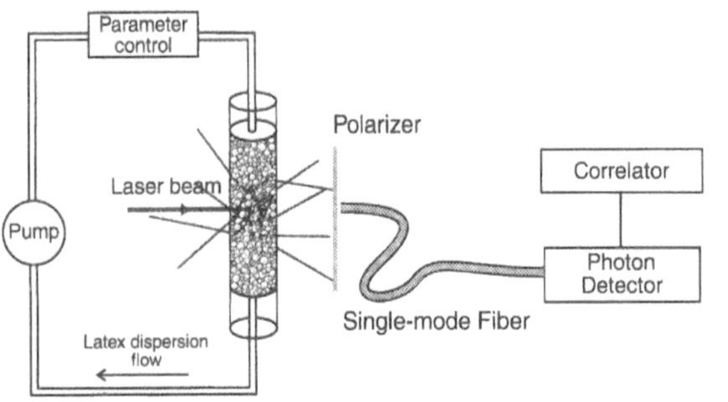

Fig. 1 Experimental setup for single-mode DLS measurements of colloidal motions in opaque glass-bead packings. See discussion in Sect. 4

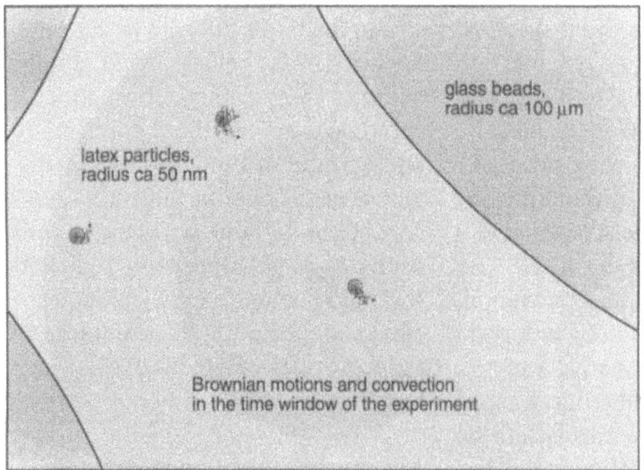

Fig. 2 Colloidal particles in a pore of the glass-bead packing. This illustration indicates the length scales and shows the expected traces of particle motions in the time window of the experiments

example closely related to our technique is Diffusive Wave Spectroscopy (DWS) [2, 3]. The key feature of our technique is the perfect selection of a single mode from the complicated light field generated by multiple scattering in the opaque glass bead packing. The basic concepts of the technique have already been presented in an earlier publication [4]. In the present contribution we put the theory on a sounder foundation and extend the analysis to include, in addition to the Brownian motion, also the convection of the colloids. Finally we present recent experimental results validating the theoretical analysis. Our theoretical approach employs the concepts of single-mode optics. The theory and practice of single-mode receivers has been extensively discussed in Refs. [5, 6]. Since, however, single-mode optics still appears to be a rather new topic, we find it advisable to begin by illustrating first the basic single-mode ideas.

Single-mode fiber receivers

A single-mode receiver consists of a single-mode fiber, a polarizer in the front of the fiber and a detector. Additional front end optics, e.g. a lens, is optional. The polarizer, which can be omitted in standard DLS, is important in our case, because in the disordered scattered field the polarization is random. Without a polarizer, we would receive 2 modes. However, the key element of the single-mode receiver is the single-mode fiber. A single-mode fiber, whose size of the light guiding core is comparable with the light wavelength λ, is actually an optical *wave-guide*. When λ exceeds a certain cutoff value, there is only one unique way to build up the light wave in the fiber, as illustrated in Fig. 3. In other words, the strict boundary condition imposed on the field by the narrow core allows the propagation of only *a single transverse mode* of the field. The impinging field may be anything, for example speckles generated by scattering in our porous medium, but the field $E_f(\rho)$ that is propagated and eventually radiated from the fiber always exhibits the same nearly Gaussian pattern in the fiber cross-section, i.e. the propagated field is patterned by the fiber mode $E_+(\rho)$. Only the complex amplitude $\mathscr{E}$ of E_f depends on the impinging light field E_s:

$$E_f(\rho) = \mathscr{E}(E_s)E_+(\rho) \, . \tag{1}$$

The propagated field exhibits a perfect transverse coherence, i.e., the fluctuations of the intensity $I(\rho, t)$ are perfectly correlated everywhere in the fibre cross-section at ρ. Consequently, the pre-detection optical signal $J(t)$, i.e., the optical power carried to the detector, is proportional to the complex square of the amplitude $\mathscr{E}(t) = \mathscr{E}[E_s(t)]$. With a proper normalization for the fiber mode E_+ one

Fig. 3 Illustration of the principle of mode selection with single-mode optical fibers. The selected mode is perfectly transverse coherent, therefore the integrated single-mode signal J relates to the amplitude $\mathscr{E}$ in the same way as a local intensity $I(r)$ relates to a local field $E(r)$

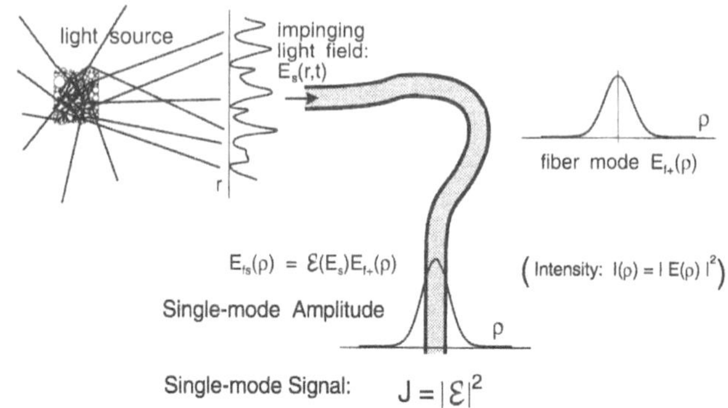

can simply write:

$$J(t) = \mathscr{E}(t)\mathscr{E}^*(t) \,. \tag{2}$$

Apparently, the signal amplitude $\mathscr{E}(t)$ relates to the pre-detection signal $J(t)$ in the same way as the conventional local field amplitude $E(\mathbf{r}, t)$ relates to the local intensity $I(\mathbf{r}, t) \propto E^*(\mathbf{r}, t)E(\mathbf{r}, t)$. However, with a single-mode receiver there is no need for an integration over finite detector area, the square of the amplitude gives the total signal directly. In the classical speckle approach to DLS this would corresponds to the "point detector limit", and this in turn offers a great simplification of theoretical analysis. Without this advantage, the present complicated situation would be prohibitively difficult. It should be noted that with the single-mode fiber the perfect transverse coherence of the selected light is achieved at finite power J, this is in contrast to the traditional pinhole technique. Consider the following example: a traditional two pinhole receiver working with a coherence factor of 0.9 receives $10 \times$ less power than a single-mode fiber receiver with the perfect coherence factor of 1.0! This is a great advantage, because due to the long light path in the strongly scattering glass packing we have to anticipate severe absorption losses. The combination of the simplicity of the theoretical analysis with the superb light gathering power makes the single-mode fiber receiver indispensable for the present experiments on porous media.

The subsequent theoretical analysis of single-mode DLS in porous media makes use of the concept of the *receiver mode*. This is an auxiliary electromagnetic field that describes completely (i.e., on the amplitude level) the receiving characteristics of the single-mode receiver. For an intuitive understanding of this concept we recommend a simple experiment: replace the detector at the output end of the fiber with an auxiliary laser. This turns the receiver into a transmitter, a radiation pattern becomes visible on an observation screen. Now observe the behavior of the pattern while bending the fiber or manipulating the coupling of the auxiliary laser. If the pattern remains unchanged (except perhaps for its intensity) during such manipulations, then the setup is single-mode. If, on the other hand, the pattern keeps changing, then the setup is multimode. The single-mode pattern does not always have to be the Gaussian radiating from a bare fiber. One can transform the single-mode beam with additional optical elements, for example collimate or focus it with a lens. But the front end optics can be something more complicated, for example a random packing of glass beads. In this case the light field is highly disordered, as suggested in Fig. 4. But once the decision on the front end optics of the transmitter is made, the light field in the single mode is fixed. In Fig. 4 the disordered single-mode field is represented in the photon-path model: a single-mode transmit-

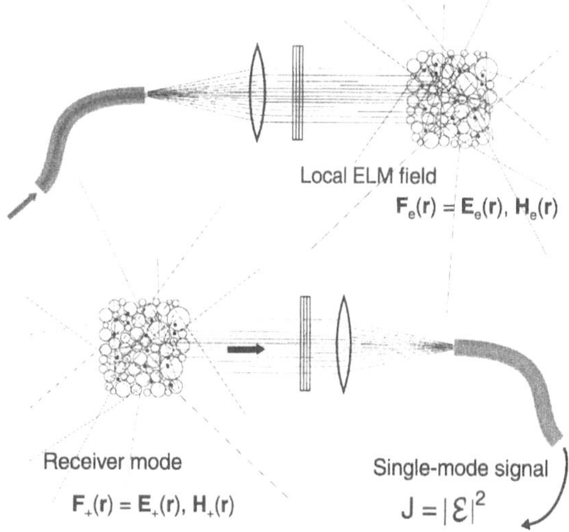

Fig. 4 Random single-mode excitation field and receiver mode. Illustration of the reciprocity principle employed in Eq. (3)

ter produces a unique set of paths (unique, but for a complex amplitude common to all paths). The photon paths model proved highly successful for the description of disordered light propagation and we will also resort to it at some occasions. However, we keep in mind that these paths are not just some particle traces. They only represent the interaction propagation in the electromagnetic vector field $\mathbf{F}(\mathbf{r}, t) \equiv \mathbf{E}(\mathbf{r}, t), \mathbf{H}(\mathbf{r}, t)$ that is rigorously given by the position and time dependence of the electric and magnetic field vectors $\mathbf{E}(\mathbf{r}, t)$ and $\mathbf{H}(\mathbf{r}, t)$.

It is important to realize that the reasoning is reciprocal, i.e., it works both ways: let us convert the transmitter back into a receiver (i.e., exchange the laser for a detector). Now one does not see any light, but one can be sure that the photons *entering* into the receiver follow the same paths as the photons flying out of the transmitter, only in reverse direction. The single-mode receiver selects a unique set of photon paths, as indicated in the lower part of Fig. 4. The key idea of the single-mode approach is to take this reciprocity very seriously and attribute to the receiver an auxiliary electromagnetic field $\mathbf{F}_+(\mathbf{r}, t) \equiv \mathbf{E}_+(\mathbf{r}, t), \mathbf{H}_+(\mathbf{r}, t)$. This is the so called *receiver mode*; the subscript + denotes the propagation *into* the receiver. The receiver mode is equivalent to a physical electromagnetic field, such as the excitation field $\mathbf{F}_e \equiv \mathbf{E}_e, \mathbf{H}_e$, except for its reversed propagation and a convenient normalization. In terms of the receiver mode one can express the signal amplitude $\mathscr{E}$ quite simply in terms of the local field quantities, without having to follow the propagation of the scattered light on its complicated paths through the packing. A quite general expression for the signal amplitude $\mathscr{E}$ in terms of the receiver mode reads [5]:

$$\mathscr{E}(t) = -\frac{1}{4} \int \mathbf{j}^*(\mathbf{r}, t) \cdot \mathbf{E}_+(\mathbf{r}, t)\, \mathrm{d}^3 r \,. \tag{3}$$

Here $\mathbf{E}_+(\mathbf{r})$ represents the electric field vector of the receiver mode. The partner of $\mathbf{E}_+(\mathbf{r})$ in the dot product $\mathbf{j}^*(\mathbf{r}, t) \cdot \mathbf{E}_+(\mathbf{r})$ is the current density distribution $\mathbf{j}$, which represents the observed light source. In our case the light source is a scattering dielectric system and $\mathbf{j}(\mathbf{r})$ is thus the dielectric displacement current excited in the medium by the electric field $\mathbf{E}_e$ of the incident laser light. Since we work with dielectrics, the magnetic field vectors $\mathbf{H}_e$ and $\mathbf{H}_+$ do not enter the calculations. We will return to Eq. (3) in Section 3.3.

Theory

The assumptions

The starting point of our analysis is a set of assumptions about the statistics of the light fields and the corresponding single-mode signal amplitudes. A first and key assumption concerns the illuminating field $\mathbf{F}_e(\mathbf{r})$ and the receiver mode $\mathbf{F}_+(\mathbf{r})$ in the *absence* of the colloid scatterers:

(M1) The multiplicity of scattering events is sufficiently high to regard the light propagation in the matrix as a diffusive process. In other words, the illuminating field $\mathbf{F}_e(\mathbf{r})$ and the receiver mode $\mathbf{F}_+(\mathbf{r})$ are completely randomized with respect to both the propagation direction and the polarization.

According to our recent experiments, reported in Ref. [7] and summarized in sub-section 4, this expectation is well fulfilled with the present porous system.

In view of the disordered light propagation we can anticipate the statistical properties of the signal selected from the scattered light with a single-mode receiver. We assume that the matrix is rigid, mobile colloidal particles are absent. Therefore the matrix signal $J_m = \mathscr{E}_m \mathscr{E}_m^*$ remains constant as long as the receiver is fixed in a certain configuration with respect to the packing. However, we also expect large signal fluctuations to appear when the receiver is moved with respect to the column. These expectations can be formally expressed as follows:

(M2) The matrix contribution $\mathscr{E}_m$ to the single-mode amplitude is constant in time, but with respect to the *ensemble* statistics, $\mathscr{E}_m$ is a complex circular Gaussian random variable. In particular $\langle \mathscr{E}_m \rangle_E = 0$. In Pusey's terminology [8, 9], we assume that $\mathscr{E}_m$ is perfectly nonergodic.

The assumption of Gaussian ensemble statistics of $\mathscr{E}_m$ is quite plausible, this in view of the disordered structure of our glass-bead packing and the disordered light propagation. For a simple test we recall that the Gaussian property of $\mathscr{E}$ implies an exponential distribution $P(J) = \exp(-J/\langle J \rangle_E)/\langle J \rangle_E$ of the ensemble values of the signal J. The first part of the assumption (M2) is more delicate. In practice, the observed signal J_m exhibits minor fluctuations. This could be due to sound waves or thermal fluctuations in the "rigid" matrix, whose effect on the scattered field is amplified by multiple scattering. (Such amplification is the principle of measurements of sub-Ångström displacements by diffusive wave spectroscopy [10–12].) However, we found that under most circumstances these fluctuations are negligible when compared with the fluctuation due to the mobile colloid particles. In addition, we profit from a time scale separation: the fluctuations of $J_m(t)$ are much slower than the fluctuations due to the colloids. Therefore we neglect the matrix fluctuations in the present paper.

When mobile colloidal particles are added to the system, the single-mode signal $J(t) = |\mathscr{E}(t)|^2$ is expected to fluctuate. The amplitude $\mathscr{E}(t)$ can be written as the sum of the constant matrix contribution $\mathscr{E}_m$ and of a fluctuating part $\mathscr{E}_p(t)$:

$$\mathscr{E}(t) = \mathscr{E}_m + \mathscr{E}_p(t) \,. \tag{4}$$

Such a decomposition is always possible, but the interpretation of the fluctuating part $\mathscr{E}_p(t)$ requires some caution and additional assumptions. The key assumption about light fields in the composed system in the *presence* of mobile colloids is the following:

(P1) The colloidal suspension within the pores is optically dilute. Light interaction with the colloidal particles therefore leads merely to single scattering, multiple scattering takes place only in the matrix. In other words, the local illuminating field $\mathbf{F}_e(\mathbf{r})$ and the receiver mode $\mathbf{F}_+(\mathbf{r})$ are only negligibly disturbed by the presence of the colloid particles (Born's first approximation).

A rigorous theoretical analysis of the validity range of this assumption would be difficult. A simple empirical criterion is $\langle J_p \rangle_E \ll \langle J_m \rangle_E$, where $\langle J_p \rangle_E = \langle J(t) - J_m \rangle_E$ is the ensemble averaged excess scattering due to the presence of colloid particles. In the strongly dilute regime it is relatively simple to justify the next assumption about the statistics of $\mathscr{E}_p(t)$:

(P2) The particle contribution $\mathscr{E}_p(t)$ is a complex circular Gaussian random variable with respect to both the time statistics and the ensemble statistics, $\mathscr{E}_p(t)$ is completely ergodic. In particular, we expect that $\langle \mathscr{E}_p(t)\mathscr{E}_p^*(t)\rangle_T = \langle \mathscr{E}_p\mathscr{E}_p^*\rangle_E$ and $\langle \mathscr{E}_p(t)\rangle_T = \langle \mathscr{E}_p\rangle_E = 0$, or, somewhat milder, $|\langle \mathscr{E}_p(t)\rangle_T|^2 \ll \langle \mathscr{E}_p\mathscr{E}_p^*\rangle_E$.

Strictly speaking, the relation $\langle \mathscr{E}_p(t)\rangle_T = 0$ is only valid for infinitely high dilution of the particle suspension: the time averaging of $\mathscr{E}_p(t)$ corresponds, in a first-order approximation, to dispersing the dielectric material of the particles uniformly in the suspension. Upon the time averaging one is left with a sponge-like rigid structure whose refractive index differs slightly, by $\Delta n \approx 3\Phi n_s[(n_p^2 - n_s^2)/(n_p^2 + 2n_s^2)]/2$, from the n_s of the solvent (Φ is the volume fraction of the latex particles, n_p their refractive index; for polystyrene latex in water $\Delta n \approx 0.25\Phi$). The amplitude of the scattering signal from this structure would correspond to $\langle \mathscr{E}_p(t)\rangle_T$. One way to check the ergodicity assumption would be to render the rigid matrix invisible by index matching with the solvent of refractive index n_s and inspect the ensemble of time averaged correlation functions from the latex suspension. Such index matching experiment is not possible in our case (the available glass beads are not perfectly homogeneous). However, instead one could simulate the time averaging $\langle \mathscr{E}_p(t)\rangle_T$ by changing the refractive index of the solvent with a suitable additive by the $\Delta n \approx 0.25\Phi$ and determine the ensemble average $\Delta J_m = \langle J_m(n_s) - J_m(n_s + \Delta n)\rangle_E$. A simple calculation shows that $\Delta J_m \approx \langle |\langle \mathscr{E}_p(t)\rangle_T|^2\rangle_E$, and thus a plausible criterion for the validity of the ergodicity assumption is $|\Delta J_m| \ll \langle J_p\rangle_E$.

Partial heterodyning

Assumptions (M2) and (P2) imply that DLS experiments with glass-bead packings can be analyzed in terms of partial heterodyning, the matrix contribution $\mathscr{E}_m$ playing the role of the local oscillator. The normalized autocorrelation function of the pre-detection signal $J(t)$ is expected to obey the usual generalized Siegert relation

$$g(\tau) = 1 + 2j_m(1 - j_m)\mathrm{Re}\,g_1(\tau) + (1 - j_m)^2|g_1(\tau)|^2 , \tag{5}$$

where $j_m = J_m/\langle J\rangle_T$ is the heterodyning parameter and $g_1(\tau) = \langle \mathscr{E}_p(0)\mathscr{E}_p^*(\tau)\rangle_T$ is the first-order correlation function of the particle contribution $\mathscr{E}_p(t)$. The strength of the local oscillator is unknown. The parameter j_m affects quite strongly the amplitude $g(0) - 1$ of the dynamic part of the correlation function $g(\tau)$. Therefore j_m can be reliably determined by standard fitting procedures, together with the parameters of $g_1(\tau)$. Thereby one can rely on the firm knowledge of the perfect single-mode selection with our fiber receiver. However, one should not be too trustful: we do not access directly the pre-detection signal $J(t)$ but the *photo-count* signal $J_c(t)$ delivered by a real detector with all

its nonidealities. Here we only consider the most important deviation from ideality, namely the dead-time of the photon-counter. A thorough analysis of the dead-time effect is one of the heritages we owe to Klaus Schätzel [13, 14]. We only extended his work to include the case of partial heterodyning. This study will be published elsewhere, here we only note the first-order results:

$$\langle J_c\rangle_T = \langle J\rangle_T[1 - \varepsilon(2 - j_m^2)] , \tag{6}$$

$$g_c(\tau) = 1 + 2j_m(1 - j_m)\mathrm{Re}\,g_1(\tau)[1 - 2\varepsilon(1 + [1 - j_m]^2)]$$
$$+ (1 - j_m)^2|g_1(\tau)|^2[1 - 2\varepsilon(2 + j_m^2)] . \tag{7}$$

Here $\varepsilon = \langle J_c\rangle_T\theta_D$ is the dead-time parameter, namely the average number of photo-counts per dead-time θ_D. The dead-time effect is larger than frequently assumed: for example, $\varepsilon = 0.02$ (dead time 100 ns, countrate 2×10^5 cps) results in $g_c(0) - 1 = 0.92$ for a homodyne experiment, whereas in a partial heterodyne experiment with an ideal detector the parameter j_m must be as large as 0.3 to have the same $g_c(0) - 1 = 0.92$.

Single-mode signal amplitude

The first task is to develop an expression for the particle contribution $\mathscr{E}_p(t)$ to the single-mode signal amplitude $\mathscr{E}(t)$. By virtue of the assumption (P1) (Born approximation) we can write $\mathscr{E}_p(t)$ as the sum of contributions from the individual scattering particles:

$$\mathscr{E}_p(t) = \sum_j \mathscr{E}_j(t) . \tag{8}$$

For the contributions $\mathscr{E}_j(t)$ we can employ Eq. (3), namely

$$\mathscr{E}_j(t) \propto \int \mathbf{j}_j^*(\mathbf{r}, t)\cdot \mathbf{E}_+(\mathbf{r}, t)\,\mathrm{d}^3r , \tag{9}$$

where $\mathbf{j}_j(\mathbf{r}, t)$ is the contribution of the particle j to the displacement current, and $\mathbf{E}_+(\mathbf{r})$ is the receiver mode. Because our colloidal scatterers are small, we extend assumption (P1) to include also the internal field in the scatterers. In other words, the latex spheres are Rayleigh–Debye scatterers, and we may write

$$\mathbf{j}_j(\mathbf{r}, t) \propto \Delta\chi_j[\mathbf{r} - \mathbf{r}_j(t)]\mathbf{E}_e(\mathbf{r}, t) . \tag{10}$$

Here $\Delta\chi_j[\mathbf{r} - \mathbf{r}_j(t)]$ denotes excess susceptibility at $\mathbf{r}$ due to a particle j that is centered momentarily at $\mathbf{r}_j(t)$ (excess susceptibility $\Delta\chi_j(\mathbf{r}, t)$ is nonzero only in the region in space which is momentarily occupied by the particle j) and $\mathbf{E}_e(\mathbf{r})$ is the exciting light field as in the absence of the colloids.

As the next step we make the assumption (M1) of disordered light propagation operational. We propose to model the light field in the packing as the superposition of plane waves with position-dependent superposition coefficients:

$$\mathbf{E}_\mathrm{e}(\mathbf{r}, t) = \sum_{\Omega_\mathrm{e}} A_{\Omega_\mathrm{e}}(\mathbf{r}) \mathbf{e}_\mathrm{e} \mathrm{e}^{-i\mathbf{k}_\mathrm{e} \cdot \mathbf{r}} \mathrm{e}^{i\omega t} ,$$

$$\mathbf{E}_+(\mathbf{r}, t) = \sum_{\Omega_\mathrm{s}} B_{\Omega_\mathrm{s}}(\mathbf{r}) \mathbf{e}_\mathrm{s} \mathrm{e}^{-i\mathbf{k}_\mathrm{s} \cdot \mathbf{r}} \mathrm{e}^{i\omega t} . \tag{11}$$

The subscripts Ω_e and Ω_s refer to the orientations of the two orthogonal tripods formed by the field vectors $\mathbf{k}_\mathrm{e}$, $\mathbf{e}_\mathrm{e}$, $\mathbf{h}_\mathrm{e}$ and $\mathbf{k}_\mathrm{s}$, $\mathbf{e}_\mathrm{s}$, $\mathbf{h}_\mathrm{s}$, where $\mathbf{k}$ is the wave-vector and $\mathbf{e}$ the unit electric polarization vector of the partial wave (the third arm of the tripod, namely the magnetic unit vector $\mathbf{h}$, does not enter the present calculations). The complex amplitudes $A_{\Omega_\mathrm{s}}(\mathbf{r})$ $B_{\Omega_\mathrm{e}}(\mathbf{r})$ of the partial waves vary randomly with the position in the sample so that the distributions of Ω_e and Ω_s are isotropic. Within the pores the amplitudes are smooth (discontinuities only occur at the glass-bead boundary), the characteristic length of their spatial fluctuations is in the order of the transport mean free path l^* of light propagation. Our l^* of 1.6 mm [7] is much shorter than the sample dimensions but much larger than the length scale l_p of approx. 50 nm that is probed in our experiments.[1] Due to the high degree of randomness we also expect that the fluctuations of the amplitudes $A_{\Omega_\mathrm{e}}(\mathbf{r})$ and $A'_{\Omega'_\mathrm{e}}(\mathbf{r})$ of two different partial waves Ω_e and Ω'_e are uncorrelated, for example we expect that $\langle A_{\Omega_\mathrm{e}}(\mathbf{r}) A^*_{\Omega'_\mathrm{e}}(\mathbf{r}) \rangle_\mathrm{E} = \delta_{\Omega_\mathrm{e} \Omega'_\mathrm{e}}$ and $\langle B_{\Omega_\mathrm{s}}(\mathbf{r}) A^*_{\Omega_\mathrm{e}}(\mathbf{r}) \rangle_\mathrm{E} = \delta_{\Omega_\mathrm{s} \Omega_\mathrm{e}}$. Upon combining Eqs. (10), (11), and (9), the contribution of the particle j to the signal amplitude becomes

$$\mathscr{E}_j(t) \propto \sum_{\Omega_\mathrm{s}} \sum_{\Omega_\mathrm{e}} A_{\Omega_\mathrm{e}}(\mathbf{r}_j) B_{\Omega_\mathrm{e}}(\mathbf{r}_j) \mathscr{F}_j(\mathbf{q}) \mathbf{e}_\mathrm{e} \cdot \mathbf{e}_\mathrm{s} \mathrm{e}^{-i\mathbf{q} \cdot \mathbf{r}_j(t)} . \tag{12}$$

The terms in this double sum each resemble the well-known approximation for the scattered field from the conventional theory of light scattering. In particular one recognizes the *scattering vector* $\mathbf{q} = \mathbf{k}_\mathrm{e} - \mathbf{k}_\mathrm{s}$. The *polarization factor* $\mathbf{e}_\mathrm{e} \cdot \mathbf{e}_\mathrm{s}$ accounts for the polarization of the incoming and the outgoing wave, whereas the *form amplitude* $\mathscr{F}(\mathbf{q})$ is related to the familiar *form factor* $P(q)$ through $P(q) = \langle \mathscr{F}(\mathbf{q}) \mathscr{F}^*(\mathbf{q}) \rangle$.

First-order correlation function

We are now ready to tackle the first-order correlation function

$$g_1(\tau) \propto \langle \mathscr{E}_\mathrm{p}(t) \mathscr{E}_\mathrm{p}^*(t + \tau) \rangle = \sum_i \sum_j \langle \mathscr{E}_i(t) \mathscr{E}_j^*(t + \tau) \rangle . \tag{13}$$

For simplicity we assume that two distinct particles i, j do not interact (we have high dilution) and therefore it is sufficient to consider only the self-terms $i = j$. Thus

$$g_1(\tau) \propto \langle \mathscr{E}_j(t) \mathscr{E}_j^*(t + \tau) \rangle , \tag{14}$$

where $\mathscr{E}_j(t)$ is given by Eq. (12). This correlation function is still a nasty four-fold sum of the contributions of two pairs of partial waves Ω_e, Ω_s and Ω'_e, Ω'_s. In order to make our life easier, we simply assume that contributions from two different pairs of partial waves are uncorrelated (not only for different $\mathbf{q}$). We only consider the contributions with $\Omega_\mathrm{e} = \Omega'_\mathrm{e}$ and $\Omega_\mathrm{s} = \Omega'_\mathrm{s}$. This appears plausible in view of the highly random nature of the fields. (The averaging in Eq. (14) involves integration over the sample, which in turn approximates the ensemble average over matrix configurations.) A rigorous justification will require much work, but we rely on the analogy with the proven photon paths model of disordered light propagation: our assumption is analogous to assuming that contributions from two different paths are uncorrelated. With this assumption, a few subsequent steps of the calculation will yield

$$g_1(\tau) \propto \langle (\mathbf{e}_\mathrm{e} \cdot \mathbf{e}_\mathrm{s})^2 P(q) F(\mathbf{q}, \tau) \rangle_{\Omega_\mathrm{e} \Omega_\mathrm{s}} . \tag{15}$$

The brackets $\langle \cdots \rangle_{\Omega_\mathrm{e} \Omega_\mathrm{s}}$ represent the averaging over the orientations of the orthogonal tripods representing the partial waves. The quantities in the averaging brackets are: $(\mathbf{e}_\mathrm{e} \cdot \mathbf{e}_\mathrm{s})^2$ the polarization factor, $P(q)$ the form factor and $F(\mathbf{q}, \tau)$ the self-intermediate scattering function, i.e., the ensemble or time average

$$F(\mathbf{q}, \tau) = \langle \mathrm{e}^{i\mathbf{q} \cdot \rho(\tau)} \rangle = \int \mathrm{e}^{i\mathbf{q} \cdot \rho} p(\rho, \tau) \, \mathrm{d}^3\rho , \tag{16}$$

where $\rho(\tau) = \mathbf{r}(t + \tau) - \mathbf{r}(t)$ is the displacement of the colloid particle during the time lag τ, and $p(\rho, \tau) \, \mathrm{d}^3\rho$ is the displacement probability. In the derivation of Eqs. (15) and (16) we took advantage of a length scale separation: the length scale l^* of the spatial fluctuations of $A(\mathbf{r})$ and $B(\mathbf{r})$ is much longer than the length scale where $F(\mathbf{q}, \tau) \neq 0$. Therefore one can set, e.g. $A(\mathbf{r}) = A(\mathbf{r} + \rho)$.

Self-intermediate scattering function

The motion of the colloid particles results from the interplay between the Brownian motion characterized by a position-dependent diffusion tensor $\mathbf{D}(\mathbf{r})$ and the convection in a random flow-field $\mathbf{v}(\mathbf{r})$. A rigorous expression for the particle displacement probability $p(\rho, \tau) \, \mathrm{d}^3\rho$ is not available, but we can employ the fact that the probing length scale of 50 nm is much shorter than the characteristic pore size of roughly 100 μm (see Fig. 2). By virtue of this length scale separation, diffusion and convection can

---

[1] As l_p we take the inverse RMS of all involved partial scattering vectors $\langle \mathbf{q}^2 \rangle^{-1/2} = 50$ nm

be de-coupled, in other words we can neglect the influence of the shear. This yields

$$F(\mathbf{q}, \tau) = \langle e^{i\mathbf{q}\cdot\rho_{B}(\tau)} \rangle \langle e^{i\mathbf{q}\cdot\mathbf{v}\tau} \rangle \,, \tag{17}$$

where $\rho_{B}(\tau)$ represents the Brownian displacement. Since the packing is isotropic (in statistical sense) we expect the Brownian factor in Eq. (17) to depend only on $q = |\mathbf{q}|$. Therefore we write

$$F_{B}(q) = \langle e^{i\mathbf{q}\cdot\rho_{B}(\tau)} \rangle = e^{-q^{2}D_{\text{eff}}\tau} \,, \tag{18}$$

where we approximately take into account the position dependence of the diffusion tensor by introducing a pre-averaged effective diffusion coefficient D_{eff}. This pre-averaging of $\mathbf{D}(\mathbf{r})$ is a short cut justified mainly by the fact that our particles are much smaller than the size of the pores. A more careful analysis would be carried out along similar lines as the subsequent case of the velocity $\mathbf{v}$. In contrast to the Brownian motion, the convection is not isotropic, we expect a net directed flow through the packing. However, the "optical" averaging, Eq. (15), involves an isotropic distribution of the orientation of the scattering vectors $\mathbf{q}$. Since the vector $\mathbf{q}$ only occurs in $\langle \exp(i\mathbf{q}\cdot\mathbf{v}\tau) \rangle_{v}$, we are free to pre-average the later expression over $\mathbf{q}/q$ and thus make the replacement

$$\langle \exp(i\mathbf{q}\cdot\mathbf{v}\tau) \rangle_{v} \rightarrow F_{v}(q, \tau) = \int \frac{\sin(qv\tau)}{qv\tau} p(v)\, \mathrm{d}v \,. \tag{19}$$

where $p(v)$ is the distribution of the convective speed $v = |\mathbf{v}|$ in the pores of the packing. The distribution $p(v)$ in Eq. (19) is unknown and therefore we content ourselves with a short time approximation:

$$F_{v}(q, \tau) \approx e^{-q^{2}\langle v^{2}\rangle\tau^{2}/6} \,. \tag{20}$$

This approximation is analogous with expanding the self-dynamic structure factor of polydisperse Brownian particles in terms of cumulants. Upon inserting Eqs. (18) and (20) into Eq. (15) and carrying out all obvious integrations (Euler angles of Ω_{e} and Ω_{s} are a convenient set of coordinates) we arrive at

$$g_{1}(\tau) \propto \int_{0}^{\pi} [1 + \cos^{2}\theta] P[q(\theta)]$$

$$\times F_{B}[q(\theta), \tau] F_{v}[q(\theta), \tau]\sin\theta\, \mathrm{d}\theta \,. \tag{21}$$

Here θ is the angle spanned by the vectors $\mathbf{k}_{e}$ and $\mathbf{k}_{s}$, whereby $q^{2} = 2k^{2}[1 - \cos\theta]$.

One more approximation, in addition to Eq. (20), is needed in order to obtain a simple explicit result: we replace the form-factor $P(q)$ of the spherical latex particles with its Guinier approximation, i.e., we set $P(q) \approx \exp[-(qR_{g})^{2}/3]$, where R_{g} is the radius of gyration of the

scatterers. This is quite harmless, because the oscillations in $P(q)$ are anyway smeared in the subsequent averaging over large range of q.

Our final expression for the first-order correlation function $g_{1}(\tau)$ reads

$$g_{1}(\tau) = f(R_{g}) \left[\frac{u^{2} + u + 1}{2u^{3}} (1 - e^{-2u}) - \frac{1}{u^{2}} \right], \tag{22}$$

where $f(R_{g})$ is a trivial normalization factor enforcing $g_{1}(0) = 1$ and

$$u = 2k^{2}(R_{g}^{2}/3 + D_{\text{eff}}\tau + \langle v^{2}\rangle\tau^{2}/6) \,. \tag{23}$$

As far as pure diffusion is concerned Eq. (22) is expected to be valid in the whole relevant range of time lags, whereas when convection is present, the validity is restricted to an initial range of τ that becomes shorter as v increases.

Experimental validation of the technique

Glass-bead packings

The investigated packings consist of thoroughly washed soda-lime glass-beads (Sovitec) with diameters ranging between 160 and 200 μm. The beads are packed into a standard chromatography column (Omni-Fit) with a diameter 24.5 mm and length approximately 100 mm. The porosity of the packing was determined by weight comparison of an empty and a water filled packing: $0.36 \pm .01$. The characteristic pore size amounts to approx. 100 μm. The packing was also examined by measuring the breakthrough curve of a neutral tracer ($NaNO_{3}$). The dispersion coefficient (see, e.g. [15, 16]) increases linearly with the seepage velocity $\bar{v}_{z}$. The slope of 0.42 indicates a good homogeneity of the packing [15, 16].

Our investigations of the optical properties of the water filled packings have been reported in our recent publication [7]. We measured the transport mean free path l^{*}, i.e., the distance over which the direction of propagation is randomized and the diffusive absorption length L_{a}, i.e., the RMS displacement of a photon before it becomes absorbed. At $\lambda = 514$ nm we obtain $l^{*} = 1.6$ mm and $L_{a} = 7$ mm. The ratio $L_{a}/l^{*} \approx 4$ is sufficiently large to regard the light propagation in the packing as statistically isotropic. This is further supported by the observed complete depolarization of the light exiting the sample.

Latex suspension

The liquid phase is a suspension of sulphonated polystyrene latex spheres of 45.5 nm radius as measured by conventional DLS. The measurements were carried out at volume fractions between $\Phi = 6 \times 10^{-5}$ and 3×10^{-6}. Only the lower value is sufficiently low to neglect multiple scattering on the mobile scatterers. The ionic strength of the suspension during the measurement was equivalent to 2×10^{-4} mol/L salt concentration and the pH amounted to 7.0. Under these conditions the screening length of the Coulomb repulsion is sufficiently short to neglect interaction among the latex particles while the repulsion is still strong enough to prevent aggregation and adsorption of the particles to the glass beads. As the pores are much larger than the effective particle size, the interactions between the glass surface and the particles can be neglected. The effects of such interactions are still poorly understood and therefore their absence is crucial in order to validate our technique.

The apparatus

The apparatus is shown schematically in Fig. 1. The suspension is circulated through the column by a pulse-free double syringe pump (skScientific) and continuously checked for aggregation, pH-value and electrical conductivity. To ensure temperature stability the column is surrounded by a cooling jacket and thermostated to $23.0 \pm 0.1°C$. The speed of flow ranged from 0 and 10 ml/min, this corresponds to a seepage velocity $\bar{v}_z$ (average velocity component in flow direction) of 0–1000 μm/s.

The optics is a part of our standard DLS apparatus, whose fiber optics allows an easy adaptation to the present needs. The illuminating beam is delivered by a polarization maintaining single-mode fiber (OZ-Optics, not shown in the picture). The laser (Lexel 3500), operating at $\lambda = 514$ nm, is equipped with an etalon in order to provide the necessary coherence length. The receiving single-mode fiber (OZ-Optics) guides two polarization modes and therefore a polarizer must be placed at its front end in order to have a perfect single-mode selection. The photocount signal, detected with an EMI-9863 photomultiplier, is processed by an ALV-5000 correlator (ALV, Langen, Ge).

Results and conclusion

The most important result, from the methodological point of view, is shown in Fig. 5. This is the dynamic part of the signal correlation function $g(\tau)$ as obtained for purely diffusive motion in the absence of flow. The solid line is the least square fit by Eq. (7) with the first-order correlation function from Eq. (22). The fit parameters are D_{eff} (the effective diffusion coefficient) and j_m (matrix contribution, strength of the local oscillator); the remaining parameters

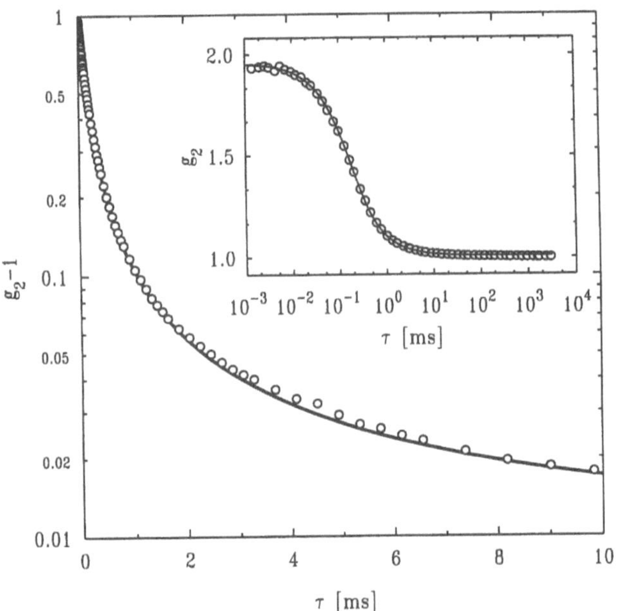

Fig. 5 Semi-log plot of signal correlation function $g(\tau)$ of monodisperse latex particles diffusing in the void space of the porous medium. The highly nonexponential shape is due to the broad range of scattering vectors q. Best fit with our theory is shown as line. Insert: logarithmic time scale. Fit parameters: $j_m = 0.224 \pm 0.002$, $D_{eff} = 6.1 \pm 0.1$, baseline due to matrix dynamics: 0.007

$R_g = 35.3$ nm (radius of gyration of the latex particles) and $\varepsilon = 3.2 \times 10^{-4}$ (dead-time parameter) are known from independent measurements. We obtain a very good agreement between the data and the fit over a dynamic range of almost two decades. (We have to admit that this high data quality is only obtained at latex fractions Φ of approx. 2×10^{-5} where double scattering already becomes apparent.) The highly nonexponential shape of the decorrelation is described by the single parameter D_{eff}. Note also that the amplitude of $g(\tau)$ is *not* normalized. We reach almost the theoretical maximum $g(0) = 2$, despite the strong scattering in the matrix. This is a merit of the mode-selective property of the fiber receiver: we are able to choose a position where the constant contribution $\mathscr{E}_m$ is small, giving j_m of only 0.22. We call this *single-mode matching*, because at this particular position the matrix appears as if index matched with the solvent. However, though $J_m = 0$ is the most probable value of the matrix scattering signal, the search for such minima would be too tedious. In practice we only seek positions where the matrix contribution is sufficiently low to avoid overloading of the detector, and rely on Eq. (7) for the analysis of the partially heterodyne data.

When the flow through the column is turned on, the limited range of the validity of Eq. (22), anticipated in

Sect. 3.5, becomes apparent. This is shown in Fig. 6. Nevertheless, from the initial slope and the initial curvature of the correlation function we can simultaneously determine the diffusion coefficient D_{eff} and the RMS-velocity $\bar{v}$, both as functions of the seepage velocity $\bar{v}_z$ (this is the average velocity in the direction of pressure gradient).

Figures 7 and 8 show the effective diffusion coefficient D_{eff} obtained for different flow velocities, $\bar{v}_z$ at two different latex volume fractions Φ. In Fig. 7 the latex volume fraction is as high as $\Phi = 2 \times 10^{-5}$. We see that D_{eff} does not depend on $\bar{v}_z$, however, D_{eff} is somewhat larger than the bulk diffusion coefficient D_0 and this indicates the presence of double scattering. At $\Phi = 2 \times 10^{-5}$ we measure $\langle J_{\text{p}} \rangle = 0.2 \ \langle J_{\text{m}} \rangle_{\text{E}}$, i.e., 20% of the particle signal originates from multiple scattering events. On the other hand, in the measurements shown in Fig. 8 the volume fraction Φ is as low as 3×10^{-6}, the scattering from the particles amounts to only 3% of the ensemble averaged matrix scattering. Despite this extremely low signal, we find that $D_{\text{eff}} \approx D_0$, as expected under the given experimental

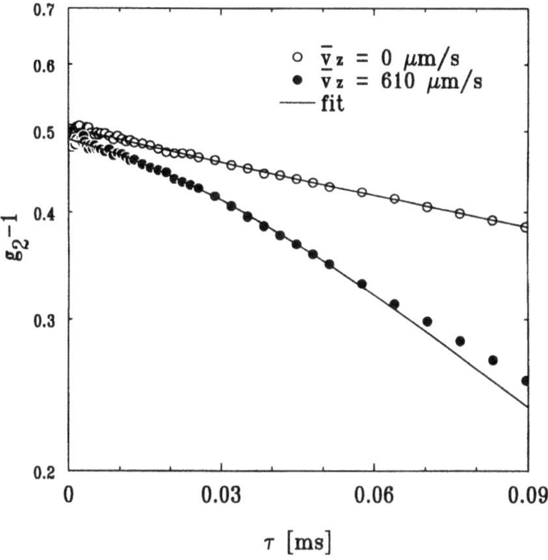

Fig. 6 Comparison of the measured signal correlation function with and without flow. The lines show the fit-function. Parameters without flow: $j_{\text{m}} = 0.71$, $D_{\text{eff}} = 5.9 \pm 0.2 \ \mu\text{m}^2/\text{s}$. Parameters with flow: $j_{\text{m}} = 0.75$, $D_{\text{eff}} = 6.1 \pm 0.4 \ \mu\text{m}^2/\text{s}$, $v = 610 \pm 60 \ \mu\text{m/s}$

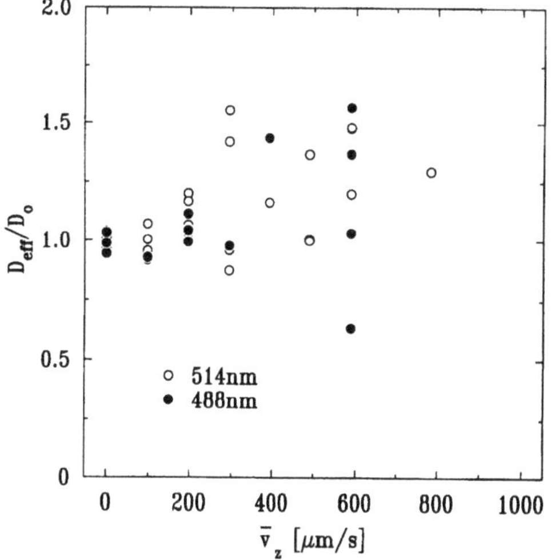

Fig. 8 The effective diffusion coefficient D_{eff} vs. the seepage velocity $\bar{v}_z$. Latex volume fraction $\Phi = 3 \times 10^{-6}$, corresponding to $\langle J_{\text{p}} \rangle = 0.03 \ \langle J_{\text{m}} \rangle_{\text{E}}$. Double scattering is absent, i.e., $D_{\text{eff}} = D_0$

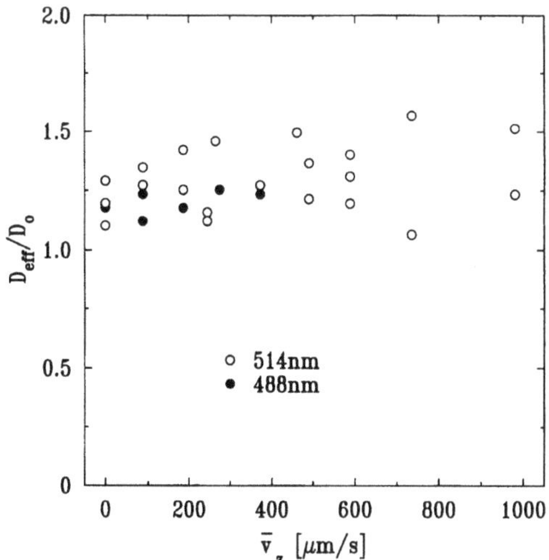

Fig. 7 The effective diffusion coefficient D_{eff} does not depend on the seepage velocity $\bar{v}_z$. Results with two excitation wavelength λ of 488 and 514 nm are shown. Latex volume fraction $\Phi = 2 \times 10^{-5}$ corresponding to $\langle J_{\text{p}} \rangle = 0.2 \ \langle J_{\text{m}} \rangle_{\text{E}}$. Double scattering causes $D_{\text{eff}} > D_0$!

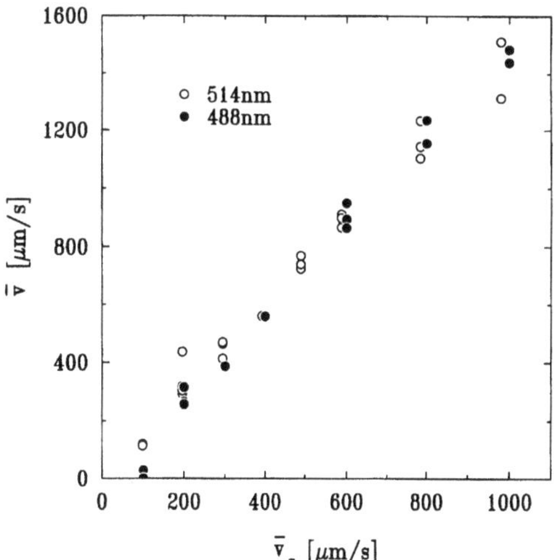

Fig. 9 The root mean square velocity $\bar{v} = \sqrt{\langle v^2 \rangle}$ increases linearly with the seepage velocity $\bar{v}_z$ (parameter D_{eff} fixed at $D_{\text{eff}} = D_0$). Latex volume fraction $\Phi = 3 \times 10^{-6}$

condition. At high flow velocities the error bars are quite large. This is because as *fitting parameters* $\bar{v}$ and D_{eff} are not entirely independent. Therefore, in a second pass of the analysis we fix $D_{\text{eff}} = D_0$ and fit only the velocity parameter $\langle v^2 \rangle$. The result is shown in Fig. 9: the RMS-velocity $\bar{v} = \sqrt{\langle v^2 \rangle}$ increases linearly with the seepage velocity. The slope θ of the line $\bar{v} = \theta \bar{v}_z$ curve is a measure of the tortuosity of the flow through the packing, a quantity hardly attainable with any other investigation method [16]. Our value of $\theta = 1.6$ is in a surprisingly good agreement with a simple classical model of flow through granular media [17]: the liquid moves in a system of equivalent cylindrical tubes. These tubes are curved, but sustain a roughly constant angle of 45° with the average flow direction z. (Another hint to the applicability of this simple tube model would be the consistency of our data with the speed distribution in Poiseuille flow. We are currently testing this conjecture.) The universality of this behavior is still questioned; our technique provides, for the first time, an adequate tool for a systematic test.

We conclude: the technique works, it is ready to provide insights in the hidden life of colloidal dispersions in opaque porous media.

Acknowledgment This work has been supported by the Swiss national foundation.

References

1. Bishop MT, Langley KH, Karasz FE (1986) Phys Rev Lett 57:1741–1744
2. Pine DJ, Weitz DA, Maret G, Wolf PE, Herbolzheimer E, Chaikin PM (1989) Dynamical Correlations of Multiply Scattered Light. World Scientific, London
3. Pine DJ, Weitz DA, Zhu JX, Herbolzheimer E (1990) J Phys France 51:2101–2127
4. Ricka J (1994) Die Makromolekulare Chemie, Macromol Symp 79:45–55
5. Ricka J (1993), Appl Opt 32:2860–2875
6. Gisler Th, Rüger H, Egelhaaf SU, Tschumi J, Schurtenberger P, Ricka J (1995) Appl Opt 34:3546–3553
7. Leutz W, Ricka J (1996) Opt Comm 126:260–268
8. Pusey PN (1989) Physica A 157:705–741
9. Pusey PN (1994) Macromol Symp 79:17–30
10. Weitz DA, Pine DJ, Pusey PN, Tough RJA (1989) Phys Rev Lett 63:1747–1750
11. Mason TG, Weitz DA (1995) Phys Rev Lett 74:1250–1253
12. Leutz W, Maret G (1995) Physica B 204:14–19
13. Schätzel K (1986) Appl Phys B 41:95–102
14. Schaetzel K, Kalström K, Stampa B, Ahrens J (1989) J Opt Soc Amer B 6:937–947
15. Charlaix E, Hulin JP, Plona TJ (1987) Phys Fluids 30:1690–1698
16. Bear J (1988) Dynamics of Fluids in Porous Media. Dover, New York
17. Carman PC (1937) Trans Inst Chem Eng London 15:150–156

Progr Colloid Polym Sci (1997) 104:59–65
© Steinkopff Verlag 1997

A. van Blaaderen

Quantitative real-space analysis of colloidal structures and dynamics with confocal scanning light microscopy

Dr. A. van Blaaderen (✉)
Van't Hoff Laboratory
Debye Research Institute
Utrecht University
Postbus 80051
3508 TB Utrecht, The Netherlands

A. van Blaaderen
FOM Institute for Atomic
and Molecular Physics
Postbus 41883
1009 DB Amsterdam, The Netherlands

Abstract Quantitative real-space coordinates of 3D structures can be obtained with confocal microscopy on concentrated dispersions of index-matched silica spheres with a fluorescent particle core. The possibilities and limitations of this technique to probe 3D colloidal particle structures (e.g., glasses, crystals, gels and electro-rheological fluids) and dynamics of colloidal processes (e.g., diffusion, glass formation and phase separation) are discussed based on examples from the literature and new preliminary work presented in this paper. Extension of the method to the study of multi-component dispersions (e.g., binary crystals and glasses) is addressed as well.

Key words Confocal microscopy – silica – colloids – real-space

Introduction

Almost a century ago Perrin studied individual colloidal particles in real space to determine Avogadro's number [1]. After this pioneering work structural and dynamic information on colloidal particle systems was mainly obtained with scattering techniques [2]. Recently, quantitative real-space techniques have become more feasible through technical advances in both data acquisition and data processing [3–6]. Usually, the approach is to image individual particles and to extract information by image analysis. However, it is possible to obtain quantitative information without visualizing colloids separately [7, 8]. Individual particles are mostly imaged in two-dimensional (2D) systems with digital video microscopy. Examples include work on phase transitions [9–11], diffusion [12, 13] and the measurement of pair potentials [14–16]. The first few layers of particles close to the container wall of three-dimensional (3D) crystals [17, 18] and crystallization [19] have also also been studied. The cause for this limitation to image only close to the glass wall is twofold. First, the imaging of unmatched relatively large colloids is hindered by significant scattering and loss of intensity deeper inside even not too concentrated dispersions. Second, with conventional microscopy, out-of-focus light contributes to the image and makes it hard to distinguish colloids deep in the bulk of concentrated dispersions even if the index is matched. There are several solutions to these problems. For instance, by only imaging tracer core-shell particles in an otherwise matched concentrated dispersion it is possible to measure diffusion of the tracer particles in a 2D plane inside the bulk [20]. Also, interactions between two particles can be measured away from the glass walls in a dilute dispersion after placing them in the focal plane with optical tweezers [15]. The solution reviewed in this paper is confocal scanning laser microscopy (CSLM) on matched dispersions of fluorescent core-shell silica spheres and the aim of the paper is to discuss the strong and weak points of this real-space technique through a number of examples taken from published, earlier work and preliminary, new results.

In confocal fluorescence microscopy the field of view is limited to a diffraction limited spot. This spot excites a fluorescent dye and the fluorescent light is imaged by the same (confocal) high numerical aperture lens onto a point detector. Light that lies outside the focal plane is imaged in front of or behind the detector pinhole and is strongly rejected. The result is the ability to take optical sections out of a 3D sample with a thickness of about 500 nm. In addition, the multiplication of the response of both a point source and point detector also leads to a higher resolution, about 250 nm in the plane of the optical section, compared to conventional light microscopy [4]. Confocal microscopy is already a well established technique for imaging biological structures and processes and also its use in characterization of integrated circuits is well documented [4]. 3D real-space imaging of colloidal particle structures in the bulk is relatively new [5]. In the first-papers the interpretation was difficult because individual particles could not be recognized [21] and/or because the colloids were not optically matched [23]. Particles with a fluorescent core and a refractive index close to that of the solvent made it possible to distinguish individual spheres even in concentrated dispersions [23] and to obtain semi-quantitative results on colloidal crystal growth rates [24] and the stacking order of close packed planes of colloidal crystals in the bulk of a dispersion [25]. The stacking order of hard-sphere crystals has also been obtained recently using phase contrast microscopy on index-matched dispersions [26]. As explained in Ref. [3] optical sectioning is possible with differential interference contrast as well. However, until now, quantitative particle positions in the bulk of 3D structures have only been obtained with confocal microscopy after a measurement of the convolution of the point spread function of the confocal microscope with the fluorescent sphere and subsequent digital image analysis [27].

In the following, the possibilities and limitations of the real-space technique to obtain quantitative data on *stationary structures*, *dynamics* and *multi-component dispersions* will be discussed after a brief experimental section.

Experimental

Particle synthesis and characterization are described in the literature [24, 28]. The preliminary experiments described here were performed with fluoresceine labeled spheres with a final radius of 525 nm (fluorescent core 200 nm, relative standard deviation in size, $\sigma = 2\%$) and two kinds of rhodamine labeled spheres with the same fluorescent core (radius 100 nm) and a final radius of 210 nm ($\sigma = 6\%$) and 460 nm ($\sigma = 2\%$).

More extensive experimental details of the preliminary experiments presented in this paper will be given elsewhere. The gel was made from an initial volume fraction of

fluoresceine labeled spheres in a matching mixture of glycerol and water containing 4 wt % tetramethylammonium hydroxide. The gellation was initiated by addition of methylformate which slowly hydrolized to methanol and formic acid causing the particles to aggregate. Because of the relatively slow initial change in pH there is time to fill a capillary before aggregation commences. The electrorheological fluid consisted of the fluoresceine labeled spheres dispersed in a matching mixture of glycerol and water. The applied alternating field strength was 1 V/μm with a frequency of 3 MHz which is much faster than the relaxation time in the double layer. The electrodes were formed by indium-tin-oxide (ITO) coated glass cover slides. The charged glass-and-crystalline sample was quenched from an initial volume fraction of 40% of the fluoresceine labeled spheres in dimethylformamide (DMF) with 0.001 M LiCl. Also the two binary mixtures were made in DMF (0.01 M LiCl) with the mixture of the spheres with almost the same size consisting of a molar ratio of 100:1, fluoresceine labeled spheres to the rhodamine spheres (total radius 460 nm), volume fraction 55%. The mixture with the larger size ratio of 0.4 consisted of 50% volume fraction of fluoresceine labeled spheres and 20% volume fraction of the smaller rhodamine labeled spheres (total radius 210 nm).

The fluorescence confocal micrographs were made on two confocal set-ups: a Multiprobe 2001 of Molecular dynamics using a 100×1.4 numerical aperture lens on an inverted Nikon Diaphot microscope and a Leica TCS 4D confocal system with a 100×1.4 numerical aperture lens on a Leitz DM IRB inverted microscope.

The experimental method to find the particle positions will be described in more detail elsewhere [29], but, briefly, consists first of a measurement of the microscope response function of a single sphere. This is done by taking a 2D arrangement of spheres in immersion oil on a coverslip and measuring the intensity of each sphere in successive optical sections that are 100 nm apart. In each image plane the sphere positions and intensities are found in a similar way as has been described by Crocker and Grier [6]. By averaging over several hundred spheres the intensity distribution as a function of the z-coordinate (that is the coordinate along the optical axis of the microscope and perpendicular to the optical section plane or, xy-plane) was obtained accurately. Subsequently, the process is repeated on a series of optical sections, typically 100 sections 100 nm apart. In each plane particles are identified and the intensity determined. Out of all the xy-coordinates and intensities the approximate z-coordinate of the centers is determined by looking for maxima. Subsequently, only those intensities with approximately the same xy-coordinate in the different planes are taken together as belonging to one individual sphere. The final xy-coordinate is determined by an intensity weighed average over the xy-positions from the different sections. To obtain accurate z-positions fits are performed using the measured functions for single spheres. If other spheres are too close above or below the sphere so that the intensities are overlapping this is taken into account.

Results and discussion

Stationary structures

3D coordinates of colloidal particles can clearly be obtained with the highest accuracy if the particles do not change their position during the time it takes to obtain a data set. Faster scan rates and consequently smaller dwell times per 3D volume element or voxel can be realized if the fluorescence intensity is increased by increasing the exciting light intensity, the amount of dye or the quantum yield of the dye used. There are several reasons however, why it is not desirable to increase the intensity of the scanning laser beam above certain levels. All organic dyes bleach and lose the ability to fluoresce after absorbing a certain amount of photons, dissipation of some of the absorbed light can cause heating and too many photons can saturate the fluorescene by overpopulating the excited state. Similarly, the concentration of the fluorophore is limited. Too high concentrations cause a decrease in fluorescent intensity by the creation of non radiative decay pathways and can also make the absorption so high it becomes difficult to probe the sample at any significant depth. Because of all these reasons, there is a lower limit to the dwell time per voxel. Roughly, this translates to about 0.25 s for a frame of 512×512 voxels and for the particles and dyes used in this study, fluoresceine and rhodamine, which have an optimized amount of dye per particle and further have a quantum yield that is close to the highest values reported for organic dyes [30]. As the determination of the z-coordinates requires several frames at different z-positions, the structure needs to be stationary for at least several seconds to obtain the coordinates without a blurring caused by motions.

There are several interesting colloidal particle structures which (almost) meet this requirement not to change on a time scale of seconds. The 3D real-space structure of glasses formed from particles interacting with a hard-sphere like potential is described in Ref. [27]. From a volume fraction of 60% until 64% there was no detectable particle movement and 10.000 sphere positions could be obtained in one measurement with high accuracy (about 15 nm in the plane of the optical section and 25 nm perpendicular to this plane). These measurements made it possible for the first time to test theories about the local structure of a hard-sphere glass quenched from a colloidal liquid in thermodynamic equilibrium on experimental real-space data.

One other example is formed by colloidal crystals made again from particles interacting with a hard-sphere like potential. These crystals were grown through a directed growth on a template in a gravitational field, a new method called colloidal epitaxy [31]. Here the almost immobility of the particles away from their lattice positions was a consequence of the almost close packed volume fraction (around 74%) which was caused by the pressure of the layers of colloid compressing the crystal in the gravitational field. Further, not only the particle positions could be obtained accurately, but also the positions of the spheres in the crystal in relation to the holes in the polymer template could be determined by confocal microscopy. This was achieved by also incorporating into the polymer template a fluorescent dye, so that it could be imaged simultaneously with the sphere centers. It is also shown in this work that by mismatching the lattice constant of the template the resulting defect structures can be studied in detail in real-space. With scattering methods the study of such local irregularities as defects in crystals is very hard.

Gels and flocs can have sufficiently immobile particles as well. In Fig. 1a the 3D structure of a gel that resulted from reaction limited aggregation is shown. The scale of the figure is such that it is just possible to see the individual particles that make up the gel. If viewed with green (right eye) and red (right eye) glasses the depth than can be seen in this figure is striking. The voxel size shown in Fig. 1a is not small enough to obtain particle positions as shown in Fig. 1b, but the 3D image on this large scale gives an insight into the gel structure that is very hard to extract from scattering data, or from an extracted parameter like a fractal dimension. For instance, it can be seen in a direct way that the volume filling gel structure is build up out of more compact particle clumps. These more compact aggregates are a result of the rearrangements early on in the aggregation process as not every collision resulted in aggregation of the spheres. Subsequently, these relatively dense sub units cluster-cluster aggregated, while also sedimenting. By necessity the number of particles of which the coordinates was determined in Fig. 1b is much smaller and the average structure as given by the radial distribution function (not shown here) is not very accurate. Although it is experimentally possible to 'patch' together several connecting data sets as shown in Fig. 1b, this is a very lengthy and difficult process. Therefore, it is difficult to probe a 3D real-space structure over several decades to test for instance for the presence of a fractal dimension. Further, these kind of structures contain inhomogeneities over different length scales making it also harder to obtain a good average of the structure. However, even without

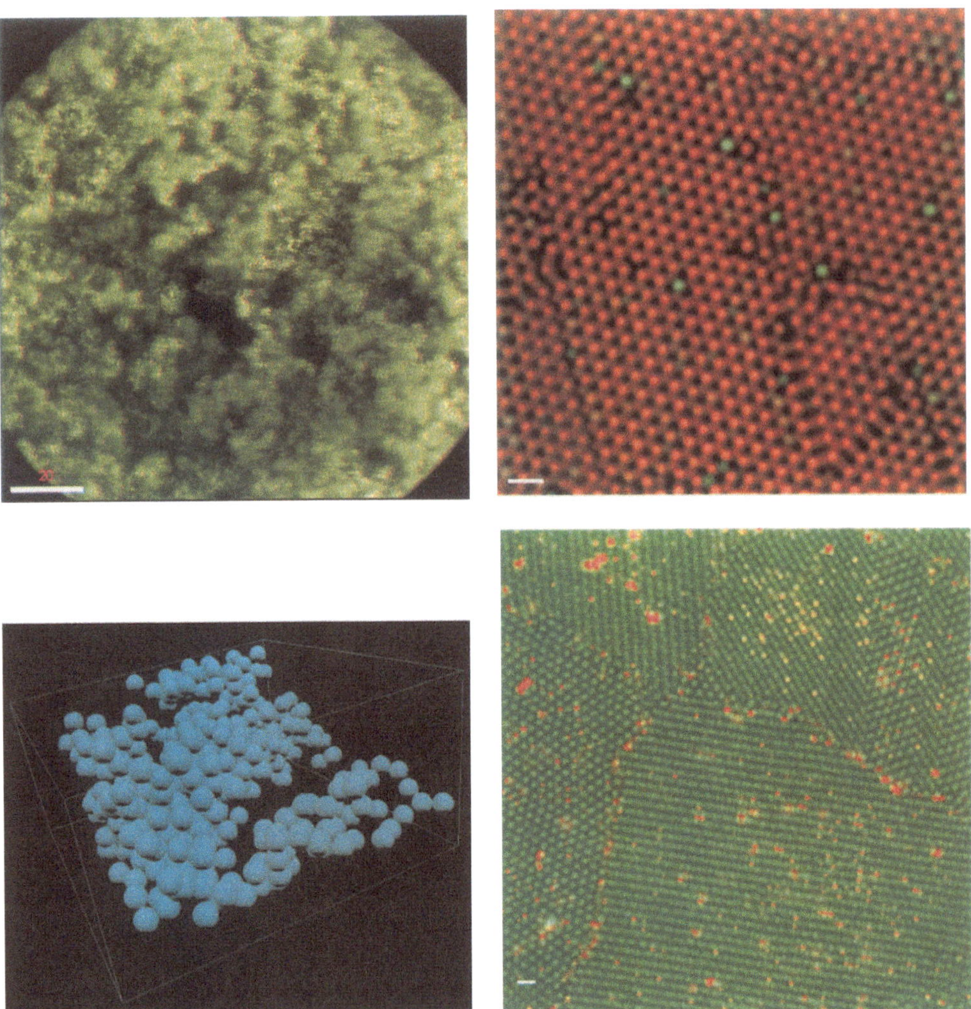

Fig. 1 Colloidal gel formed through reaction-limited aggregation of fluorescent-core (radius 200 nm) silica-shell particles (total radius 525 nm); volume fraction 20% in index-matched mixture of glycerol/water. Bar is 20 μm. (a) 3D confocal micrograph of the gel. (b) Computer generated picture of part of the gel after the particle coordinates were obtained from a 3D experimental data set (33 μm × 33 μm × 9 μm)

Fig. 4 Confocal micrographs of binary mixtures of fluoresceine (total radius 525 nm) and rhodamine labeled spheres dispersed in DMF with 0.01 M LiCl. (a) Rhodamine labeled spheres with total radius of 460 nm, molar ratio of rhodamine to fluoresceine labeled spheres: 100 : 1, total volume fraction 55%. (b) Rhodamine labeled spheres with total radius of 210 nm, volume fraction of rhodamine labeled spheres 20% of fluoresceine labeled spheres 50%. Bars are 2 μm

good statistical averages the real-space measurements can provide for insights in the structure that can for instance be used to interpret and model scattering data.

A final example is presented in Fig. 2 where a confocal micrograph taken halfway between two ITO electrodes conveys the structure of an electro-rheological fluid. The relatively high alternating electric field (1 V/μm, 3 MHz) is perpendicular to the optical section and the "strings" of particles that are visible in this section are in reality just 2D slices of sheets of particles where the normal of the sheets is perpendicular to the electric field. The sheets itself are built up from true, straight strings of spheres that are

parallel to the field lines. The curvature of the sheets, which is not changing in time, is what appears as the worm-like "strings" in the section of Fig. 2. The field strength, 1 V/μm, at which these sheets were formed was so high that the dipolar interaction energy between two spheres was many times kT. Immobilization, and therefore a relative easy structure determination is made possible in this example because of this strong restraining, external field. Therefore, these sheets are very likely non-equilibrium structures, because it is generally accepted that a tetragonal crystalline arrangement of strings of spheres packed close together is the equilibrium phase at high fields. It is

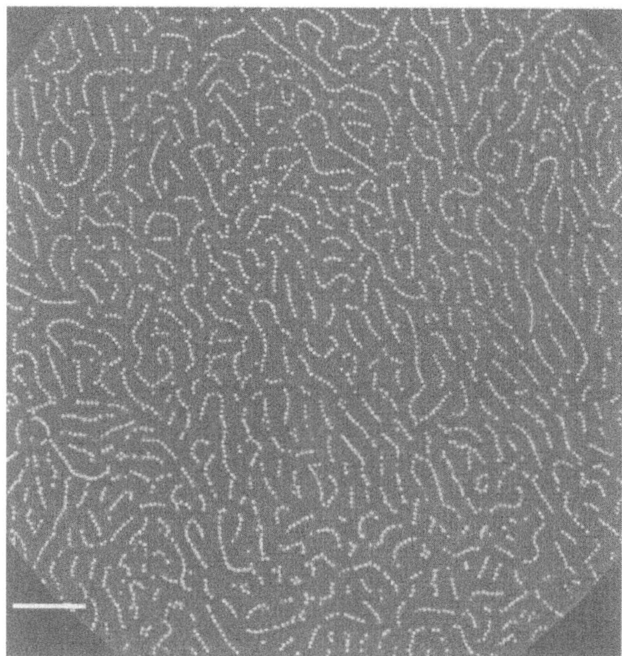

Fig. 2 Confocal micrograph of an electro-rheological fluid consisting of fluoresceine labeled spheres (total radius 512 nm) in a matching mixture of glycerol and water in between two ITO electrodes (10 μm apart). Field strength of 1 V/μm (frequency 3 MHz) is perpendicular to the optical section which was taken halfway in between the electrodes. Bar is 20 μm

interesting to notice that exactly the same kind of sheets were observed in magneto-rheological fluids only if the magnetic field was increased to high values quickly so that no equilibrium phase could be reached [32]. Applications, like a variable transmission, will operate under high fields, because this will give the fastest and largest increase in viscosity. This one brief example demonstrates that a quantitative real-space analysis of the resulting (out of equilibrium) structures can be very revealing and helpful to find the most optimal conditions.

In the above-mentioned examples the structures were more or less stationary during a measurement. There is of course no reason to limit the structural measurements to these relatively rare occasions. For instance in the case of crystals made out of charged particles with extended double layers or hard-sphere crystals at volume fractions significantly below 74% the particles will diffuse around on the average lattice position and this will diminish the accuracy of the particle positions. The average structure or type of lattice can still be determined accurately.

Dynamics

It has been shown that by imaging a single particle in a small field of view coupled with fast image analysis and a feedback mechanism the colloidal particle can be traced in 3D and its diffusion measured [33]. For a whole collection of particles in a concentrated dispersion this is much harder and has not yet been done. However, if the dynamics of the process, like gas–liquid like phase separation, or if the particle mobilities are isotropic, then in principle following x–y displacements in a thin optical section could give all necessary information. For tracer or (semi) 2D systems this has already been shown to give diffusion coefficients with an accuracy comparable to that of e.g., dynamic light scattering [20].

In Fig. 3a a section can be seen that was taken inside the bulk of a colloidal dispersion of charged particles (double layer thickness of 15 nm) that was quenched by centrifugation after crystallization had started. During twenty time steps of 5 s the particle positions in this 2D slice were followed as far as they did not go too far above or below the section. It can be seen that the crystalline regions did not grow, but also that in the glassy regions in between the crystals there still was substantial particle movement (Fig. 3b). From a movie made of the 20 sections it became obvious that these particle movements are not random but have a clear localized and "collective" component both in time and direction, unfortunately this cannot be seen in the figure. In other words, particles move together in more or less the same direction in a kind of bursts. Localized motions in "dynamic pockets" has also been observed by us in completely glassy, quenched dispersions of charged particles. If these first results are no artifacts and can be reproduced and quantified, this would be very important for an understanding of mobility in very condensed liquids or glasses. For instance, in mode coupling theory these very localized regions of particle mobility are not considered. The resemblance of Fig. 3 to a figure in Ref. [34] where also the dynamics in a 2D liquid has been described as kinetic structural inhomogeneous is striking.

Multi-component dispersions

With multiple labeling the real-space measurements can be extended to multi-component systems if multi-line excitation and multi-channel detection are possible on the confocal set-up. In Fig. 4a a binary crystal of fluoresceine labeled spheres (green, total radius 525 nm) and 13% smaller rhodamine labeled spheres (red, total radius 460 nm) can be seen. The minority fluoresceine labeled spheres form a solid-solution (although in this case "solid-gas" might be a better term) on the crystal lattice of smaller spheres. The crystal consists of a random stacking of hexagonal close-packed layers as is usually found for

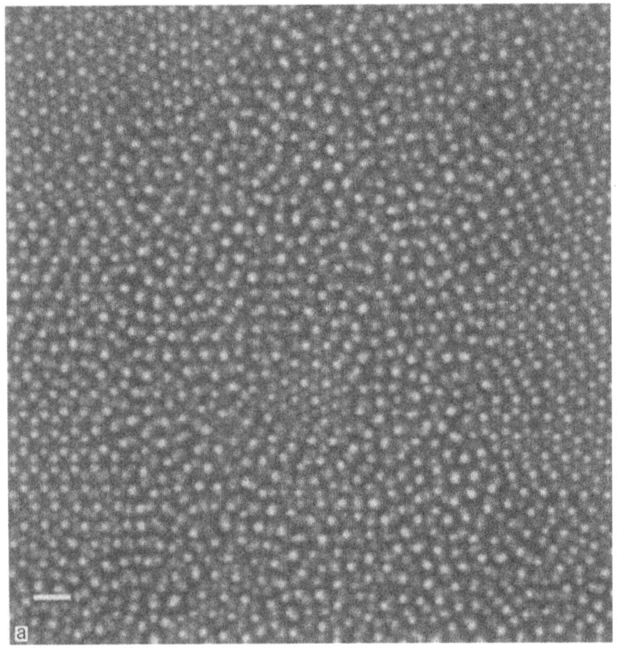

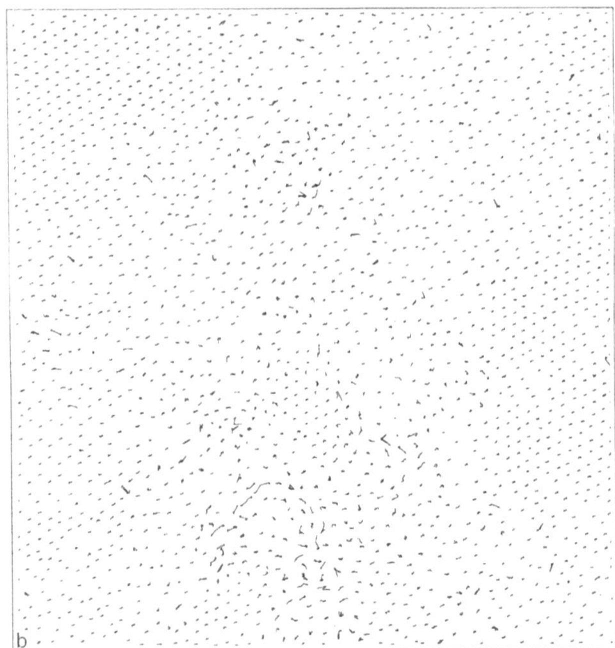

Fig. 3 Dynamics in the bulk of a crystalline/glassy charged colloidal system (DMF with 0.001 M LiCl). (a) Optical section at time = 0 s. (b) Particle trajectories of (most of) the particles present in (a) 20 time steps, 5 s apart. Bar 2 μm

hard-sphere like particles [31]. It will be very interesting to analyze the glasses that can be quenched from this binary system, because the size difference is such that, close to an optimal mixing ratio, icosahedral fragments can be formed

more easily than in a dispersion with just one size of spheres.

In the other example of a binary dispersion, Fig. 4b, one of the two components, the rhodamine labeled spheres (red, total radius 200 nm), cannot be resolved as separate spheres while the crystalline fluoresceine labeled spheres (green, total radius 525 nm) can be distinguished easily. There seems to be a preference of the small spheres for the grain boundaries and defects in the crystals of the larger spheres. Here and there the small spheres can be seen on lattice positions, but a definite AB or AB_2 crystalline structure has not (yet) been formed. In a fully developed binary crystalline structure, e.g., AB_2, the average positions of the smaller spheres could still be determined. And even if this is not possible, like in the grain boundaries, the fluorescent intensity of the small spheres can be used to determine its local volume fraction. Similar measurements would be possible in a binary mixture consisting of (large) labeled spheres and labeled polymers.

Conclusions

The importance of real-space analysis of colloidal structures and dynamics is increasing, on the one hand because of the development of new and more quantitative methods, but also because the structures and processes under study are becoming more complex. Through examples it was shown that with confocal microscopy on matched dispersions of fluorescent-core silica-shell particles structures can be measured accurately inside bulk if the structure is stationary on a time scale of seconds. With a frame refresh rate of about 0.25 s (512^2 voxels) it is also possible to follow the dynamics of micron size particles in the bulk. If the motion of many particles needs to be followed, the analysis is at the moment limited to a (semi) 2D section of a section situated in the bulk. The examples shown in this paper have made it clear that although it is in some cases hard to get statistically good averages, the real-space information is of great use for a detailed description of complex structures and processes. The information can subsequently be used to model and interpret scattering data that generally give better averaged information. Finally, the extension of the method to multi-component systems that are labeled with fluorophores is straightforward and can even provide valuable information if one of the components cannot be resolved on the particle level.

Acknowledgments Pierre Wiltzius and Cherry Murray (Bell Laboratories, Lucent Technologies, where most of the measurements were done) are thanked for discussions and encouragement.

Progr Colloid Polym Sci (1997) 104:59–65
© Steinkopff Verlag 1997

References

1. Perrin J (1910) Brownian Motion and Molecular Reality. Taylor and Francis, London
2. Pusey PN (1990) In: Levesque D, Hansen J-P, Zinn-Justin J (eds) Liquids, Freezing and the Glass Transition. Elsevier, Amsterdam
3. Inoué S (1986) Video Microscropy. Plenum, New York
4. Wilson T (ed) (1995) Confocal Microscopy. Academic Press, London
5. van Blaaderen A (1993) Adv Mater 5(1):52–55
6. Crocker JC, Grier DG (1996) J Colloid Interface Sci 179:298–310
7. Monovoukas Y, Gast AP (1991) Langmuir 7:460–468
8. Aastuen DJ, Clark NA, Cotter LK, Ackerson BJ (1986) Phys Rev Lett 57 (14):1733–1736
9. Murray CA (1992) In: Strandburg KJ (ed) Bond-Orientational Order in Condensed Matter Systems, Ch. 4. Springer, New York
10. Helgesen G, Skjeltorp AT (1991) Physica A 170:488
11. Marcus AH, Rice SA (1996) Phys Rev Lett 77 (12):2577–2580
12. Schaertl W, Sillescu H (1993) J Colloid Interface Sci 155:313
13. Marcus AH, Lin B, Rice SA (1996) Phys Rev 53:1765–1776
14. Kepler GM, Fraden S (1994) Phys Rev Lett 73:356–360
15. Crocker JC, Grier DG (1996) Phys Rev Lett 73:352–355
16. Vondermassen K, Bongers J, Mueller A, Versmold H (1994) Langmuir 10: 1351
17. Hachisu S, Yoshimura S (1980) Nature 283:188–189
18. Grier DG, Murray CA (1994) J Chem Phys 100(12):9088–9095
19. Bongers J, Versmold H (1996) J Chem Phys 104(4):1519–1523
20. Kasper A, Kirsch S, Renth F, Bartsch E, Sillescu H (1996) Progr Colloid Polym Sci 100:151–155
21. Bremer LGB (1992) Thesis, University of Wageningen, Wageningen
22. Yoshida H, Ise N, Hashimoto T (1995) J Chem Phys 103:10146–10151
23. van Blaaderen A, Imhof A, Hage W, Vrij A (1992) Langmuir 8:1514–1517
24. Verhaegh NAM, van Blaaderen A (1994) Langmuir 10:1427–1438
25. Verhaegh NAM, van Duijneveldt JS, van Blaaderen A, Lekkerkerker HNW (1995) J Chem Phys 102:1416
26. Elliot MS, Bristol BTF, Poon WCK (1997) Physica A 235:216–223
27. van Blaaderen A, Wiltzius P (1995) Science 270:1177–1179
28. van Blaaderen A, Vrij A (1993) Langmuir 8:29221–2931
29. van Blaaderen A, Lubachevsky B, Wiltzius P, in preparation
30. Imhof A (1995) Thesis, Utrecht University, Utrecht
31. van Blaaderen A, Ruel R, Wiltzius P (1997) Nature 385:321–324
32. Lawrence EM, Ivey GA, Flores A, Liu J, Bibette J, Richard J (1994) Int Mod Phys B 8:2765–2777
33. Schätzel K, Neumann W-G, Müller J, Merzok B (1992) Appl Opt 31:770–778
34. Hurley MM, Harrowell P (1995) Phys Rev E 52:1694–1698

Progr Colloid Polym Sci (1997) 104:66–75
© Steinkopff Verlag 1997

J.K.G. Dhont

Spinodal decomposition of colloids in the intermediate stage

Dr. J.K.G. Dhont (✉)
Van't Hoff Laboratory
University of Utrecht
Padualaan 8
3584 CH Utrecht, The Netherlands

Abstract Spinodal decomposition in the initial and intermediate stage is described on the basis of the Smoluchowski equation, "the Liouville equation for Brownian systems". For the intermediate stage, where a dominant length scale exists, a general scaling relation for the static structure factor is derived: the scattered intensity divided by the cubed dominant length and the second moment of the intensity should be time independent. The corresponding dynamic scaling function is found from the non-linear equations of motion for the structure factor as derived from the Smoluchowski equation. It turns out that this dynamic scaling function is universal in the sense that it is independent of the kind of colloid and the quench parameters. The scaling function in the intermediate stage is much sharper than the well known Furukawa scaling function which applies to the late stage. The time dependence of the wavevector where the scattered intensity is maximum is found to follow power law behaviour, with an exponent in (0.2, 1.1), depending on the relative importance of hydrodynamic interaction, which is set by the quench parameters.

Key words Spinodal decomposition – scaling – light scattering

Introduction

The aim of this paper is to present a microscopic description of the initial and intermediate stages of spinodal decomposition of colloidal systems. An equation of motion for the density is obtained from the Smoluchowski equation with an appropriate closure relation for the pair-correlation function, as discussed in the next section. The closure relation may be regarded as the statistical mechanical analogue of thermodynamic local equilibrium. Initial decomposition kinetics is then discussed where the well known Cahn–Hilliard equation is rederived from microscopic considerations. In the initial stage, a dominant length scale develops, which leads to dynamic scaling of the static structure factor in the subsequent, so-called intermediate stage, where non-linear coupling is important. General considerations on dynamic scaling are also given. Next, the non-linear equation of motion for the structure factor is derived from the Smoluchowski equation from which the dynamic scaling function is found. Furthermore, power law behaviour of the time dependence of the wavevector where the structure factor exhibits a maximum is predicted, with an exponent that depends on the quench parameters. Comparison with experiments is made in the last section.

Before entering into quantitative considerations, let us discuss the several stages during spinodal decomposition.

After immersion of the cuvette into a thermostating bath with a temperature pertaining to an unstable state of the suspension, first of all, the solvent and the core material of the colloidal particles attain the thermostat temperature. The correlations between the colloidal particles, in an ideal quench, do not change yet, and will be the same as in the stable larger temperature state prior to quenching. The system will therefore find itself out of equilibrium, and correlations will change with time in order to adjust to the low temperature of the solvent. Correlations will adjust in the homogeneous suspension up to a point where these render the system unstable. From that time on, inhomogeneities develop, ultimately leading to phase separation. The time where phase separation sets in will be referred to in the sequel as time $t = 0$.

The kind of density variations that become unstable after adjustment of correlations are sinusoidal density variations (also referred to as density waves) with an arbitrary small amplitude. These density waves increase their amplitude without any further time delay. As the creation of gradients in the density increases the free energy of the system, only long wavelength density waves are the unstable ones, since these give rise to smaller gradients. Thus, in the initial stage, right after a quench into the unstable part of the phase diagram, long wavelength density waves grow in amplitude. Large gradients are formed only in later stages of the phase separation process, where sharp interfaces separate large regions with differing density. Furukawa [1] distinguished between four different stages during phase separation before coexistence is achieved: the initial stage, the intermediate stage, transition stage and the final stage. In order to describe these different stages, it is convenient to write the spatially and temporally varying macroscopic density $\rho(\mathbf{r}, t)$ as

$$\rho(\mathbf{r}, t) = \bar{\rho} + \delta\rho(\mathbf{r}, t) , \tag{1}$$

where $\bar{\rho} = N/V$ is the number density of colloidal particles of the homogeneous system, and $\delta\rho(\mathbf{r}, t)$ is the change in density due to the ongoing phase separation. The four stages can now be specified as follows:

Initial stage:
$\delta\rho/\bar{\rho}$ is small,
gradients are small ("diffuse interfaces"),

Intermediate stage:
$\delta\rho/\bar{\rho}$ is not small,
gradients are small ("diffuse interfaces"),

Transition stage:
$\delta\rho/\bar{\rho}$ is large,
gradients are not small ("sharp interfaces"),

Final stage:
$\delta\rho/\bar{\rho}$ is large,
gradients are large ("very sharp interfaces").

Here, "small" means that equations of motion for the density may be expanded to leading order, "not small" means that the first higher order term should also be taken into account, and "large" means that all terms should be considered, that is, the full equation of motion must be analysed. Equations of motion for the density in the initial and intermediate stage can therefore be expanded to leading order with respect to gradients of the density, while the leading non-linear contribution in $\delta\rho/\bar{\rho}$ must be included in the intermediate stage. The first higher order terms in an expansion with respect to gradients of the density, which must be included in the transition stage, are referred to here as describing of *sharp interfaces*, while even higher order terms describe the dynamics of *very sharp interfaces* in the final stage. These very sharp interfaces have a width of the order of a few particle diameters, except in case of quenches very close to the critical point, where the equilibrium interfacial thickness may be large. It is a formidable task to describe decomposition kinetics in the transition and final stage from microscopic considerations. Probably the only feasible way to go about here is to hint at the mechanism that drives the phase separation, and to device a phenomenological theory that describes the essentials that are believed to be important [2–5]. The present paper is aimed at the description of the initial and intermediate stages.

The above-described subdivision of the phase separation process into different stages is an idealization, and the distinction may not be that rigorous in reality. Crossover from one stage to the other may be very smooth, and, for example, in the intermediate stage sharp interfaces may already start to form to some extent.

Equation of motion for the macroscopic density

The equation of motion for the macroscopic density is derived from the Smoluchowski equation, which is the equation of motion for the probability density function (pdf) $P \equiv P(\mathbf{r}_1, \mathbf{r}_2, \ldots, \mathbf{r}_N, t)$ of the position coordinates $\mathbf{r}_j$, $j = 1, 2, \ldots, N$, of all N colloidal particles in the system [6, 7]. This equation reads, with the neglect of hydrodynamic interaction between the colloidal particles,

$$\frac{\partial}{\partial t} P = D_0 \sum_{j=1}^{N} \nabla_{r_j} \cdot [\beta[\nabla_{r_j}\Phi]P + \nabla_{r_j}P] , \tag{2}$$

where D_0 is the Stokes–Einstein diffusion coefficient, $\beta = 1/k_B T$ (with k_B the Boltzmann constant and T the temperature), and $\Phi \equiv \Phi(\mathbf{r}_1, \mathbf{r}_2, \ldots, \mathbf{r}_N)$ the potential energy of the assembly of colloidal particles. The role of hydrodynamic interaction is discussed in the subsection titled "Contribution of hydrodynamic interaction". Since the macroscopic density is related to the pdf P as,

$$\frac{1}{N}\,\rho(\mathbf{r}_1,t) = \int d\mathbf{r}_2 \cdots \int d\mathbf{r}_N\, P(\mathbf{r}_1,\mathbf{r}_2,\ldots,\mathbf{r}_N,t)\,, \tag{3}$$

the equation of motion for the macroscopic density can be obtained from the Smoluchowski equation by integration with respect to all the position coordinates, except for $\mathbf{r}_1$. In order to integrate the Smoluchowski equation, a pair-wise additive interaction potential is assumed, that is (with $r_{ij} = |\mathbf{r}_i - \mathbf{r}_j|$),

$$\Phi(\mathbf{r}_1,\mathbf{r}_2,\ldots,\mathbf{r}_N) = \sum_{i,\,j=1,\,i<j}^{N} V(r_{ij})\,, \tag{4}$$

with V the pair-interaction potential. Further introducing the (time dependent non-equilibrium) pair-correlation function g as

$$\int d\mathbf{r}_3 \cdots \int d\mathbf{r}_4\, P(\mathbf{r}_1,\mathbf{r}_2,\mathbf{r}_3,\ldots,\mathbf{r}_N,t)$$
$$= \frac{1}{N^2}\,\rho(\mathbf{r}_1,t)\rho(\mathbf{r}_2,t)g(\mathbf{r}_1,\mathbf{r}_2,t)\,, \tag{5}$$

the integrated Smoluchowski equation reads, for identical Brownian particles (with $\mathbf{r}_1$ renamed as $\mathbf{r}$ and $\mathbf{r}_2$ as $\mathbf{r}'$),

$$\frac{\partial}{\partial t}\rho(\mathbf{r},t) = D_0\left[\nabla^2\rho(\mathbf{r},t) + \beta\nabla\cdot\rho(\mathbf{r},t)\int d\mathbf{r}'\,\nabla V(|\mathbf{r}-\mathbf{r}'|)\right]$$
$$\times \rho(\mathbf{r}',t)g(\mathbf{r},\mathbf{r}',t)\Bigg]\,, \tag{6}$$

where ∇ is the gradient operator with respect to $\mathbf{r}$. There are two terms to be distinguished on the right hand-side: the first term between the square brackets describes the effect of Brownian motion, while the second term describes the effect of direct interactions. The combination,

$$\mathbf{F}^{\text{int}}(\mathbf{r},t) \equiv -\int d\mathbf{r}'\,[\nabla V(|\mathbf{r}-\mathbf{r}'|)]\rho(\mathbf{r}',t)g(\mathbf{r},\mathbf{r}',t)\,, \tag{7}$$

is the direct force on a colloidal particle at $\mathbf{r}$ due to particles in a volume element with position $\mathbf{r}'$, averaged with respect to the position of the latter. The Brownian term always tends to homogenize the system, whereas the direct force term can lead to a growth of inhomogeneities. In case the pair-potential has an attractive component, the direct interaction force $\mathbf{F}^{\text{int}}$ can, at low enough temperatures, "pull" colloidal particles toward regions of higher density, so that "uphill diffusion" occurs, which is the mechanism that renders the system unstable.

To obtain a closed equation for the density, the pair-correlation function g must be expressed in terms of the density itself. Such a closure relation may be obtained as follows. The important feature here is that the pair-correlation function in the integral in the Smoluchowski equation (6) is multiplied by the pair-force $\nabla V(|\mathbf{r}-\mathbf{r}'|)$, so that a closure relation is only needed for small distances

$|\mathbf{r} - \mathbf{r}'| \leq R_V$, with R_V the range of the pair-interaction potential. Correlations over such small distances establish much faster than the demixing rates of the very long unstable wavelengths, simply because it takes more time for a colloidal particle to diffuse over larger distances. The dynamics of short wavelength density waves is much faster than the dynamics of long wavelength density waves. Therefore, on a coarsened time scale that is much larger than relaxation times of density waves with wavevectors $k \geq 2\pi/R_V$, but which still resolves the phase separation process, the pair-correlation function in the integral in the Smoluchowski equation may be replaced by the equilibrium pair-correlation function. The separation of time scales between the short and long wavelength dynamics is even enhanced by the fact that the effective diffusion coefficient is much smaller for large wavelengths than for small wavelengths (this will be established in the following section, and is usually referred to as "critical slowing down"). Furthermore, for the diffuse interfaces at hand in the initial and intermediate stage, the equilibrium pair-correlation function may be taken equal to that of a homogeneous system with a density equal to that inbetween the points $\mathbf{r}$ and $\mathbf{r}'$: $\rho((\mathbf{r} + \mathbf{r}')/2,t)$. The closure relation where the pair-correlation function is simply replaced by the equilibrium pair-correlation function g^{eq} for a homogeneous system evaluated at the local density can be regarded as the statistical equivalent of thermodynamic local equilibrium. Hence,

$$g(\mathbf{r},\mathbf{r}',t) = g^{\text{eq}}(|\mathbf{r}-\mathbf{r}'|)|_{\text{density}=\rho((\mathbf{r}+\mathbf{r}')/2,t)}\,. \tag{8}$$

The equation of motion (6) and the above closure relation (8) can only be used to describe decomposition kinetics in the initial and intermediate stages (and possibly in the transition stage). In the final stage it makes no sense to replace the pair-correlation function by its equilibrium form at the local density, since the (non-equilibrium) interfaces generally have a thickness of the order R_V.

Initial spinodal decomposition kinetics

Consider the initial stage of the phase separation, where the change $\delta\rho$ of the macroscopic density is small. The closure relation (8) can now be Taylor expanded to leading order as

$$g(\mathbf{r},\mathbf{r}',t) = g^{\text{eq}}(|\mathbf{r}-\mathbf{r}'|) + \frac{dg^{\text{eq}}(|\mathbf{r}-\mathbf{r}'|)}{d\bar{\rho}}\,\delta\rho\left(\frac{\mathbf{r}+\mathbf{r}'}{2},t\right) \tag{9}$$

where g^{eq} is understood to be the equilibrium pair-correlation function at the initial density $\bar{\rho} = N/V$. Linearization of the equation of motion (6) with respect to $\delta\rho$ yields (renaming $\mathbf{R} = \mathbf{r} - \mathbf{r}'$),

$$\frac{\partial}{\partial t} \delta\rho(\mathbf{r}, t) = D_0 \left[\nabla^2 \delta\rho(\mathbf{r}, t) + \beta\bar\rho\nabla \cdot \int d\mathbf{R} \left[\nabla_R V(R) \right] \right.$$

$$\left. \times \left(g^{\mathrm{eq}}(R)\delta\rho(\mathbf{r} - \mathbf{R}, t) + \bar\rho \frac{dg^{\mathrm{eq}}(R)}{d\bar\rho} \delta\rho(\mathbf{r} - \tfrac{1}{2}\mathbf{R}, t) \right) \right], \quad (10)$$

with ∇_R the gradient operator with respect to $\mathbf{R}$. Since the density varies only little over distances of the order of the range R_V of the pair-potential, we may use the Taylor expansion,

$$\delta\rho(\mathbf{r} - \mathbf{R}, t) = \delta\rho(\mathbf{r}, t) - \mathbf{R} \cdot \nabla\delta\rho(\mathbf{r}, t) + \tfrac{1}{2}\mathbf{R}\mathbf{R} : \nabla\nabla\delta\rho(\mathbf{r}, t)$$

$$- \tfrac{1}{6}\mathbf{R}\mathbf{R}\mathbf{R} \vdots \nabla\nabla\nabla\delta\rho(\mathbf{r}, t) + \cdots , \quad (11)$$

and,

$$\delta\rho(\mathbf{r} - \tfrac{1}{2}\mathbf{R}, t) = \delta\rho(\mathbf{r}, t) - \tfrac{1}{2}\mathbf{R} \cdot \nabla\delta\rho(\mathbf{r}, t) + \tfrac{1}{8}\mathbf{R}\mathbf{R} : \nabla\nabla\delta\rho(\mathbf{r}, t)$$

$$- \tfrac{1}{48}\mathbf{R}\mathbf{R}\mathbf{R} \vdots \nabla\nabla\nabla\delta\rho(\mathbf{r}, t) + \cdots . \quad (12)$$

Using these expressions as written, the leading order expansion in gradients of the density, as referred to in the introduction, is implemented. Substitution of these Taylor expansions into Eq. (10), and subsequent Fourier transformation finally leads to,

$$\frac{\partial}{\partial t} \delta\rho(\mathbf{k}, t) = - D^{\mathrm{eff}}(k)k^2 \delta\rho(\mathbf{k}, t) , \quad (13)$$

and hence,

$$\delta\rho(\mathbf{k}, t) = \delta\rho(\mathbf{k}, t = 0)\exp\{- D^{\mathrm{eff}}(k)k^2 t\} , \quad (14)$$

where the effective diffusion coefficient is equal to,

$$D^{\mathrm{eff}}(k) = D_0\beta \left[\frac{d\Pi}{d\bar\rho} + \Sigma k^2 \right] , \quad (15)$$

where,

$$\Pi = \bar\rho k_B T - \frac{2\pi}{3} \bar\rho^2 \int_0^\infty dR\, R^3 \frac{dV(R)}{dR} g^{\mathrm{eq}}(R) , \quad (16)$$

and,

$$\Sigma = \frac{2\pi}{15} \bar\rho \int_0^\infty dR\, R^5 \frac{dV(R)}{dR} \left(g^{\mathrm{eq}}(R) + \frac{1}{8} \bar\rho \frac{dg^{\mathrm{eq}}(R)}{d\bar\rho} \right) . \quad (17)$$

The expression in Eq. (16) for Π is precisely the osmotic pressure of a homogeneous system in equilibrium with a density equal to $\bar\rho$.

The effective diffusion coefficient is small in the neighbourhood of the spinodal, since there $|\beta d\Pi/d\bar\rho|$ is small. This enhances the separation in time scales of the dynamics of long and short wavelength density waves as discussed in the previous section.

Those density waves for which $D^{\mathrm{eff}}(k) < 0$ are unstable, since for such wavevectors, according to Eq. (14), density waves increase in amplitude exponentially with time. This results in what Cahn and Hilliard referred to as "uphill diffusion", where colloidal particles diffuse toward regions of higher concentration [8, 9]. The effective diffusion coefficient can become negative only when $d\Pi/d\bar\rho < 0$, which is the classic thermodynamic criterion for instability. Density waves with wavevectors which are smaller than the "critical wavevector",

$$k_{\mathrm{c}} = \sqrt{ - \frac{d\Pi}{d\bar\rho}/\Sigma } , \quad (18)$$

are unstable. The wavevector of the most rapidly growing density wave is found from Eq. (14) to be equal to,

$$k_{\mathrm{m}} = k_{\mathrm{c}}/\sqrt{2} = \sqrt{ - \frac{d\Pi}{d\bar\rho}/2\Sigma } . \quad (19)$$

A deeper quench, where $- \beta\, d\Pi/d\bar\rho$ is relatively large, results in a larger value for k_{m}, and gives rise to faster demixing.

The above results were derived for the first time by Cahn and Hilliard [8, 9] on the basis of thermodynamic arguments. The present Smoluchowski equation approach [10] allows for a microscopic rederivation of these now classic results for colloids.

Scaling of the static structure factor

As seen in the previous section, there is a particular density wave (with wavevector k_{m}) that grows most rapidly in the initial stage. This gives rise to the existence of a dominant length scale of density variations in the subsequent intermediate stage. As a consequence, the static structure factor exhibits dynamic scaling behaviour, the generic features of which are discussed in the present section.

In the intermediate stage, where a non-linear equation of motion for the density must be considered, in contrast to the above-described initial stage, the wavevector where the structure factor exhibits a maximum will be time dependent. Let us denote this wavevector as $k_{\mathrm{ms}}(t)$. This wavevector is larger than the most rapidly growing wavevector $k_{\mathrm{m}}(t)$, as $k_{\mathrm{ms}}(t)$ shifts to smaller values with time (see the next section). Both wavevectors are equal and time-independent only during the initial stage. The dominant length scale,

$$L(t) = 2\pi/k_{\mathrm{ms}}(t) , \quad (20)$$

is a function of time during the intermediate stage.

For the small (unstable) wavevectors under consideration, the static structure factor can be written in terms of the macroscopic density as [7],

$$S(k, t) = \frac{1}{N} \langle |\delta\rho(\mathbf{k}, t)|^2 \rangle_{\text{init}} , \qquad (21)$$

where the subscript "init" refers to ensemble averaging over all initial realizations of the density. The initial state is the state of the suspension when it developed long ranged correlations that render it unstable, but before significant phase separation has occurred (see the discussion in the introduction).

This is the quantity that is measured in a small-angle time-resolved static light scattering experiment.

The dominance of a single length scale implies that distances can only be measured in units of that single length scale, so that,

$$\frac{\langle \delta\rho(\mathbf{r}, t)\delta\rho(\mathbf{r}', t) \rangle_{\text{init}}}{\langle \delta\rho^2(\mathbf{r}, t) \rangle_{\text{init}}} \equiv F\left(\frac{|\mathbf{r} - \mathbf{r}'|}{L(t)}\right) . \qquad (22)$$

The normalizing denominator on the left hand-side fixes the value of the *scaling function* $F(x)$ to unity at $x = 0$ for all times. It follows that (with $x = |\mathbf{r} - \mathbf{r}'|/L(t)$),

$$S(k, t) = \frac{1}{N} \int d\mathbf{r} \int d\mathbf{r}' \, \langle \delta\rho(\mathbf{r}, t)\delta\rho(\mathbf{r}', t) \rangle_{\text{init}} \exp\{i\mathbf{k} \cdot (\mathbf{r} - \mathbf{r}')\}$$

$$= \frac{1}{N} \langle \delta\rho^2(\mathbf{r}, t) \rangle_{\text{init}} \int d\mathbf{r} \int d\mathbf{r}' \, F\left(\frac{|\mathbf{r} - \mathbf{r}'|}{L(t)}\right)$$

$$\times \exp\{i\mathbf{k} \cdot (\mathbf{r} - \mathbf{r}')\}$$

$$= \frac{4\pi}{\bar{\rho}} \langle \delta\rho^2(\mathbf{r}, t) \rangle_{\text{init}} \int\limits_0^\infty d|\mathbf{r} - \mathbf{r}'||\mathbf{r} - \mathbf{r}'|^2 \, F\left(\frac{|\mathbf{r} - \mathbf{r}'|}{L(t)}\right)$$

$$\times \frac{\sin\{k|\mathbf{r} - \mathbf{r}'|\}}{k|\mathbf{r} - \mathbf{r}'|}$$

$$= \frac{4\pi}{\bar{\rho}} L^3(t) \langle \delta\rho^2(\mathbf{r}, t) \rangle_{\text{init}} \int\limits_0^\infty dx \, x \, F(x) \frac{\sin\{kL(t)x\}}{kL(t)} .$$

We used here that $\langle \delta\rho^2(\mathbf{r}, t) \rangle_{\text{init}}$ is independent of position. By integration of the first equation above with respect to the wavevector it is easily found that,

$$\langle \delta\rho^2(\mathbf{r}, t) \rangle_{\text{init}} = \frac{1}{2\pi^2} \, \bar{\rho} \int\limits_0^\infty dk \, k^2 S(k, t) . \qquad (23)$$

Substitution of this result into the last of the above equations yields the following relation,

$$\frac{S(k, t)L^{-3}(t)}{\int_0^\infty dk' \, k'^2 S(k', t)} = \frac{2}{\pi} \int\limits_0^\infty dx \, x \, F(x) \frac{\sin\{kL(t)x\}}{kL(t)} . \qquad (24)$$

The right-hand side of this relation is a function of $kL(t) \sim k/k_{\text{ms}}(t)$ only. Therefore, plots of the quantity on the left-hand side of Eq. (24) vs. $k/k_{\text{ms}}(t)$ for various times must collapse onto a single curve. This is why Eq. (24) is referred to as a *dynamic scaling relation*.

Notice that it follows from the scaling relation (24), together with Eq. (20) for the dominant length scale, that plots of $S(k, t)/S(k_{\text{ms}}(t), t)$ vs. $k/k_{\text{ms}}(t)$ for various times should also collapse onto a single curve. This scaling means that the structure factor peaks have the same form, and differ only in the location of their maxima. One might call this scaling *dynamic similarity scaling*.

Demixing kinetics in the intermediate stage

Beyond the initial stage of the phase separation, linearization of the equation of motion is no longer allowed. Non-linear terms must be taken into account to describe demixing in the intermediate stage. What the initial stage and the intermediate stage have in common, however, is that the density varies smoothly on the length scale of the order of the range R_V of the pair-interaction potential.

In this section the static structure factor in the intermediate stage is analysed. To begin with, the evolution of density waves with the neglect of hydrodynamic interaction is considered, after which the hydrodynamic interaction is treated in an approximate way.

Decomposition kinetics without hydrodynamic interaction

The equation of motion for the static structure factor is obtained from that of the macroscopic density through the expression:

$$\frac{\partial}{\partial t} S(k, t) = 2 \frac{1}{N} \int d\mathbf{r} \int d\mathbf{r}' \left\langle \delta\rho(\mathbf{r}', t) \frac{\partial \delta\rho(\mathbf{r}, t)}{\partial t} \right\rangle_{\text{init}}$$

$$\times \exp\{i\mathbf{k} \cdot (\mathbf{r} - \mathbf{r}')\} . \qquad (25)$$

The equation of motion (6), together with the closure relation (8), is substituted into this expression, and subsequently expanded with respect to $\delta\rho$, including higher order terms. We will assume here that $\delta\rho(\mathbf{r}, t)$ for a fixed position and time is approximately a Gaussian variable. This is certainly wrong in the transition and late stage, where the probability density function of the density is peaked around two concentrations, which ultimately become equal to the two binodal concentrations. In the initial and intermediate stage, such a splitting of the pdf is assumed not to occur, and the pdf is approximately "bell-shaped" like a Gaussian variable [11]. When one is willing to accept the Gaussian character of the macroscopic density, averages $\langle \cdots \rangle_{\text{init}}$ of odd products of changes in the density are zero, while averages of products of four density changes can be written as a sum of products containing only two density changes, using Wick's theorem. Furthermore, as discussed in the introduction, the gradient expansions (11), (12) which were used for the initial stage are also sufficient in the intermediate stage. Extending the Taylor expansion (9) of the closure relation (8) to third order, yields, after a considerable amount of mathematics [12, 7],

$$\frac{\partial}{\partial t} S(k,t) = -2D_0\beta k^2 S(k,t)\left[\frac{d\Pi}{d\bar{\rho}} + \Sigma k^2\right]$$

$$- D_0\beta k^2 S(k,t)\left[\frac{d^3\Pi}{d\bar{\rho}^3} + \frac{d^2\Sigma}{d\bar{\rho}^2} k^2\right]\langle\delta\rho^2(\mathbf{r},t)\rangle_{\text{init}}$$

$$+ 2D_0\beta k^2 S(k,t)[\Sigma^\circ\langle\delta\rho(\mathbf{r},t)\nabla^2\delta\rho(\mathbf{r},t)\rangle_{\text{init}}$$

$$+ \Sigma^\bullet\langle|\nabla\delta\rho(\mathbf{r},t)|^2\rangle_{\text{init}}\bigg], \tag{26}$$

where,

$$\Sigma^\circ = \frac{4\pi}{15}\int_0^\infty dR\, R^5\, \frac{dV(R)}{dR}$$

$$\times\left(\frac{5}{8}\frac{dg^{\text{eq}}(R)}{d\bar{\rho}} + \frac{5}{6}\bar{\rho}\frac{d^2 g^{\text{eq}}(R)}{d\bar{\rho}^2} + \frac{5}{48}\bar{\rho}^2\frac{d^3 g^{\text{eq}}(R)}{d\bar{\rho}^3}\right), \tag{27}$$

and,

$$\Sigma^\bullet = \frac{4\pi}{15}\int_0^\infty dR\, R^5\, \frac{dV(R)}{dR}$$

$$\times\left(\frac{5}{8}\bar{\rho}\frac{d^2 g^{\text{eq}}(R)}{d\bar{\rho}^2} + \frac{1}{16}\bar{\rho}^2\frac{d^3 g^{\text{eq}}(R)}{d\bar{\rho}^3}\right). \tag{28}$$

Notice that averages like $\langle\delta\rho^2(\mathbf{r},t)\rangle_{\text{init}}$ are independent of position (since on average there is no preferred position in the system) but are still time dependent. In fact, these averages can be expressed in terms of wavevector integrals of the structure factor. The average $\langle\delta\rho^2(\mathbf{r},t)\rangle_{\text{init}}$ has been calculated in terms of the structure factor in Eq. (23). The other averages in Eq. (26) can be likewise expressed as [12, 7].

$$\langle|\nabla\delta\rho(\mathbf{r},t)|^2\rangle_{\text{init}} = -\langle\delta\rho(\mathbf{r},t)\nabla^2\delta\rho(\mathbf{r},t)\rangle_{\text{init}}$$

$$= \frac{1}{2\pi^2}\bar{\rho}\int_0^\infty dk\, k^4 S(k,t). \tag{29}$$

Substitution of this expression into Eq. (26) yields a closed equation of motion for the static structure factor.

It is important to note that the *static structure factor that is integrated with respect to the wavevector in the above equations, is only that part of the static structure factor that relates to the demixing process*: the expression (21) for the static structure factor describes scattering due to inhomogeneities that exist during phase separation, and does

not include any contribution that one would measure in a homogeneous equilibrium system. The integration therefore does not actually extend to infinity, but really goes up to some finite wavevector of the order of a few times $k_{\text{ms}}(t)$. The "molecular contribution" to the static structure factor, where $k \approx 2\pi/R_V$ or larger, is understood not to be included in any of the above equations. In an experiment, the integrals over the static structure factor in the above equations can be obtained by numerically integrating the intensity peak at small scattering angles that emerges during demixing.

Not all terms in the equation of motion (26) are equally important. Neglect of the irrelevant terms simplifies the equation of motion and reduces the number of independent parameters.

The wavevector dependent contribution $\sim\Sigma k^2$ in the very first term on the right-hand side of Eq. (26) is essential, even though the wavevectors of interest are small. This is due to the fact that near the spinodal $d\Pi/d\bar{\rho}$ is small and negative. The wavevector dependent contribution $\sim k^2 d^2\Sigma/d\bar{\rho}^2$ to the second term, however, is not essential, since $d^3\Pi/d\bar{\rho}^3$ is not small, except possibly for quenches close to the critical point. For the small wavevectors under consideration one may neglect the contribution $\sim k^2 d^2\Sigma/d\bar{\rho}^2$ in the second term on the right-hand side in Eq. (26). Physically this means that the local density dependence of the contribution of gradients in the density to the Helmholtz free energy is neglected, that is, the density dependence of the Cahn–Hilliard square gradient coefficient is neglected.

Furthermore, the dimensionless numbers $\beta\bar{\rho}^2 d^3\Pi/d\bar{\rho}^3$, $\beta\bar{\rho}^2\Sigma^\circ/R_V^2$, and $\beta\bar{\rho}^2\Sigma^\bullet/R_V^2$ are probably not of a different order of magnitude. The ratio of the third and second term on the right-hand side of Eq. (26) is then of the order

$$\frac{\text{third term}}{\text{second term}} = O\left(\int_0^\infty dk'\, k'^2 (k'R_V)^2 S(k',t)\bigg/\int_0^\infty dk'\, k'^2 S(k',t)\right).$$

This ratio is small since $k'R_V \ll 1$, so that the third term may be neglected against the second term.

The equation of motion (26) thus reduces to,

$$\frac{\partial}{\partial t} S(k,t) = -2D_0\beta k^2 S(k,t)\left[\frac{d\Pi}{d\bar{\rho}} + \Sigma k^2\right]$$

$$- D_0\beta k^2 S(k,t)\frac{d^3\Pi}{d\bar{\rho}^3}\frac{\bar{\rho}}{2\pi^2}\int_0^\infty dk'\, k'^2 S(k',t). \tag{30}$$

This equation contains the relevant features of phase separation in the intermediate stage, for quenches away from the critical point.

In the initial stage, the wavevector k_{m} of the most rapidly growing density wave is independent of time, and coincides with the wavevector k_{ms} where the static structure factor exhibits its maximum. This is no longer true beyond the initial stage. The wavevector of the most rapidly growing sinusoidal density variation is easily found from Eq. (30),

$$k_m(t) = \sqrt{ -\left[\frac{\mathrm{d}\Pi}{\mathrm{d}\bar{\rho}} + \frac{\mathrm{d}^3\Pi}{\mathrm{d}\bar{\rho}^3} \frac{\bar{\rho}}{4\pi^2} \int_0^\infty \mathrm{d}k'\, k'^2 S(k',t) \right] \Big/ 2\Sigma } \; . \quad (31)$$

Since $\mathrm{d}^3\Pi/\mathrm{d}\bar{\rho}^3 > 0$, and $\int_0^\infty \mathrm{d}k'\, k'^2 S(k',t) = (2\pi^2/\bar{\rho}) \times \langle \delta^2 \rho(\mathbf{r},t) \rangle_{\mathrm{init}}$ evidently increases with time, $k_m(t)$ shifts to smaller wavevectors with time. This result is identical to that derived in Ref. [11] for molecular systems.

The wavevector $k_{ms}(t)$, where the static structure factor is maximal, does not coincide with $k_m(t)$ beyond the initial stage. Since $k_m(t)$ shifts to smaller wavevectors, and since necessarily $k_m(t) < k_{ms}(t)$, the maximum $k_{ms}(t)$ will also shift towards smaller wavevectors.

The critical wavevector $k_c(t)$, beyond which density waves are stable, is easily seen to be equal to,

$$k_c(t) = \sqrt{2}\, k_m(t) \, , \quad (32)$$

just as in the initial stage.

For numerical purposes and to reduce the number of parameters, the equation of motion (30) is rewritten in dimensionless form. To this end we introduce the dimensionless wavevector K and time τ,

$$K = k/k_{m,0} \, , \quad (33)$$

$$\tau = -2D_0\beta \frac{\mathrm{d}\Pi}{\mathrm{d}\bar{\rho}} k_{m,0}^2 t \, , \quad (34)$$

where $k_{m,0} = k_m(t=0)$ is the wavevector of the most rapidly growing density wave during the initial stage. The dimensionless variable τ is the time that a particle with an effective diffusion coefficient $-D_0\beta\, \mathrm{d}\Pi/\mathrm{d}\bar{\rho}$ requires for diffusion over a distance $\sim k_{m,0}^{-1}$. The equation of motion (30) in the desired dimensionless form reads,

$$\frac{\partial}{\partial \tau} S(K,\tau) = \left[\left(\frac{k_m(\tau)}{k_{m,0}}\right)^2 K^2 - \frac{1}{2} K^4 \right] S(K,\tau) \, . \quad (35)$$

The ratio $k_m(\tau)/k_{m,0}$ is similarly written in dimensionless form, using Eq. (31), as

$$\left(\frac{k_m(\tau)}{k_{m,0}}\right)^2 = 1 - C \int_0^\infty \mathrm{d}K'\, K'^2 S(K',\tau) \, , \quad (36)$$

with,

$$C = \sqrt{ -\frac{\mathrm{d}\Pi/\mathrm{d}\bar{\rho}}{2\Sigma} \frac{\mathrm{d}^3\Pi/\mathrm{d}\bar{\rho}^3}{2\Sigma} \frac{\bar{\rho}}{4\pi^2} } > 0 \, , \quad (37)$$

and $K' = k'/k_{m,0}$. The number of parameters is thus reduced to the single dimensionless constant C.

Contribution of hydrodynamic interaction

In the above description of spinodal decomposition kinetics we have neglected the hydrodynamic interaction between colloidal particles. In the present subsection the effect of hydrodynamic interaction is considered in an approximate way. It is not feasible to tackle this problem by simply starting with the Smoluchowski equation with the inclusion of hydrodynamic interaction. On integrating the Smoluchowski equation to obtain an equation of motion for the macroscopic density, integrals containing three particle correlation functions are encountered. Moreover, these integrals probe the long ranged non-equilibrium part of correlation functions. A sensible closure relation then requires a separate analysis of the Smoluchowski equation for the three particle correlation functions. These equations are extremely complicated and not amenable to further analysis.

The following reasoning allows for an approximate evaluation of the effect of hydrodynamic interaction. Consider a subdivision of the entire system into small volume elements. The linear dimensions of these volume elements are small in comparison to the unstable wavelengths but should contain many colloidal particles. There are now two contributions from hydrodynamic interactions to be distinguished: hydrodynamic interaction between colloidal particles within a volume element and long ranged hydrodynamic interaction between different volume elements.

The short ranged hydrodynamic interaction between particles within single volume elements is simply accounted for by replacing the Stokes–Einstein diffusion coefficient D_0 in the equation of motion (6) by a "renormalized diffusion coefficient", which is denoted by $D_0^{(\mathrm{rn})}$. This expresses the change of the mobility of the entire assembly of colloidal particles within a volume element due to their mutual hydrodynamic interaction. The renormalized diffusion coefficient is virtually wavevector independent for the small wavevectors of interest here. Furthermore, the change of concentration during the intermediate stage is assumed to be small enough in order to neglect the concentration dependence of this diffusion coefficient.

The long ranged hydrodynamic interaction of colloidal particles in distinct volume elements may be treated as follows. The additional velocity that particles in a certain volume element attain is equal to the solvent velocity $\mathbf{u}(\mathbf{r},t)$ that is induced by the motion of the colloidal particles in the other volume elements, with $\mathbf{r}$ the position of the volume element under consideration. The solvent velocity is in turn related to the forces $\mathbf{F}^h$ that the fluid exerts on each colloidal particle as,

$$\mathbf{u}(\mathbf{r},t) = -\int \mathrm{d}\mathbf{r}'\, \mathbf{T}(\mathbf{r}-\mathbf{r}') \cdot \rho(\mathbf{r}',t) \mathbf{F}^h(\mathbf{r}',t) \, , \quad (38)$$

where $\mathbf{T}$ is the Oseen matrix. On the Brownian time scale there is a balance of the hydrodynamic, Brownian and direct forces. The Brownian force is equal to $-k_B T \nabla' \ln\{\rho(\mathbf{r}',t)\}$, while the direct force is given in Eq. (7). Hence,

$$\mathbf{F}^h(\mathbf{r}',t) = k_B T \nabla' \ln\{\rho(\mathbf{r}',t)\}$$
$$+ \int d\mathbf{r}'' [\nabla' V(|\mathbf{r}' - \mathbf{r}''|)] \rho(\mathbf{r}'',t) g(\mathbf{r}',\mathbf{r}'',t) . \quad (39)$$

The additional contribution to the equation of motion for the macroscopic density now follows by substitution of Eqs. (38) and (39) into the continuity equation,

$$\left.\frac{\partial \delta\rho(\mathbf{r},t)}{\partial t}\right|_{\text{hydro}} = -\nabla \cdot [\rho(\mathbf{r},t)\mathbf{u}(\mathbf{r},t)]$$

$$= k_B T [\nabla \delta\rho(\mathbf{r},t)] \cdot \int d\mathbf{r}' \mathbf{T}(\mathbf{r} - \mathbf{r}')$$

$$\cdot \left\{ \nabla'\rho(\mathbf{r}',t) + \beta\rho(\mathbf{r}',t) \right.$$

$$\left. \times \int d\mathbf{r}'' [\nabla' V(|\mathbf{r}' - \mathbf{r}''|)] \rho(\mathbf{r}'',t) g(\mathbf{r}',\mathbf{r}'',t) \right\}, \quad (40)$$

where $\nabla \cdot \mathbf{T}(\mathbf{r}) = 0$ is used. The subscript "hydro" refers to the additional contribution due to hydrodynamic interaction.

Using the same closure relation for the pair-correlation function g as before and expanding up to fourth order in $\delta\rho$'s, yields, with some effort [12,7],

$$\left.\frac{\partial S(K,\tau)}{\partial \tau}\right|_{\text{hydro}} = C'K^4 S(K,\tau) \int_0^\infty dK' f(K'/K) S(K',\tau) , \quad (41)$$

where C' depends on the quench parameters,

$$C' = -\frac{3}{40} \frac{\bar{\rho}}{\Sigma} (k_{m,0}a) \frac{D_0}{D_0^{(\text{rn})}} \int_0^\infty dR R^5 \frac{dV(R)}{dR}$$

$$\times \left(\frac{2}{3} g^{\text{eq}}(R) + \bar{\rho} \frac{dg^{\text{eq}}(R)}{d\bar{\rho}} \right), \quad (42)$$

and the function f is equal to,

$$f(z) = z(1 - z^2) \left[2z + (1 + z^2) \ln\left| \frac{1-z}{1+z} \right| \right]. \quad (43)$$

The constant C' is most likely positive, due to the large positive values of $g^{\text{eq}}(R)$ and $dg^{\text{eq}}(R)/d\bar{\rho}$ at contact. Here, τ is given by Eq. (34), except that D_0 is replaced by the renormalized diffusion coefficient $D_0^{(\text{rn})}$.

The expressions for $k_m(t)$ and $k_c(t)$ with the inclusion of hydrodynamic interaction are much more involved than the simple expressions (31), (32) in the absence of hydrodynamic interaction. For details see Refs. [12, 7].

Solution of the equation of motion

The equation of motion for the static structure factor with the inclusion of hydrodynamic interaction is the sum of Eqs. (35) and (41),

$$\frac{\partial}{\partial \tau} S(K,\tau) = K^2 S(K,\tau) \left[1 - C \int_0^\infty dK' K'^2 S(K',\tau) - \frac{1}{2} K^2 \right.$$
$$\left. + K^2 C' \int_0^\infty dK' f(K'/K) S(K',\tau) \right], \quad (44)$$

This equation of motion is easily solved numerically, where the wavevector integration extends up to the non-zero wavevector where the actual structure factor becomes equal to the initial structure factor.

A remarkable feature of the equation of motion is that, *except for the very early times, the structure factor is insensitive to the precise initial wavevector dependence $S(K, \tau = 0)$ of the static structure factor.*

Furthermore, when the equation of motion (44) is solved numerically, it turns out that the scaling function in Eq. (24) is independent of the values of both C and C' in the intermediate stage. Hence, *dynamic scaling is universal in the sence that scaling functions do not depend on the quench parameters.*[1] Although the evolution of the static structure factor itself may be very different for different choices of the quench parameters, the dynamic scaling functions are identical in each case. The scaling function in Eq. (24) and the similarity scaling function are depicted in Fig. 1. The dynamic similarity scaling function is described, to within a fraction of a percent, by the simple function,

$$\frac{S(k,t)}{S(k = k_{\text{ms}}(t),t)} = \exp\left\{ -30\left(\frac{k}{k_{\text{ms}}(t)} - 1\right)^2 \right.$$
$$\left. - 25\left(\frac{k}{k_{\text{ms}}(t)} - 1\right)^3 \right\}. \quad (45)$$

The late stage off-critical similarity dynamic scaling function of Furukawa [13] is also plotted in Fig. 1, and is seen to be much broader than the above scaling function which applies in the intermediate stage (for more recent developments on scaling during the late stages, see Refs. [14]).

The numerical solution of the equation of motion (44) shows power law behaviour of the time dependence of the

---

[1] This is true, provided that the quench parameters $S(K, \tau = 0)$, C ad C' are such that the non-linear terms are negligible at time $\tau = 0$, i.e. non-linear terms should become important solely due to the growth of the static structure factor.

Fig. 1 (a) The scaling function in Eq. (24) and (b) the similarity scaling function $S(K,\tau)/S(K_{\mathrm{ms}}(\tau),\tau)$ in Eq. (45) as functions of $K/K_{\mathrm{ms}}(\tau)$. The dotted line in (b) is the off-critical Furukawa similarity scaling function, which applies to the late stage of demixing

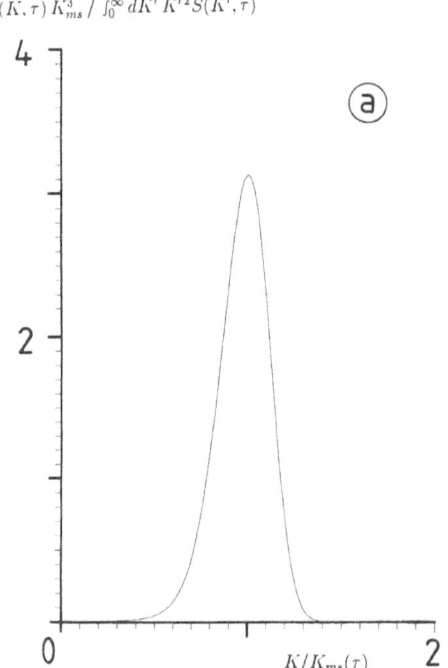

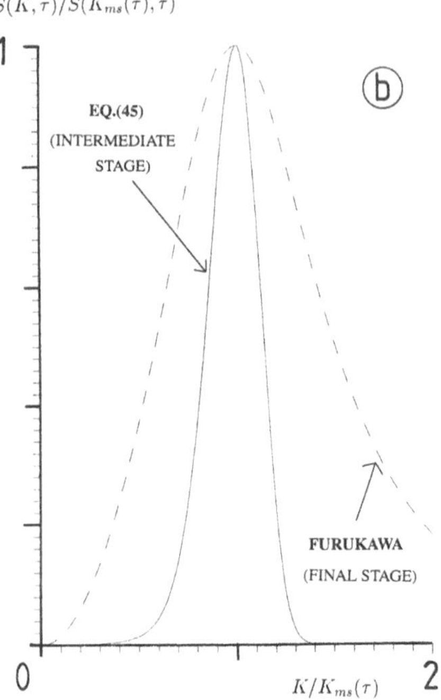

Fig. 2 (a) Dynamic similarity scaling for a microemulsion and (b) for a binary polymer melt. The dashed line is the theoretical dynamic similarity scaling function. Data are taken from Refs. [15] and [16], respectively

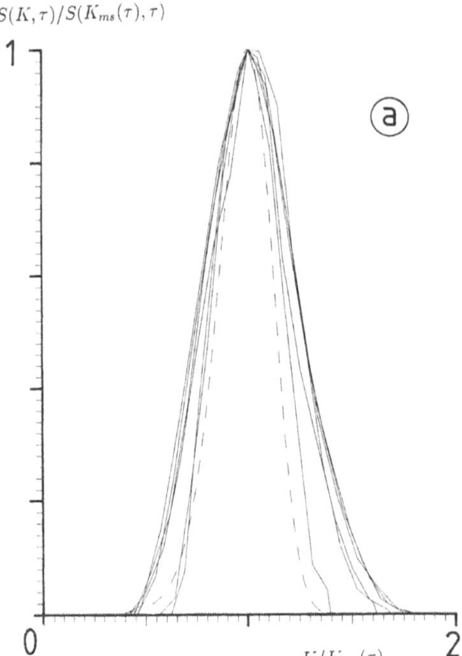

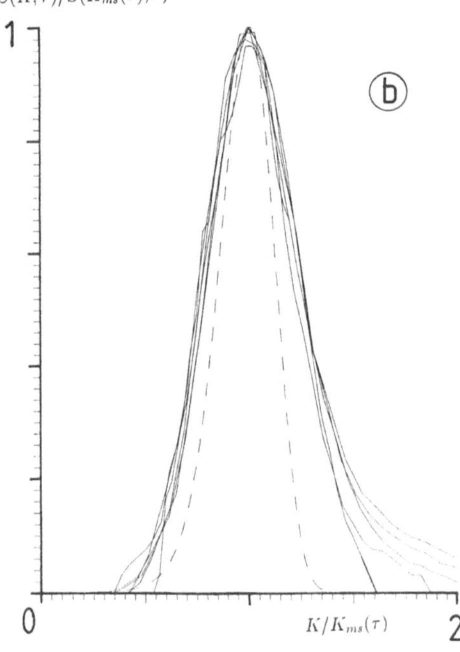

position k_{ms} of the maximum in the structure factor,

$$k_{\mathrm{ms}} \sim t^{\alpha} , \tag{46}$$

where the exponent α varies between 0.2 (when hydrodynamic interaction is neglected) and 1.1 (when hydrodynamic interaction is dominant, that is, $C' \gg C$).

Comparison with experiment

Two sets of experimental results are discussed here: data on a microemulsion system, from Ref. [15], and on a binary polymer melt, from Ref. [16]. When scaling the data, pertaining to the intermediate stage, as given in Eq. (24), it turns out that experimental curves do not scale, but decrease their amplitude with time. There are two possible reasons for this behaviour. First of all, in the integral over the structure factor in the denominator in Eq. (24), only the demixing contribution should be included. In a scattering experiment it might be that part of the larger wavevector scattered intensity is due to stable modes. Secondly, it could be that in the intermediate stage sharp interfaces are already beginning to form, giving rise to scattering at somewhat larger wavevectors. Since intensities are weighted in the wavevector integral in the scaling relation (24) with k^2, both contributions lead to an overestimation of the integral as compared to the assumed situation where only unstable modes scatter, and only diffuse interfaces exist. What does apply is the dynamic similarity scaling, where the integral is divided out. Only at somewhat larger wavevectors the dynamic similarity scaling is somewhat off, as a result of the above-mentioned scattering contributions. Dynamic similarity scaling is indeed seen to hold in Fig. 2, except at somewhat larger wavevectors. The dotted curve is the predicted similarity scaling function in Eq. (45).

Power law behaviour of k_{ms} in the intermediate stage is found in both Refs. [15, 16], and the exponent is indeed always in the theoretically predicted range $\alpha \in (0.2, 1.1)$. Moreover, in Ref. [16], the exponent α in Eq. (46) is found to be dependent on the quench parameters, in accordance with the theoretical finding that α depends on the parameters C and C' in the equation of motion (44), which in turn depend on the quench parameters.

References

1. Furukawa H (1984) Physica A 123:497; (1985) Adv Phys 34:703
2. Binder K, Stauffer D (1974) Phys Rev Lett 33:1006
3. Lifshitz IM, Slyozov VV (1961) J Phys Chem Solids 19:35
4. Wagner C (1961) Z Electrochem 65:581
5. Siggia ED (1979) Phys Rev A 20:595
6. Murphy TJ, Aguirre JL (1972) J Chem Phys 57:2098
7. Dhont JKG (1996) In: Möbius D, Miller R (eds) An Introduction to Dynamics of Colloids. Elsevier, Amsterdam
8. Cahn JW (1968) Trans Metall Soc AIME 242:166
9. Hilliard JE (1970) In: Aronson HI (ed) Phase Transformations, Ch 12. American Society for Metals. Metals Park, Ohio
10. Dhont JKG, Duyndam AFH, Ackerson BJ (1992) Physica A 189:503
11. Langer JS, Bar-on M, Miller HD (1975) Phys Rev A 11:1417
12. Dhont JKG (1996) J Chem Phys 105:5112
13. Furukawa H (1984) Physica A 123:497
14. Yeung C (1988) Phys Rev Lett 61:1135; Koga T, Kawasaki K (1991) Phys Rev A 44:817; (1993) Physica A 196:389
15. Mallamace F, Micali N, Trusso S, Chen SH (1995) Phys Rev E 51:5818
16. Wiltzius P, Bates FS, Heffner WR (1988) Phys Rev Lett 60:1538

Progr Colloid Polym Sci (1997) 104:76–80
© Steinkopff Verlag 1997

A. Vailati
M. Giglio

Very low-angle static light scattering from steady-state and time-dependent nonequilibrium fluctuations

Dr. A. Vailati (✉) · M. Giglio
Dipartimento di Fisica and Istituto
Nazionale per la Fisica della Materia
Università di Milano
via Celoria 16
Milano 20133, Italy

Abstract We investigate both steady-state and time-dependent nonequilibrium fluctuations by means of very low-angle static light scattering. The system is a thin layer of a binary liquid mixture close to a critical consolution point and placed in a temperature gradient. Due to this choice, the Soret driven nonequilibrium fluctuations are large and their time evolution is slow enough to be followed at ease. We discuss the fast q divergence at steady state and its low q frustration induced by the presence of gravity. We also report measurements taken during the build-up of the concentration gradient. The data indicate that no additional contributions are present during the time-dependent part of the process beyond those calculated in the quasi-steady-state approximation.

Key words Nonequilibrium fluctuations – static light scattering

While equilibrium fluctuations have been studied very extensively, nonequilibrium ones have received much less attention so far. From an experimental point of view, nonequilibrium fluctuations are rather elusive. They grow large at very long wavelengths, and therefore the excess scattering above the equilibrium value can be appreciated only by going to extremely small scattering angles, in a range where the conventional scattering instruments become unreliable.

Some very specialised set-ups have been recently employed to study the nonequilibrium contributions from a thin layer of fluid under a temperature gradient in steady-state conditions. Dynamic light scattering has been used to discriminate between equilibrium and nonequilibrium fluctuations, and the data nicely confirm the theoretical prediction that the mean square of nonequilibrium fluctuations exhibits a strong q^{-4} divergence [1–5]. Recently it has also been predicted [6, 7] that under the influence of gravity this divergence should be frustrated at very small q vectors, since the vertical velocity fluctuations that are responsible of the large nonequilibrium fluctu-

ations in a stratified medium are actually discouraged by the presence of gravity (the temperature gradient is applied from above, and the system is free of buoyancy driven fluctuations).

In this paper we will present some low-angle static light scattering from nonequilibrium fluctuations in a thin layer of a binary liquid system under a (stabilising) temperature gradient. Both steady-state and time-dependent data will be presented. The steady-state data have already been presented in a recent publication [8], but we will discuss them here in connection with hydrodynamic data obtained on a convective instability (heating from below-destabilizing gradient). Also, the discussion of the steady-state data is quite convenient to introduce the dynamic data.

While the steady-state data are collected once the temperature-induced concentration gradient (Soret effect) is established, the time-dependent ones are taken during the build-up of the gradient. In spite of the fact that there is no theory at present for nonsteady-state nonequilibrium fluctuations, the data can be interpreted on the basis of a simple model that takes into account the evolution of the concentration profile during its growth.

The experimental set-up has already been briefly described [8, 9], and will also be described at some length in a forthcoming work [10]. We will therefore simply recall here that the scattering set-up allows to collect scattered light over 31 angles which cover almost two decades in scattering wave vectors. The scattering wavevector range can be selected anywhere inside the interval 20–50 000 cm^{-1}. The heart of the instrument is a monolithic multielement solid state sensor and the data acquisition is controlled via a PC. The scattering cell has two thick sapphire windows that confine vertically a 1.00 mm thin, horizontal slab of fluid. The scattering beam is directed vertically across the slab. The temperature difference of the plates is controlled via an electronic servo acting on two annular shaped Peltier elements. The temperature difference between the sapphire plates can be kept constant within a few millidegrees indefinitely.

The choice of the sample is a peculiar one. We have chosen a binary liquid mixture of aniline and cyclohexane near the critical consolution point. The concentration is that of the critical isochore, and the temperature anywhere in the sample is above the critical point so that the system never undergoes a phase separation. The very particular choice was made because nonequilibrium fluctuations in such a system grow particularly large.

A typical set of data is presented in Fig. 1. The scattered intensity is plotted as a function of the scattering wave vector q on a log scale. One can immediately appreciate the strong q^{-4} divergence at larger q vectors. To compare the contribution due to nonequilibrium fluctu-

Fig. 1 Nonequilibrium (boxes) and equilibrium (dotted line) scattered intensity plotted as a function of q

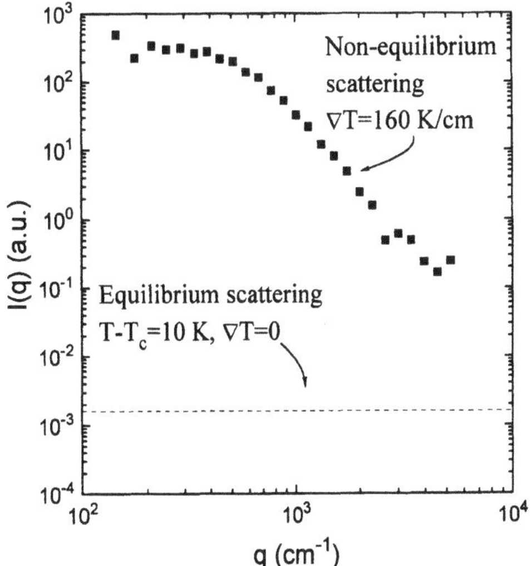

ations, we have also added in Fig. 1 the estimated scattered intensity due to ordinary fluctuations (shown with the dotted line) at a temperature equal to the temperature in the plane half a way between the top and bottom liquid boundaries. The intensity scattered from equilibrium fluctuations is too weak to be appreciated with the instrument. Indeed its level is too small compared with the stray light contributions. In order to bypass this difficulty, we have performed scattered intensity measurement from a static scatterer. The beam attenuation was also determined. Since the sample was scattering power well within the sensor solid angle of collection, we could then determine on an absolute scale the levels of the scattered intensity, since the integral of the scattered intensity over all angles should match the beam attenuation. We have then used the Reference data for the turbidity of the binary liquid mixture, and also the data for the long correlation length [11]. We are thus in the position of determining the equilibrium contribution to the scattered intensity, and these are reported in Fig. 1.

One can notice the many orders of magnitude increase in the scattered intensity due to the nonequilibrium fluctuations.

At shorter wavelengths the divergence is frustrated and the data level off to a plateau.

Let us see how these results compare with the theoretical predictions. We have first to recall some definitions related to the Soret effect. As it is well known, a mass flow (a concentration flow) can be induced by relaxing a concentration gradient. The flow will continue until the original gradient is eventually destroyed by the flow. Although it is not widely appreciated, a mass flow can be induced in a mixture by the presence of a temperature gradient. This is the Soret effect [12]. The overall mass flow equation then becomes

$$J = -\rho D \left(\nabla c + \frac{k_{\mathrm{T}}}{T} \nabla T \right) \tag{1}$$

where we have introduced the Soret thermal diffusion ratio k_{T} that actually determines the steady-state concentration gradient one observes when the overall flow is set to zero:

$$\nabla c = -\frac{k_{\mathrm{T}}}{T} \nabla T . \tag{2}$$

The customary definition of the concentration c is the mass fraction of the heavier component with respect to the total mass. The sign of the Soret thermal diffusion ratio can be both positive and negative, depending on the mixture. The most common case is that it is positive, thus implying that the heavier component migrates to the cold plate.

The intensity distribution of the light scattered from nonequilibrium fluctuations is given by [5–7]:

$$I(q) \propto \frac{1}{vD} \frac{1}{1 - \dfrac{R(q)}{R_{\mathrm{c}}}} \frac{(\nabla c)^2}{q^4} \qquad (3)$$

where we have introduced the mass diffusion coefficient D, the kinematic viscosity v and the q-dependent Rayleigh number ratio $R(q)/R_{\mathrm{c}}$ given by

$$\frac{R(q)}{R_{\mathrm{c}}} = -\frac{\beta \mathbf{g} \cdot \nabla c}{vDq^4} \qquad (4)$$

where $\beta = \rho^{-1}(\partial \rho / \partial c)$, $\mathbf{g}$ is the gravity acceleration and R_{c} is the critical Rayleigh number that determines the onset of a convective instability when heating from below. The threshold condition for the convective instability is

$$R(q)/R_{\mathrm{c}} = 1 \ . \qquad (5)$$

Under these condition, and assuming that the thermal diffusion ratio is positive, when the temperature gradient attains the critical value indicated by Eq. (5), a buoyancy-driven convective instability is established. The physical origin of the instability and the structure of the threshold equation (Eq. (5)), can be easily understood.

Let us consider a binary liquid mixture under a temperature gradient and at steady state. Let us assume that a velocity fluctuations displaces a small parcel of fluid in the vertical direction, for example upwards. This small parcel will have both a density and concentration at variance with the neighbouring volumes in the same horizontal plane. Of course these differences will be greater for larger temperature gradients and larger Soret coefficients. The density variations associated with the thermal expansion of the fluid will relax very fast, while the changes in concentration will last longer, since for liquids the mass diffusion coefficient D is typically two orders of magnitude smaller than the thermal diffusivity. So even after the temperature of the parcel has become equal to that of the neighbouring volumes, a net buoyancy force is still present because of the long lived concentration difference. As a result the small parcel will continue his motion upwards. Whether it will make it to reach the top plate (and thus promote the convective planform) will depend on the competition between buoyancy and the two adversing effects of diffusion, that tend to equalize the concentration (and thus kill the buoyancy) and viscosity, that impedes the vertical motion. This is why both viscosity and the diffusion coefficient appear at the denominator in the threshold condition given by Eq. (5), while the temperature gradient and the thermal diffusion ratio appear in the numerator. Alternatively, the threshold can be seen as the point at which the nonequilibrium fluctuations diverge, since $R(q)/R_{\mathrm{c}} = 1$ and therefore the denominator vanishes in Eq. (3).

In the case under study here, the reverse happens. Heating is from above so that the ratio $R(q)/R_{\mathrm{c}}$ is negative and therefore as the ratio becomes comparable with unity, the amplitude of the fluctuations is actually diminished. So gravity plays here a quenching effect, and this leads to the levelling of the intensity scattered at low wave vectors.

A crude quantitative comparison between the observed position of the q_{ro} at which $R(q_{\mathrm{ro}})/R_{\mathrm{c}} = 1$ can be attempted on the basis of the Eq. (4). All the thermodynamic quantities are known from literature for the system under study here. Of course the dependence of q_{ro} on the fluid parameters is rather insensitive due to the 1/4 power dependence, but at least we can check if the experimental value falls reasonably close to the theoretical value. We find $q_{\mathrm{ro(exp)}} = 537 \ \mathrm{cm}^{-1}$, while from reference data we estimate that the value should be $q_{\mathrm{ro(estim)}} = 410 \ \mathrm{cm}^{-1}$. We conclude that the agreement is fair.

We can also compare the relative magnitude of the nonequilibrium contributions with respect to the equilibrium one (dashed line in Fig. 1). Again, a quantitative comparison is difficult, because all the parameters are temperature and concentration dependent, since we are close to a critical consolution point. We can, however, estimate the ratio between the two contributions by assuming that the parameters have a fixed value equal to that corresponding to the critical isochore and at the temperature of the midplane of the sample. In this way the estimated ratio between nonequilibrium and equilibrium contributions should be $r_{\mathrm{estim}} = 790$ at $q = 1500 \ \mathrm{cm}^{-1}$, to be compared with the experimental value $r_{\mathrm{exp}} = 3800$. In considering the crude assumptions made, the agreement is more than satisfactory.

Let us now turn to the time-dependent nonequilibrium contributions. The measurements have been performed during the build-up of the Soret-driven concentration gradient. The temperature difference is switched on at $t = 0$ and the scattered intensity distributions are collected as a function of time. Of course it is essential to compare the time constant for the establishment of the temperature gradient across the sample with the time constant associated with the concentration gradient build-up. The first is essentially controlled by the finite maximum power provided by the Peltier elements and the heat mass of the sapphire plates and associated parts. It turns out that the time evolution of the temperature gradient follows rather nicely an exponential growth with a time constant $\tau_{\mathrm{thermal}} = 60 \ \mathrm{s}$. The build-up of the concentration gradient is given by $\tau_{\mathrm{conc}} = a^2/(\pi^2 D)$, where a is the thickness of the sample. Since the temperature in the midplane is roughly $T = 313 \ \mathrm{K}$, $T - T_{\mathrm{c}} = 10 \ \mathrm{K}$, we have from reference data [13] that $D = 10^{-6} \ \mathrm{cm}^2 \mathrm{s}^{-1}$ and consequently $\tau_{\mathrm{conc}} =$

1000 s. Notice that this is larger than the thermal time constant (although they differ by an order of magnitude only). The choice of working at an average temperature closer to the critical point was explicitly made so that the concentration relaxation is slower than the thermal relaxation, and the process is genuinely a time-dependent nonequilibrium process. Should the time constant be comparable, or even worse, should the temperature relaxation be the longest, then one would follow a sequence of quasi-steady-state conditions and this would be of little interest.

We present in Fig. 2 a sequence of scattered intensity distributions at various instants during the build-up. The very early curves have been discarded since their intensity levels are so low that the spurious stray light contributions dominate and cannot be accounted for adequately by blank subtraction procedures. Moreover, we had to restrict the useful q vector range to about one decade, since in the early stages the intensity scattered at the smaller and larger q is dominated by stray light.

One can immediately notice that the scattered intensity profiles show little change in the rolloff position, while the scattered intensity level increases as a function of time. To compare the various curves we show in the inset the scaling of the curves by readjusting their amplitudes only. Let us see which kind of conclusions we can draw from these data.

It is important to understand how the concentration changes as a function of height in the cell and as a function of time. Again, the correct calculation should take into account the fluid parameters changes as a function of both the concentration and the temperature. This would be

a rather formidable task, especially if one takes into account the fact that we are close to a critical point. To simplify, we assume as done before that the fluid parameters are fixed and equal to those corresponding to the critical isochore and at a temperature equal to that of the midplane. Under these conditions the concentration profiles are described by the following expression [14]:

$$c(z, t) = c_\infty(t) - 2\pi \frac{k_\mathrm{T}}{T} \Delta T \left\{ \sum_{k=1}^{\infty} \frac{[1 - (-1)^k]}{k^2 \pi^3} \cos\left(k\pi \frac{z}{a}\right) \right.$$
$$\left. \times \exp\left(-\frac{k^2 \pi^2 D t}{a^2}\right)\right\} \tag{6}$$

where ΔT is the applied temperature difference and $c_\infty(t)$ is the steady-state concentration profile:

$$c_\infty(t) = c_\mathrm{crit} + \frac{k_\mathrm{T}}{T} \Delta T \left(\frac{1}{2} - \frac{z}{a}\right) \tag{7}$$

c_crit is the critical aniline weight fraction.

A plot of the concentration profiles at various times is (qualitatively) shown in Fig. 3. There are few important things to be noticed in these curves. First, it is immediate to notice that at steady state a uniform concentration gradient is established throughout the entire cell height. At any instant however (even at the shortest times) the gradient at the boundaries attains the same value as the steady-state one. Indeed at the boundary this gradient is attained instantaneously. This has always been well known to the practitioners of analytical centrifuges as the Archibald

Fig. 2 Nonequilibrium scattered intensity distributions plotted as a function of q at various times ($t = 0$ corresponds to the switching of the temperature gradient). The inset shows the distributions rescaled by their forward scattered intensity

Fig. 3 Time evolution of the concentration profile described by Eq. (6). Here $k_\mathrm{T} = 3.5$, $D = 1.3 \times 10^{-6}\,\mathrm{cm^2\,s^{-1}}$, $\Delta T = 16\,\mathrm{K}$, $a = 1$ mm, $c_\mathrm{crit} = 0.47$ and $T = 313\,\mathrm{K}$

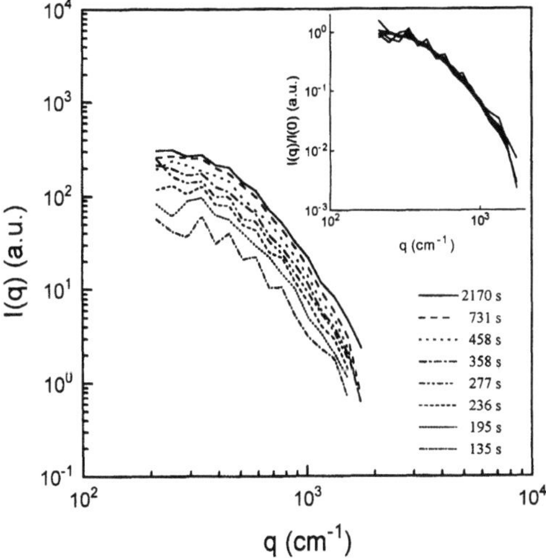

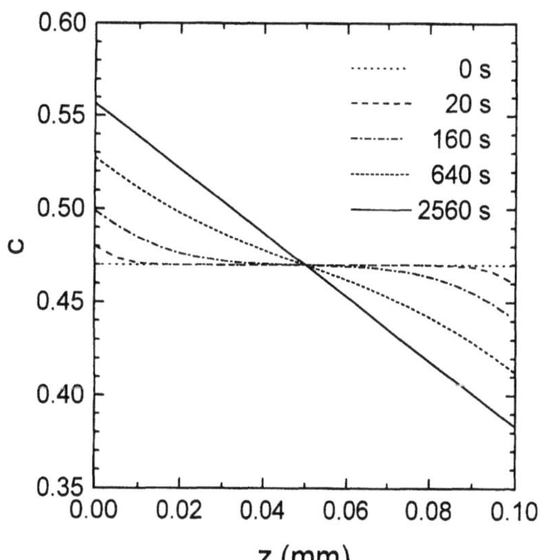

method [15]. Looking at Eq. (1), since exactly at the boundary the mass flow J is always zero (no solute is drained from the boundaries or moves into them), we have that instantaneously the gradient is the steady-state one. As the time progresses, the thickness of the layers adjoining the boundaries where the steady-state gradient is attained gets larger and larger and eventually the whole height of the sample has the same gradient.

The basic hypothesis we want to test is whether during transients, effects are present other than those described by the theory for the steady-state behaviour. In other words, if at a given instant a layer of fluid has a given concentration gradient, is the scattered intensity the same one would expect as if that layer was in steady-state conditions or are there additional effects? If we assume that the former hypothesis is valid, then from a qualitative point of view it is easy to understand why the curves show little change in the rolloff position and exhibit only a variation of the scattered intensity level. We can crudely divide the cell height into regions where we have the (same) steady-state gradient and regions where the gradient is zero. Then things would make sense, since the rolloff is dictated by the magnitude of the gradient (which as we said is the same), while the intensity is actually controlled by the height over which the steady gradient is attained, and this grows as a function of time, until it eventually attains a terminal, steady-state value.

Let us see if we can test the hypothesis above on a more quantitative basis. We can estimate the scattered intensity by assuming that the overall effect is the sum of the contributions, layer by layer, taking into account the various gradients in the layers as described by Eq. (6), the calculation being carried out as a function of time. We have to integrate Eq. (3) between the boundaries. When that is done, we find that asymptotically the value of the scattered intensity extrapolated at zero q vectors should behave as an exponential:

$$I(0) = c(0) - c(a) \propto 1 - \exp\left(-\frac{a^2}{\pi^2 D}t\right). \qquad (8)$$

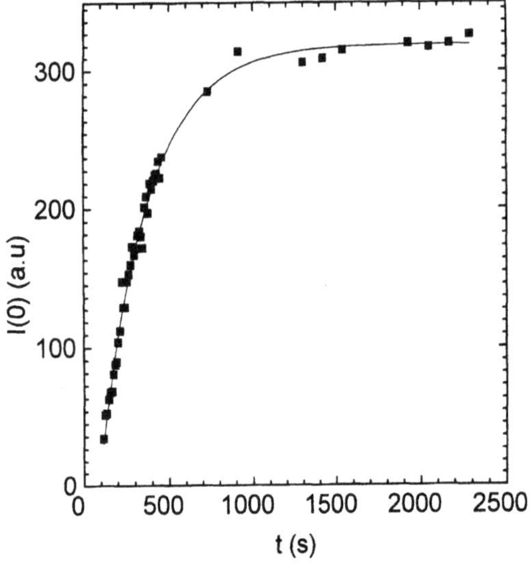

Fig. 4 Time evolution of the nonequilibrium forward scattered intensity. The solid line is the best fit of the data with Eq. (8)

When the actual data for this quantity are plotted as a function of time, a fairly good fit with an exponential is obtained (see Fig. 4). The experimentally derived estimate for the diffusion coefficient D that appears in Eq. (8) turns out to be $D = 3.6 \times 10^{-6}$ cm^2/s to be compared with the estimated value from literature data $D = 1.3 \times 10^{-6}$ cm^2/s [13]. We consider the agreement as fair, since there are so many approximations, especially those stemming from neglecting the concentration and temperature dependence of D and k_T over the cell height.

In conclusion, the preliminary data we present here on time-dependent nonequilibrium fluctuations indicate that no additional effects other than those associated to the presence of a stress induced gradient and described by the steady-state theory are present.

Acknowledgment This work has been partially supported by the Italian Space Agency (ASI) and by the Italian Ministry of University and Research.

References

1. Law BM, Gammon RW, Sengers JV (1988) Phys Rev Lett 60:1554
2. Law BM, Segrè PN, Gammon RW, Sengers JV (1990) Phys Rev A 41:816
3. Segrè PN, Gammon RW, Sengers JV, Law BM (1992) Phys Rev A 45:714
4. Segrè PN, Gammon RW, Sengers JV (1993) Phys Rev E 47:1026
5. Li WB, Segrè PN, Gammon RW, Sengers JV (1994) Physica A 204:399
6. Segrè PN, Schmitz R, Sengers JV (1993) Physica A 195:31
7. Segrè PN, Sengers JV (1993) Physica A 198:46
8. Vailati A, Giglio M (1996) Phys Rev Lett 77:1484
9. Carpineti M, Ferri F, Giglio M, Paganini E, Perini U (1990) Phys Rev A 42:7347
10. Vailati A, Giglio M, to be submitted
11. Calmettes P, Lagues I, Laj C (1972) Phys Rev Lett 28:478
12. de Groot SR, Mazur P (1962) Nonequilibrium Thermodynamics. North-Holland Amsterdam
13. Giglio M, Vendramini A (1975) Phys Rev 34:561
14. Bierlein JA (1955) J Chem Phys 23:10
15. Tanford C (1961) Physical Chemistry of Macromolecules. Wiley, New York

Progr Colloid Polym Sci (1997) 104:81–89
© Steinkopff Verlag 1997

H. Löwen
M. Schmidt

Freezing in confined suspensions

H. Löwen[1] (✉) · M. Schmidt
Institut für Theoretische Physik II
Heinrich-Heine-Universität Düsseldorf
Universitätsstraße 1
40225 Düsseldorf, Germany

[1] also at:
Institut für Festkörperforschung
Forschungszentrum Jülich
52425 Jülich, Germany

Abstract The freezing transition of hard sphere colloids confined between two parallel hard plates is studied for different plate distances ranging from one to two particle diameters. Using Monte Carlo simulations and free volume theory, the full phase diagram is obtained exhibiting solid-to-solid transitions between buckled, rhombic and layered crystals involving several triangular or square layers. While the fluid freezing transition is always strongly first order, both strong and extremely weak transitions occur between different crystalline structures. These predictions should be experimentally observable in confined suspensions of sterically stabilized or highly salted charge-stabilized colloidal particles.

Key words Confined fluids – freezing transitions – colloidal suspensions

Introduction

If a fluid is confined on a microscopic scale, the location of its phase transitions can be significantly shifted with respect to that of the bulk system. While this effect is well-known and well-studied for the liquid-vapour transition [1] where capillary condensation can stabilize the liquid phase at the expense of the gas it is much less clear how the freezing transition is affected by a confinement. Recent studies indicate [2, 3] that the direction of the shift of the fluid freezing line depends delicately on the range and nature of the wall–fluid interaction. Another significant shift is expected for the (dynamical) glass transition in a confined system, see e.g. Ref. [4] for a computer simulation study and a compilation of literature.

Different kinds of confinement are conceivable: the liquids can be inside a porous material (like vycor glass or silica gels), inside a spherical or cylindrical cavity or in between two parallel smooth plates. In the following we shall mainly study a system confined between two parallel walls but we add also some qualitative remarks for other confinements. In between two walls, the effective dimensionality of the confined fluid may be continuously interpolated between three and two by varying the plate distance from macroscopic towards molecular spacings. This may also help to explain why the freezing transition in such a confining geometry is difficult to understand: In strictly three- or two-dimensional fluids it was found that the freezing and melting transition can be quite different. While it is a usual first-order transition in 3D, it may be a two-stage continuous transition in 2D with an intermediate hexatic phase possessing long-ranged bond-orientational order [5]. Hence it is a priori unclear which of these two situations is realized in between two and three dimensions although there are certain indications that an intermediate phase can persist between two and three dimensions provided it occurs in the pure 2D case. The orientational symmetry of such an intermediate phase may be sixfold ("hexatic phase") as well as fourfold ("tetratic phase"), threefold ("triatic phase") or twofold ("duatic phase").

Colloidal suspensions represent excellent model liquids on a mesoscopic length scale. They possess many advantages over molecular (microscopic) liquids: First, they can be easily confined between parallel glass plates. On a mesoscopic scale these plates are much smoother than in any confinement of microscopic fluids. The larger size of colloidal spheres allows one to watch their positions in real space using video microscopy, see e.g. Refs. [6–13]. What is known from this work is that many different crystalline phases can become stable as one varies the plate separation. In the experimental work [7], the following cascade of solid-to-solid transitions was found

$$\ldots n\triangle \rightarrow (n + 1)\square \rightarrow (n + 1)\triangle \ldots . \tag{1}$$

This implies an alternation of crystals involving n square layers ($\square$) with crystals consisting of $n + 1$ layers of stacked two-dimensional triangular lattices ($\triangle$).

Theoretical work, on the other hand, is much less comprehensive and was mainly done in the framework of a hard-sphere model confined between hard walls: Pieranski and coworkers have calculated the close packing density [14] and used a cell model to calculate some solid-to-solid transitions [15]. The structure of the confined hard sphere fluid was investigated by Percus [16] and Wertheim et al. [17] without addressing the freezing transition. Finally, within a Landau approach, a transition from a crystalline monolayer to buckled solid phase was recently pointed out by Chou and Nelson [18].

In this paper we present calculations of the full phase diagram for the confined hard-sphere model for arbitrary density and moderate plate distances lying between one and two sphere diameters. The phase diagram exhibits a rich structure with a fluid phase and many different solid phases including buckled, rhombic and layered crystalline structures. We find that the sequence (1) suggested by the experiments is in fact more complicated since additional buckled and rhombic phases may also occur. We have also addressed the question after the order of the solid-to-solid transitions. In fact they can be strongly first order as well as very weakly first order. The results are obtained using extensive Monte Carlo (MC) simulations. We also present a simple theory for the phase diagram, combining free volume theory of the crystalline phase with an effective-diameter theory of the fluid phase, which yields qualitative and semi-quantitative agreement with our exact simulation data. At least for our model studied we do not find an indication for an intermediate phase with algebraically decaying bond-orientational order. A part of this work has been published already elsewhere [19].

The paper is organized as follows: In the second section we propose and define our model. Then we describe our Monte-Carlo simulation technique in the third section. Then, in the fourth section, we summarize our results and discuss our simple theory in terms of a cell model in the fifth section. We conclude in the sixth section. One peculiar emphasis of our paper is to give a survey on other possible fascinating transitions in confining geometry. This is finally done in the seventh section.

The model: hard spheres between hard plates

Our model consists of N hard spheres of diameter σ confined between parallel hard plates with area A and gap thickness $H = (h + 1)\sigma$, such that $h = 0$ corresponds to the 2D limit of hard disks. Since temperature is irrelevant for excluded-volume interactions, the only thermodynamic quantities are the reduced particle density $\rho_H = N\sigma^3/(AH)$ and the effective reduced plate separation h. The particle coordinate perpendicular to the plates is z, with $-h\sigma/2 \leq z \leq h\sigma/2$. In the limit $h \rightarrow \infty$, the effect of the confining plates vanishes and the 3D bulk hard sphere system is recovered.

Monte-Carlo simulation

In our Monte-Carlo simulations, we use the canonical ensemble with particle numbers ranging from $N = 192$ to $N = 4608$ in order to check systematically for finite-size effects. Careful attention is paid to the boundary conditions, which are crucial in a system exhibiting structural phase transitions. To allow any periodically ordered structure to fit into the simulation box for a suitable particle number N, the box is allowed to change its shape in the course of the simulation while its volume is fixed. The area in the lateral plane is a parallelogram with periodic boundary conditions. MC moves concerning its angles and aspect ratio are performed, so that the system can relax to equilibrium via shearing and squeezing. Between 10 and 100 million MC steps per particle were computed to determine the equation of state for fixed h in the region of a phase transition. Phase transitions are detected by looking for van der Waals loops in the equation of state for fixed h. By performing a Maxwell construction, the corresponding density jump is calculated by equating both the lateral pressures $p_{\text{lat}} = -H^{-1}\mathrm{d}F/\mathrm{d}A$ (F denoting the Helmholtz free energy) and the chemical potentials of the coexisting phases. As a consistency check, we have also used the single occupancy cell method [20] for $h = 0.85$ finding the same phase boundaries. In addition, we have monitored the behavior of suitably defined order parameters in order to characterize the emerging crystalline phases. We introduce a set of double-indexed complex order parameters Ψ_{mn}, defined via

$$\Psi_{mn} \equiv \left\langle N^{-1} \sum_{\alpha=1}^{N} |\Psi_n(\alpha)| \exp(im \arg \Psi_n(\alpha)) \right\rangle . \tag{2}$$

Here $\langle \ldots \rangle$ denotes a canonical average and $\Psi_n(\alpha) \equiv N_\alpha^{-1} \sum_\beta \exp(in\Theta_{\alpha\beta})$, where the sum is over N_α neighbours of particle α possessing lateral distances smaller than 1.2σ and having opposite signs in their z-coordinates and $\Theta_{\alpha\beta}$ is the angle between the bond of particles α and β and an arbitrary axis. The quantity Ψ_{mn} tests for solid structures with an $m \cdot n$-fold rotational symmetry. By calculating the order parameter set $\{\Psi_{mn}\}$ during the simulation, one can readily distinguish between different local surroundings of particles and thus identify the crystalline structure. An abrupt change in the order parameters signals a phase transition. The fluctuations of the order parameters are measured by means of an order parameter susceptibility

$$\chi_{mn}(N) = N(\langle |\Psi_{mn}|^2 \rangle - \langle |\Psi_{mn}| \rangle^2) \,,$$

depending on the particle number N. It will be used to investigate the order of weak phase transitions. (The dependence of Ψ_{mn} and χ_{mn} on the thermodynamic variables ρ_H and h is suppressed in the notation.) The following two scenarios are conceivable: In the case of a continuous phase transition, diverging fluctuations are observed at the critical point, and $\chi_{mn}(\rho_H \to \rho_H^{\text{crit}}) \to \infty$. Of course, diverging fluctuations are only encountered in the thermodynamic limit $N \to \infty$, which is not directly accessible in computer simulations. Hence we study the dependence of χ_{mn} on the system size N. In the discontinuous case, on the other hand, entering the coexistence region would simply mix the susceptibilities of the coexisting phases according to the relative weight of both phases $\chi_{mn}(\rho_H) = \alpha\chi_{mn}(\rho_H^{(1)}) + (1 - \alpha)\chi_{mn}(\rho_H^{(2)})$, $\alpha = (\rho_H^{(1)}/\rho_H)(\rho_H^{(2)} - \rho_H)/(\rho_H^{(2)} - \rho_H^{(1)})$, (here the superscripts 1 and 2 stand for the low- and high-density coexisting phases). Consequently, no divergence is encountered as $N \to \infty$.

Results

In Fig. 1, the resulting phase diagram is shown in the plane spanned by the reduced particle density ρ_H and the effective plate spacing h. The region in phase space is naturally limited by the close-packed density (dashed line). Different symbols represent different system sizes showing that the dependence on system size is only weak. For $h = 0$, in agreement with recent simulations of Weber et al. [21], we recover the first-order freezing transition of hard disks into a triangular lattice. As h is increased, the fluid freezes first into one triangular layer and with increasing density subsequently undergoes a further first-order transition into a crystalline structure of buckled lines (b). For intermediate h, the fluid freezes into two layers of a square lattice ($2\square$) via a strong first-order transition and then transforms into the buckling phase (b). The latter transition is marked by $\triangle$-symbols indicating phase boundaries, where the

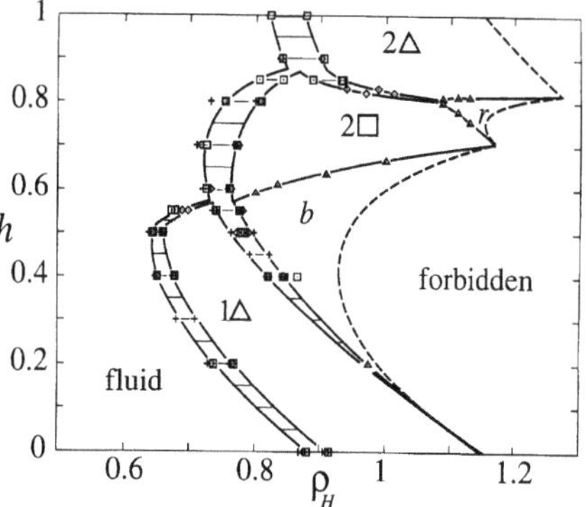

Fig. 1 Simulated phase diagram for hard spheres of reduced density ρ_H between parallel plates with effective reduced distance h. Symbols indicate different system sizes: $N = 192(+)$; $N = 384, 512$ ($\diamond$); $N = 576(\triangle)$; $N = 1024, 1156$ ($\square$). Six phases occur (fluid, $1\triangle$, b, $2\square$, r and $2\triangle$). The closed-packed density is marked by a dashed line. Solid lines are guides to the eye. Thin horizontal lines represent two-phase coexistence

equation of state shows no van der Waals loop, but the order parameter Ψ_{21} exhibits anomalous behaviour. The order parameter Ψ_{21} is shown in Fig. 2 for fixed plate separation distance $h = 0.62$ as a function of density ρ_H. The function $\Psi_{21}(\rho_H)$ increases in a relative small density interval $\rho_H = 0.87 - 0.9$ from a value close to zero to a finite value of about 0.04. We find that the fluctuations on the low-density side vanish with a law $\propto 1/\sqrt{N}$ and tend to zero in the thermodynamic limit $N \to \infty$. We thus conclude, that the $2\square$-phase is thermodynamically stable for low densities. In contrast, on the high-density side, no decrease of Ψ_{21} as a function of particle number N is observed; from the finite-size systems' data we can conclude that a value clearly larger than zero is reached as $N \to \infty$, and the buckling structure with only two-fold rotational symmetry is stable. We thus conclude that a phase transition happens between $2\square$ and b, signalled by a rapid increase of the order parameter Ψ_{21} as a function of density ρ_H. As a function of N, the increase happens more rapidly (with a larger slope) and is shifted towards higher densities.

Having demonstrated the existence of $2\square - b$ phase transition, we are now concerned with its order. Therefore the order parameter susceptibility is considered. In Fig. 3 the susceptibility χ_{21} as a function of density ρ_H is shown for plate separation distance $h = 0.62$. The susceptibility has small, but finite values in the low-density $2\square$- and high-density buckling phase. In between the pure phases a pronounced maximum occurs that can be interpreted in terms of fluctuations driving the system from one phase to the other. In contrast to the finite-size dependence of the order parameter itself, its susceptibility shows only weak dependence on system size. Although statistical errors are present, there is no indication up to $N = 4608$ of a

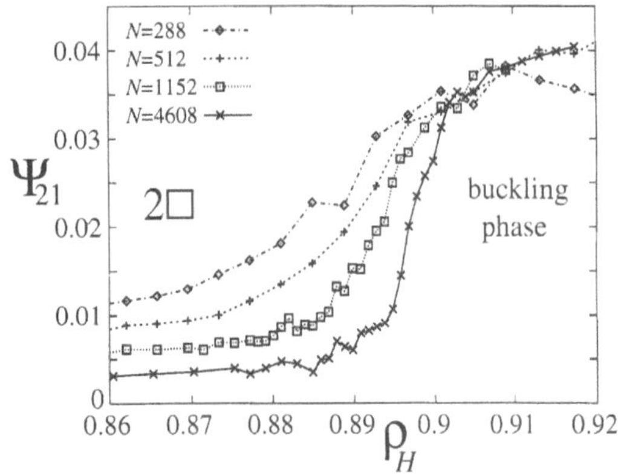

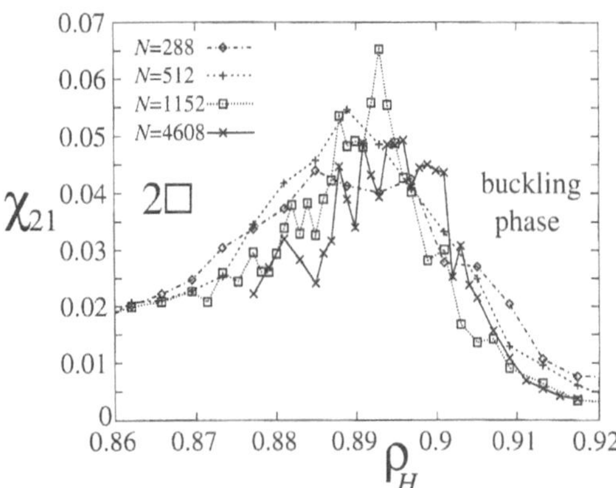

Fig. 2 Behaviour of the order parameter Ψ_{21} across the $2\square$–b phase boundary as a function of density ρ_H for plate separation distance $h = 0.62$. There are four different system sizes shown $N = 288, 512, 1152, 4608$

Fig. 3 Finite-size dependence of the order parameter susceptibility χ_{21} for particle numbers $N = 288, 512, 1152, 4608$ and plate separation $h = 0.62$. The maximum of the curves remains finite as a function of N

divergence in the thermodynamic limit. We thus conclude, that although we cannot resolve a probably miniscule density jump from the equation of state, the phase transition between the $2\square$ and the buckling structure is of first order. However, for $h = 0.6$, for example, we can exclude a density jump $\Delta\rho_H$ larger than 0.0004. Hence the $2\square \rightarrow b$ transition is extremely weak.

For even higher h, a transition occurs from the $2\square$ phase into a crystal with two triangular layers ($2\triangle$). Finally there is a new stable crystalline phase which we call rhombic (r) since its unit cell is a rhombus. It is a close-packed structure but its stability also extended towards slightly smaller densities. The $2\square \rightarrow r$ and $2\triangle \rightarrow r$ transitions are again very weak.

Let us add some more details for the two less common phases b and r. Typical particle configurations and the corresponding unit cells are depicted in Figs. 4a and b.

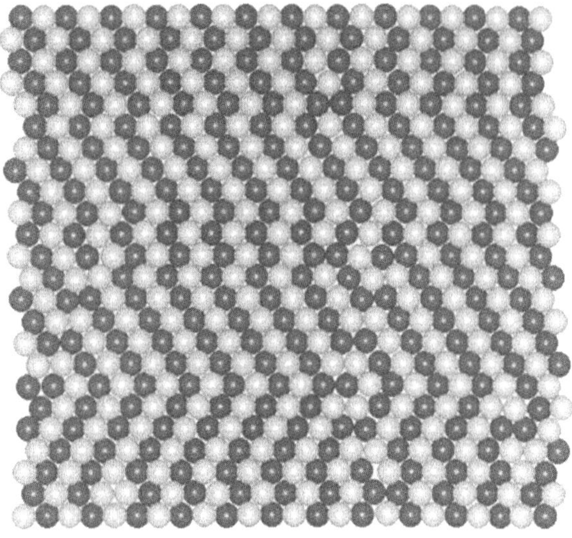

(a)

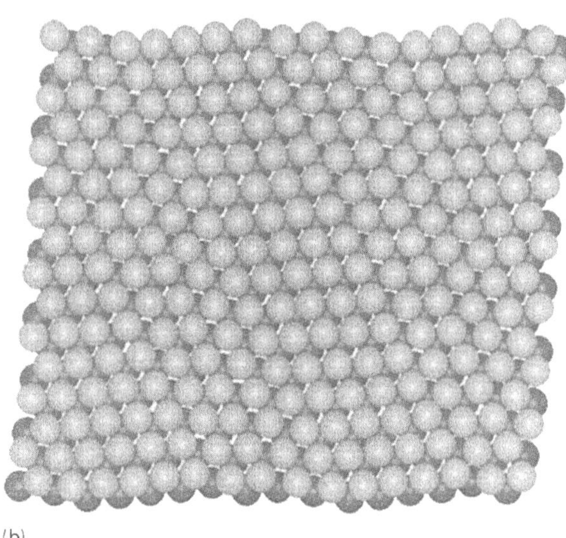

(b)

Fig. 4 Typical configurations of (a) the buckling phase, and (b) the rhombic phase

Both structures possess twofold rotational symmetry. Interestingly enough both the buckling and the rhombic phase are *highly degenerate*. For the buckling phasely, there is linear buckling (constituted by a single rectangular unit cell), periodic zig-zag buckling (built up from left and right kites) and a random succession of both (shown in Fig. 4a). Still, in the horizontal direction, there is strict periodicity. Likewise, the rhombic phase can appear to be linear rhombic (with a single rhombic elementary cell), zig-zag rhombic (with an alternating succession of the two rhombi) and again a radom succession of both as shown in Fig. 4b. All of these structures are close packed at the corresponding values of h. Away from close packing we cannot distinguish within our simulation which of these phases is the thermodynamically stable one since the free-energy differences are too miniscule. Let us remark that this is quite similar to the 3D hard-sphere crystal, where the three close-packed structures fcc, hcp, and random stacking are extremely close in free energy and the actual crystalline structure depends on the history of the sample.

We emphasize that the fluid freezing transition is first order. In fact we checked that bond-angle orientational correlation functions with two-, four- and six-fold symmetry decay exponentially with distance in the fluid phase and reach a finite plateau value in the solid phase. At least for a system size of $N \leq 1156$, we never found an intermediate "duatic", "tetratic" or "hexatic" phase characterized by an algebraic decay of the corresponding orientational correlation. This fact, of course, does not exclude the occurrence of such phases in larger systems and in systems that are governed by softer interactions.

Free volume theory

Cell model for crystalline phases

The cell model exploits the physical picture of a solid with particles being located around given lattice sites [15, 22, 23]. It enables one to determine the thermodynamically stable crystalline structure and its equation of state approximately. Furthermore, it provides an exact upper bound on the free energy.

In principle, the theoretical problem of hard spheres in confined geometry consists of computation of the configurational integral over N 3D spatial coordinates, the integrand being the Boltzmann factor, which is in the case of hard bodies either unity for allowed configurations or zero if at least two particles overlap. Adopting the cell model, we first impose a candidate lattice structure with given average density ρ_H and plate separation H. In general, this structure will depend on a set of free geometric parameters $\{a_i\}$, e.g. angles, ratios of lattice constants, etc. Second, the integration regime of each particle in the configurational integral is restricted from the total volume V to a smaller region in space around each lattice site, called the free volume cell. Thereby the cell size is chosen small enough, so that any two neighbouring spheres restricted within their respective cells cannot penetrate each other, hence the integrand of the restricted configurational integral is unity, the N integrals decouple and yield v_f^N, where v_f is the spatial volume of one cell. Clearly, the expression $- k_B T \ln(v_f/\Lambda^3)$ provides an upper bound on the exact (Helmholtz) free energy per particle, Λ denoting the thermal de Broglie wave length. To optimize this bound, we minimize it with respect to the set of free parameters $\{a_i\}$.

We checked the available analytical expression for v_f for the $1\triangle$-, b-, $2\square$-phases against an exact numerical procedure [24]. A modification is done for the $1\triangle$-phase [15] where we inserted a different effective diameter $\sigma_{1\triangle}^* = \sigma\sqrt{1 - h^2/6}$ into the expression for the $1\triangle$-free volume in order to enlarge the free volume for two touching spheres with different z-coordinates.

As the deviation of the cell model result from the exact free energy is expected to depend only weakly on the explicit crystal structure, and only free energy *differences* enter in the calculation of the phase diagram, phase boundaries between different crystalline phases are expected to be reasonably reproduced within the cell model.

Effective-diameter fluid theory

The fluid in slab geometry is approximately treated as a strictly two-dimensional hard-disk system with an effective diameter σ^* obtained from the implicit relation

$$\sigma^{*2} = \sigma^2 - \int\limits_{-h\sigma/2}^{h\sigma/2} dz_1 \int\limits_{-h\sigma/2}^{h\sigma/2} dz_2 (z_1 - z_2)^2 \, \rho(z_1, \sigma^*)$$
$$\times \rho(z_2, \sigma^*)/(\rho_H(h + 1)\sigma^{-2})^2 \tag{3}$$

where the one-particle density profile $\rho(z, \sigma^*)$ is given by $\rho_H(h + 1) \exp(\alpha(\sigma^*)z^2)/\mathcal{N}(\alpha(\sigma^*))$ with the normalization $\mathcal{N}(\alpha) = \sqrt{\pi/\alpha}\sigma^2 \, \text{erfi}(\sqrt{\alpha}h\sigma/2)$ and $\alpha(\sigma^*) = \pi\rho_H(h + 1)g(\sigma^*)/\sigma^2$, $\text{erfi}(x)$ denoting the imaginary error function. Here, $g(\sigma^*) = (1 - \eta(\sigma^*)/2)/(1 - \eta(\sigma^*))^2$ is the contact value of the 2D pair correlation function within scaled particle theory, and $\eta(\sigma^*) = (\pi/4)\rho_H(h + 1)\sigma^{*2}/\sigma^2$ is the effective area fraction of the 2D system. The expression (3) takes into account the fact that two spheres can be laterally closer than σ if they differ in their z-coordinates. Finally the Helmholtz free energy is obtained via integration of $dF/dA = - k_B T\rho_H(1 + h)(1 + \pi\rho_H(h + 1)\sigma^{*2}g(\sigma^*)/2\sigma^2)/\sigma^2$ guaranteeing the correct second virial coefficient in the low-density limit. The integration constant F_0 is empirically chosen to fit the location of the hard-disk freezing transition.

Phase diagram

The theoretical phase diagram is shown in Fig. 5. It looks similar to the exact simulation data reproducing the stability of the six different phases found in the simulation. All transitions are first order. The density gap between $b \rightarrow 2\square$ and $r \rightarrow 2\triangle$ is extremely small, for instance, $\Delta\rho_H = 0.00047$ at $h = 0.72$ for the $b \rightarrow 2\square$ transition. Also the fluid $\rightarrow 1\triangle$ and $1\triangle \rightarrow b$ transitions are quantitatively correct. Another interesting property concerns the above-mentioned

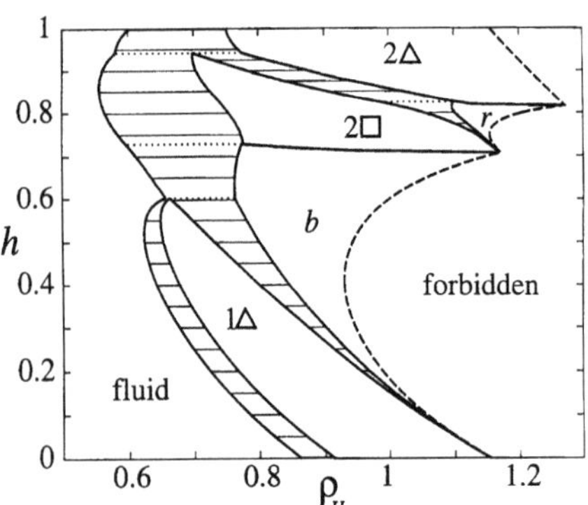

Fig. 5 Same as Fig. 1 but now obtained from effective-diameter and cell theory. The dotted lines represent a situation with three coexisting phases

degeneracy of the b and r phases. Indeed, also away from close packing, the free volumes are identical for different realizations of the b and r phases and hence the cell model cannot distinguish between the subspecies. One can thus conclude that the cell model, which requires much less numerical effort than a direct simulation, gives reliable results as far as the topology of the phase diagram of confined hard-body systems is concerned. Some details of the phase diagram, however, are not reproduced. The slope of the $2\square \rightarrow b$ line is positive in simulation but negative in cell theory. Hence the cell theory overestimates the stability of the buckled phase. Furthermore, the agreement of the fluid–solid coexistence region grows worse as h increases since within effective-diameter theory we map a multi-layer fluid onto a single-layer fluid.

Conclusions

In conclusion we have calculated the phase diagram of hard spheres confined between two parallel hard plates for small plate distances H ranging from one to two sphere diameters. Although solely excluded volume interactions enter in our model, it may also be used for softer interactions as long as they can be mapped onto an effective hard sphere system confined between effective hard plates. The interaction of a charged suspension is Yukawa-like [25–28] but image charges are also important if the dielectric constant of the solvent and the confining plates differ [25]. The screening length decreases with increasing salt concentration of the solvent. Therefore a charged suspension between charged plates could be mapped onto an effective hard-sphere system if the salt content is high. However, one should be careful in details of the freezing transition as far as an intermediate phase is concerned. It is known that the softness of a pretty harsh interaction $\propto r^{-12}$ is already sufficient to produce an intermediate

hexatic phase in 2D [29, 30] which is absent for hard disks [21].

An experimental verification of the full phase diagram is highly desirable in confined sterically stabilized colloidal suspensions or in charged but highly salted dispersions. Using video microscopy or scattering methods one should be able to resolve the order of the solid–solid transitions in order to verify our predictions. Let us also mention that similar solid-to-solid transitions were recently found [31, 32] for bilayer Wigner crystals in a double-quantum-well system exposed to a strong magnetic field.

Outlook and open problems

There are still many related questions open which we summarize in the following:

Larger plate distances: multilayer systems

The first problem obviously concerns the extension of our results to larger plate separations H. Here the situation becomes more and more complicated due to two reasons. First there are much more structures conceivable in packing spheres between larger spaced plates. Second all these phases compete in free energy and the free energy differences can become tiny. Hence it is difficult and practically impossible to resolve these small differences by a computer simulation. For a multilayer system, only the location of the fluid–solid transition has been simulated [33] without resolving the different solid phases. What is also known is that for infinite plate separation a layer of finite thickness of the metastable bulk crystal covers the plates. This precrystallization effect was demonstrated by van Swol and coworkers [34].

In our work, we have left out any possible transition structure to three layers, which emerge at $H > 1.82\sigma$ for high densities. Even the density of close packing is not known in this case. Recent experiments [11] suggest that again quite exotic structures can be generated in three and more layer systems. For instance, one strongly expects the stability of a "super-buckled phase". This phase consists of straight rows of triangles. In a two-dimensional cut perpendicular to the walls and normal to the row direction, there are alternating triangles of different orientation. Such a phase can be understood by packing Swiss "Toblerone" chocolates into an intersecting buckled structure. The corresponding two-dimensional cut is sketched in Fig. 6. Phases involving more layers can possess an even more exotic structure. After all, the simple sequence (1) is only true for a path well in between the crystal melting density and close-packed density but is not valid in general.

Fig. 6 A sketch of the "Toblerone" structure. The structure is periodic in the direction perpendicular to the drawing plane

Ensemble of constant wall pressure

As a second problem, one can also think in terms of a different thermodynamic ensemble where the pressure perpendicular to the walls is fixed instead of a fixed wall separation H. The choice of the ensemble clearly depends on the kind of experiment one is doing. In the new ensemble of constant wall pressure, the phase transition lines will shift a bit although the overall topology of the phase diagram should not change. However, one should bear in mind that transition between phases belonging to different wall separations are possible. Therefore additional phase transitions are conceivable. One example of these are the so-called stratification transitions found in Ref. [35]. We finally remark that the Gibbs phase rule is not violated in this ensemble since the relative portion of the three coexisting phases is uniquely determined by the wall pressure exerted on the system.

Another complication is the wedge geometry which is actually used in experiments [7]. Only if the relative tilt angle is small one can view the sample as one with parallel plates.

Density functional theory for the phase diagram

The third problem is to do a more sophisticated theoretical treatment involving classical density functional theory of 3D bulk freezing [36]. Many approximations for the free energy density functional which describe 3D bulk freezing are now available. However, most of these approaches suffer from the fact that there are unphysical divergencies if one shrinks the three-dimensional functional towards two dimensions by applying an external potential of two parallel hard walls. This was particularly investigated by Rosenfeld, Tarazona and the present authors [37]. It was found that a geometrically based density functional for hard spheres as originally proposed by Rosenfeld [38] does survive the 2D (and even the 1D and quasi-zero-dimensional) limit. This functional can be modified to describe very accurately the bulk freezing transition of hard spheres [37]. Hence one should take

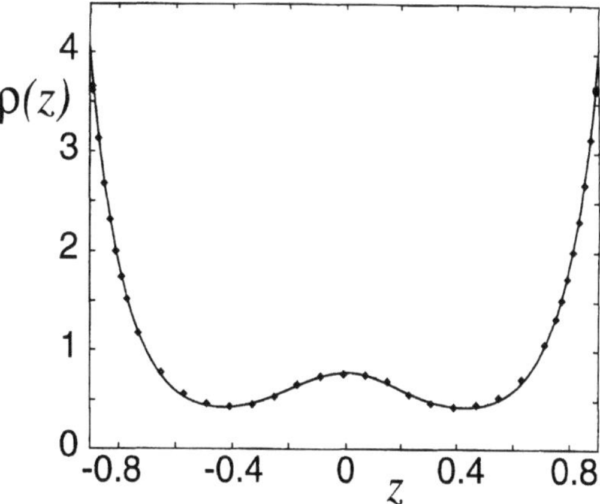

Fig. 7 Density profile of the hard sphere fluid in a planar slit of reduced separation $h = 1.8$. Symbols are Monte-Carlo results, the solid line is predicted by the fundamental-measure free energy density functional [38]

Rosenfeld's functional in order to investigate the phase diagram of hard spheres between hard plates. The problem of density parameterization can be avoided by adopting the free minimization scheme as developed and applied in Ref. [39]. In order to demonstrate the ability of Rosenfeld's functional to describe accurately the density profile we have confronted the density profiles in the fluid phase as resulting from density functional theory with the results from computer simulation, see Fig. 7. The agreement is excellent. One should bear in mind, that theories relying on weighted densities reproducing the direct correlation function in 3D do fail in describing the correct number of maxima in the density profile [40]. This gives a strong indication that Rosenfeld's functional is superior to other approximations, at least in a situation of heavily confined geometry.

Effect of gravity

Real suspensions are not completely density matched, i.e. the mass density of the solvent and that of the colloidal particles differ in general resulting in a non-vanishing buoyant mass M of the colloidal particles. This implies that gravity comes into the game which typically acts normal to the walls and leads to sedimentation. Gravity can be modelled by adding an external potential linear in the z-coordinate perpendicular to the plates. Including gravity, the phase diagram becomes more complicated depending also on the reduced coupling $\alpha \equiv Mg\sigma/k_{\mathrm{B}}T$ where g is the gravitational acceleration.

One may ask whether the complicated superlattice structures found in recent experiments [11] which do not occur in our "free" phase diagram ($\alpha = 0$) are due to gravity. In order to check this one can study the case $\alpha \to \infty$ implying that gravity becomes dominant. In this case, entropy becomes irrelevant and the stable configuration simply minimizes the gravitational potential energy. Even in this limit, the whole phase diagram depending on h and ρ_H, is not known. It is, however, expected that the typical structure is a phase separation involving dense-packed triangular layers residing on the lower plate with a coexisting part of the close-packed structure at the given H. This statement can at least be proved in the range where the buckled phase is closed packed. This is illustrated in Fig. 8. From this consideration one can conclude that gravity does not induce complicated superstructures. Still results for finite α are missing.

Hard cylinders between parallel hard plates

Replacing the hard spheres by hard cylinders or hard spherocylinders one obtains an even more complicated system which can also be realized in the context of colloidal suspensions or microscopic anisotropic fluids [41]. Cylinders carry an additional orientational degree of freedom and the 3D bulk phase diagram involves different mesophases, namely isotropic, nematic, smectic A, plastic crystalline and other crystalline phases [42]. Confining the system by parallel hard plates, the aspect ratio p of the cylinders is an additional parameter (for spheres p is unity). Even the closed-packed density is not known as a function of p and H. There may be a multitude of possible phases which are stable. One interesting phase transition is depicted in Fig. 9 which happens if the plate separation is a bit larger than the length of the cylinders. Then, due to simple packing arguments, a transition occurs from crystalline layers containing cylinders oriented parallel to the walls to one smectic or crystalline layer with rod orientations perpendicular to the walls. This transition is expected to possess a large hysteresis since the structures are completely different.

   In order to resolve all these interesting transitions one should perform a simple free volume theory as well as an extensive Monte-Carlo simulation for different aspect ratios p.

Hard spheres in cylindrical and spherical cavities

If the cavity is spherical or cylindrical the kind of transitions are quite different. For a *cylindrical* cavity of fixed diameter, the system is one dimensional. Then the configurations have a helix structure, but there are no phase transitions as the density is varied. Varying the

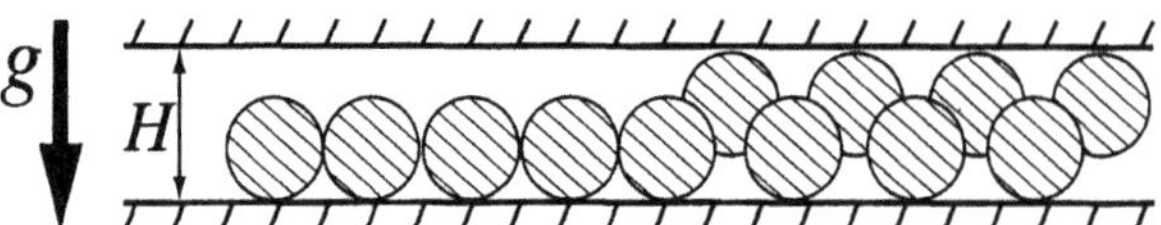

Fig. 8 In the limit of strong gravity g (heavy particles) there occurs a phase separation into triangular layers (left side) and close-packed regions (right side)

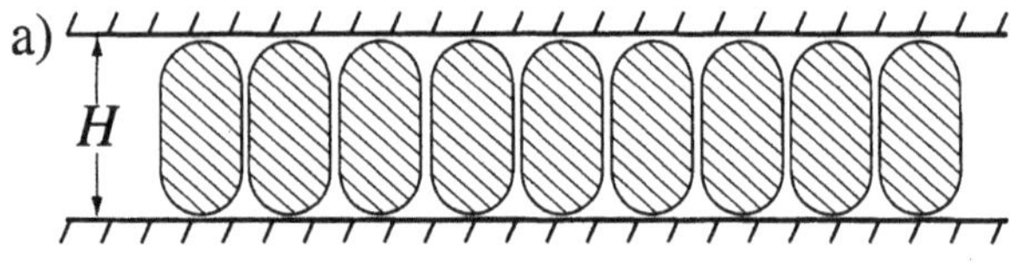
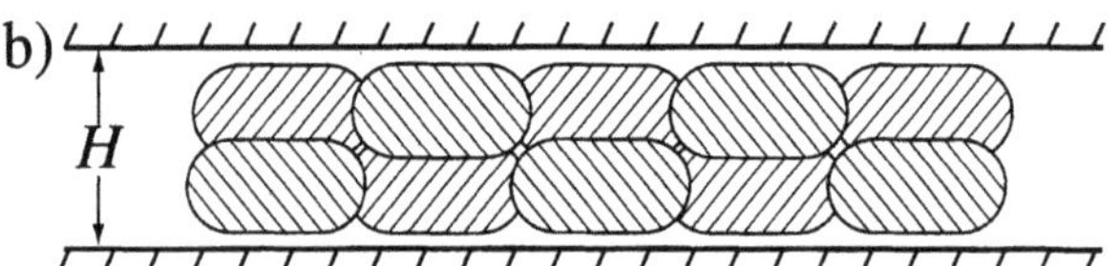

Fig. 9 Two possible phases for hard spherocylinders between parallel plates with plate distance comparable to particle length: (a) one smectic or crystalline layer with rod orientations perpendicular to the walls, (b) two-layered close-packed structure with spherocylinders aligned parallel to the confining walls

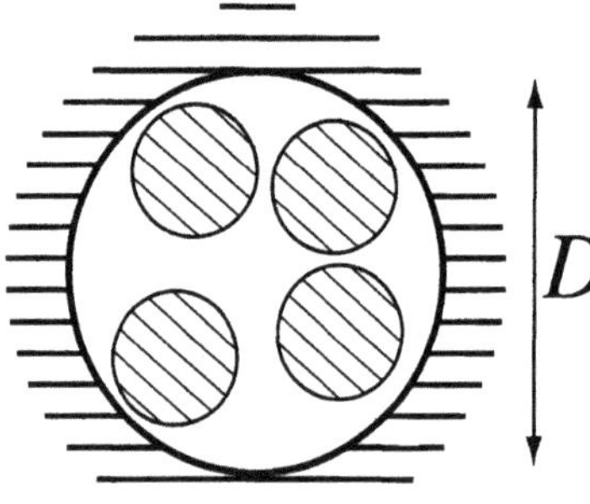

Fig. 10 Four hard spheres confined in a spherical cavity of diameter D

cylinder radius R^*, however, one can reach certain points where an anomalous behaviour can be found; the easiest case is $R^* = \sigma$ which is the border where two neighbouring spheres can exchange their position by hopping over each other. As far as we know, the close-packed density has not yet been calculated for $R^* > \sigma$.

   In a *spherical* cavity of diameter D, the number N of hard spheres which can be packed into the big sphere depends crucially on the ratio D/σ, see Fig. 10. The close-packed configuration is a dumb-bell for $N = 2$, a triangle for $N = 3$ and a tetrahedron for $N = 4$. For $N > 4$, however, this configuration is non-trivial and not yet known. The simpler two-dimensional problem of packing circles into a circle is better understood [43]. Also, recently, computer codes were developed and applied to a similar problem of how to pack two-dimensional circles into a square [44]. The solution for $1 \leq N \leq 20$ was presented in Ref. [44] exhibiting really some unexpected configuration for "odd" numbers N. A similar interesting result should be obtained in three dimensions. These dense-packed configurations should then be observable in porous materials.

Acknowledgments We thank H. Wagner, Y. Rosenfeld, A. Hüller, P. Leiderer, T. Palberg and S. Neser for helpful discussions. This work was supported by the Deutsche Forschungsgemeinschaft within the Gerhard-Hess-Programm.

References

1. Rowlinson JS, Widom B (1982) Molecular Theory of Capillarity. Clarendon, Oxford
2. Christenson HK (1995) Phys Rev Letters 74:4675
3. Duffy JA, Clarke AP, Fretwell HM, Alam MA, Evans R (1996) J Phys Condens Matter, in press
4. Fehr T, Löwen H (1995) Phys Rev E 52:4016
5. Strandburg KJ (1988) Rev Mod Phys 60:161
6. Pieranski P, Strzelecki L, Pansu B (1983) Phys Rev Lett 50:900
7. Van Winkle DH, Murray CA (1986) Phys Rev A 34:562; Murray CA, Sprenger WO, Wenk RA (1990) Phys Rev B 42:688; Murray CA (1992) In: Strandburg KJ (ed) Bond-orientational Order in Condensed Matter Systems. Springer, New York
8. Weiss J, Oxtoby DW, Grier DG, Murray CA (1995) J Chem Phys 103:1180
9. Schaertl W, Sillescu B (1994) J Stat Phys 77:1007
10. Marcus AH, Lin B, Rice SA (1996) Phys Rev E 53:1765
11. Neser S, Palberg T, Leiderer P, to be published
12. Larsen AE, Grier DG (1996) Phys Rev Lett 76:3862
13. Kepler GM, Fraden S (1994) Langmuir 10:2501
14. Pansu B, Pieranski P, Pieranski P (1984) J Phys 45:331
15. Bonissent A, Pieranski P, Pieranski P (1984) Phil Mag A 50:57
16. Percus JK (1980) J Stat Phys 23:657
17. Wertheim MS, Blum L, Bratko D (1990) In: Chen S-H, Rajagopalan R (eds) Micellar Solutions and Microemulsions: Structure, Dynamics, and Statistical Thermodynamics. Springer, New York
18. Chou T, Nelson DR (1993) Phys Rev E 48:4611
19. Schmidt M, Löwen H (1996) Phys Rev Lett 76:4552
20. Hoover WG, Ree FH (1967) J Chem Phys 47:4837; see also, Frenkel D, Ladd AJC (1984) J Chem Phys 81:3188
21. Weber H, Marx D (1994) Europhys Lett 27:593; Weber H, Marx D, Binder K (1995) Phys Rev B 51:14636
22. Kirkwood JG (1950) J Chem Phys 18:380
23. Wojciechowski KW (1987) Phys Lett A 122:377
24. Schmidt M, Löwen H, to be published
25. Chang E, Hone DW (1988) J Phys 49:25
26. Löwen H (1992) J Phys Condensed Matter 4:10105
27. Aranda-Espinoza H, Carbajal-Tinoco M, Urrutia-Banuelos E, Arauz-Lara JL, Medina-Noyola M, Alejandre J (1994) J Chem Phys 101:10925
28. Chávez-Páez M, Méndez-Alcaraz JM, Arauz-Lara JL, Medina-Noyola M, (1996) J Coll Interface Sci 179:426
29. Chen K, Kaplan T, Mostoller M (1995) Phys Rev Lett 74:4019
30. Bagchi K, Anderson HC, Swope W (1996) Phys Rev Lett 76:255
31. Zheng L, Fertig HA (1995) Phys Rev B 52, 12282; Narasimhan S, Ho T-L (1995) Phys Rev B 52:12291
32. Goldoni G, Peeters FM (1996) Phys Rev B 53:4591
33. Hug JE, van Swol F, Zukoski CF (1995) Langmuir 11:111
34. Courtemanche DJ, van Swol F (1992) Phys Rev Lett 69:2078; Courtemanche DJ, Pasmore TA, van Swol F (1993) Mol Phys 80:861
35. Schoen M, Diestler DJ, Cushman JH (1994) J Chem Phys 101:6865
36. For a review see: Löwen H (1994) Phys Rep 237:249
37. Rosenfeld Y, Schmidt M, Löwen H, Tarazona P (1996) J Phys Condes Matter 8:L577
38. Rosenfeld Y (1993) J Chem Phys 98:8126
39. Ohnesorge R, Löwen H, Wagner H (1994) Phys Rev E 50:4801
40. Choudhury N, Ghosh SK (1996) J Chem Phys 104:9563
41. Idziak SHJ, Koltover I, Davidson P, Ruths M, Li Y, Israelachvili JN, Safinya CR (1996) Physica B 221:289
42. Bolhuis P (1996) PhD thesis, University of Utrecht
43. Kravitz S (1967) Math Mag 40:65; (1969) U Pirl Math Nachr 40:111; Goldberg M (1971) Math Mag 44:134
44. Peikert R (1994) El Math 49:16

Progr Colloid Polym Sci (1997) 104:90–96
© Steinkopff Verlag 1997

W. Richtering

Investigation of shear-induced structures in lyotropic mesophases by scattering experiments

PD Dr. W. Richtering (✉)
Albert-Ludwigs-Universität Freiburg
Institut für Makromolekulare Chemie
Stefan-Meier-Straße 31
79104 Freiburg, Germany

Abstract The influence of shear on the structure of lyotropic liquid crystalline phases was investigated by small angle light and neutron scattering (SALS, SANS). Three different systems were studied involving nonionic and ionic surfactants: a hexagonal phase, a lamellar phase and a defective lamellar phase. Three different regimes are discussed in the hexagonal phase. A polydomain sample with dominating elastic properties, an aligned sample with the rodlike micelles being aligned in flow direction and characterized by a stripe texture. This texture can be disrupted at high shear rates leading to a further decrease of the moduli. Different regimes were also found with the defective lamellar phase. Multilamellar vesicles were formed at intermediate shear rates leading to an increase of viscosity. At higher shear rates these vesicles were destroyed and oriented lamellae were obtained. The lamellae were aligned with the layer normal parallel to the neutral (vorticity) axis. Vesicles were also obtained with a classical lamellar phase. Here the vesicles could not be destroyed at high shear rates but a butterfly pattern was observed in light and neutron scattering.

Key words Lamellar phase – hexagonal phase – vesicle – viscosity – neutron scattering – light scattering – SALS – SANS

Introduction

Micelles are formed when surfactant molecules are dissolved in water. These aggregates can be of different shape, e.g. spherical, wormlike or disklike. At high concentration ordered structures that consist of micelles are observed [1]. Cubic phases are known that are comparable with crystals formed by rigid colloidal particles. Anisotropic mesophases are formed when nonspherical micelles are present, e.g. the hexagonal phase of rodlike micelles and the lamellar phase which consists of surfactant bilayers of infinite extension. Since these mesophases are characterized by a long-range orientation correlation between anisotropic micelles, they are called lyotropic liquid crystals. Similar structures are known from block copolymer melts.

Liquid crystalline samples of macroscopic size usually display a polydomain structure and a typical texture is observed in polarizing microscopy. The texture strongly influences the rheological properties and often complex flow behavior is found because the flow field itself can alter the structure of the sample [2]. Consequently, it is necessary to monitor the structure under flow in order to understand rheological properties and various techniques have been developed for this purpose. Here we will focus on scattering experiments, namely small angle light and neutron scattering, SALS and SANS, respectively. Both methods provide similar information but on a different length scale. Therefore it is possible to detect shear-induced structural changes on the microscopic (micellar) and the mesoscopic (texture) dimensions. This aspect is especially important in surfactant systems. Micellar aggregates are not rigid in contrast to the rigid particles present in colloidal dispersions. A mechanical deformation can therefore influence both the suprastructure of the micelles and the shape of the surfactant aggregate itself.

In the following, we will describe some effects of shear on the structure of lyotropic hexagonal and lamellar mesophases. In the quiescent state all samples showed a texture typical of a polydomain structure, i.e., there were no external fields present which could induce a monodomain and wall effects could also be neglected. Consequently, effects of shear on both the texture and the micellar orientation have to be taken into account.

Experimental

Synthesis and purification of the nonionic surfactants were performed as described in the literature. Sodium dodecyl sulfate and decanol were purchased from Merck. Solutions were prepared with deionized H_2O (milliQ plus) and D_2O, respectively, as described previously [3–5]. Rheo-SALS experiments have been performed with an apparatus described in [6]. It consists of a Bohlin controlled stress rheometer equipped with a cone-plate shear geometry made out of quartz glass. A laser beam is guided through the tools parallel to the direction of the velocity gradient. The scattered light is detected by a two-dimensional detector below the lower plate, i.e., in the plane given by the flow direction and the neutral (vorticity) direction. Polarized and depolarized scattered light can be detected. SANS measurements have been performed with the instrument D11 at the Institut Laue-Langevin. A shear cell of two concentric cylinders was used and again the incident beam was parallel to the direction of the velocity gradient. A detailed description of the experimental setup has been given previously [7]. Schematic diagrams of the shear cells are shown in Fig. 1.

Results and discussion

Hexagonal phase

The hexagonal phase of different surfactant/water systems has been studied. Since many phase diagrams of aqueous mixtures of nonionic surfactants have been reported in the literature, it was possible to use systems with a hexagonal to isotropic phase transition near room temperature. This is very important because the viscosity of the hexagonal phase is extremely high and it is not possible to avoid effects of preshearing when the rheometer is loaded with the sample being in the hexagonal state. Therefore, the rheometer was loaded with an isotropic sample in all experiments described here. The hexagonal phase was then reached by cooling the sample into the hexagonal phase. Here we discuss the results obtained with a branched nonionic surfactant $H_3C(CH_2)_{13}OCH[CH_2O(CH_2CH_2O)_4CH_3]_2$ and with the common C12E6 surfactant $CH_3[CH_2]_{11}[OCH_2CH_2]_6OH$ [8].

The rheological properties of the polydomain sample can be probed by small amplitude oscillatory shear experiments. However, the linear viscoelastic region can be extremely small and a quantitatively correct measurement of the modulus is often difficult. Nevertheless, it could be shown that elastic properties dominate in the polydomain sample, i.e. the storage modulus G' is larger than the loss modulus G'' [9].

In small angle light scattering, the isotropic to hexagonal phase transition can be detected by the observation of depolarized scattering intensity when the sample is in the anisotropic phase. The scattered intensity is symmetrically distributed around the primary beam in the case of a polydomain sample. The scattering pattern changed strongly when the sample was sheared, as can be seen in Fig. 2. At first scattering was observed along the flow direction and then scattering perpendicular to the direction of flow

Fig. 1 Scattering geometries used in SALS and SANS

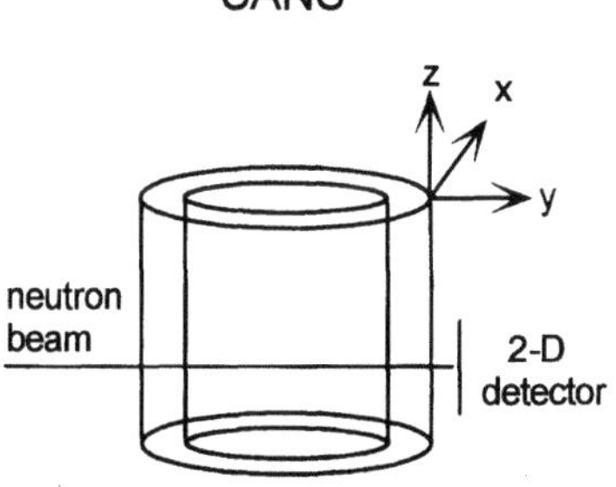

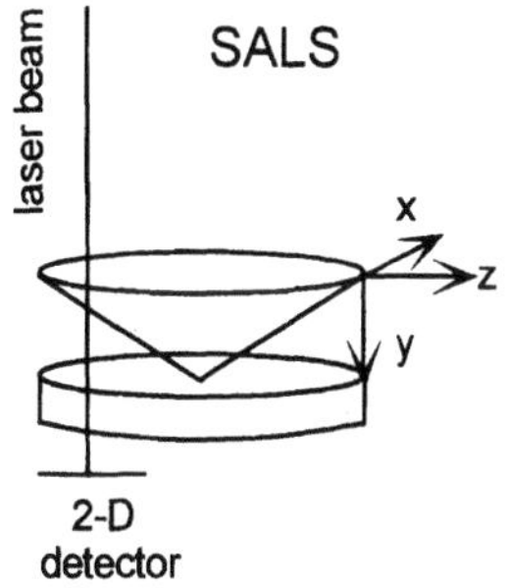

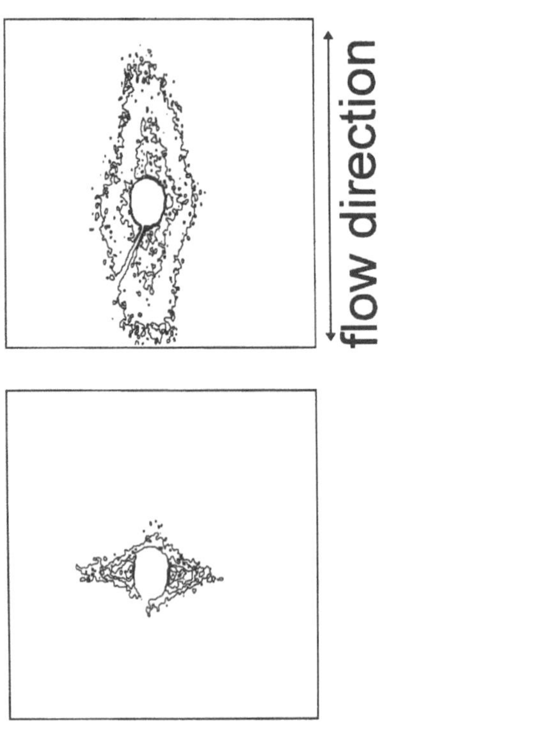

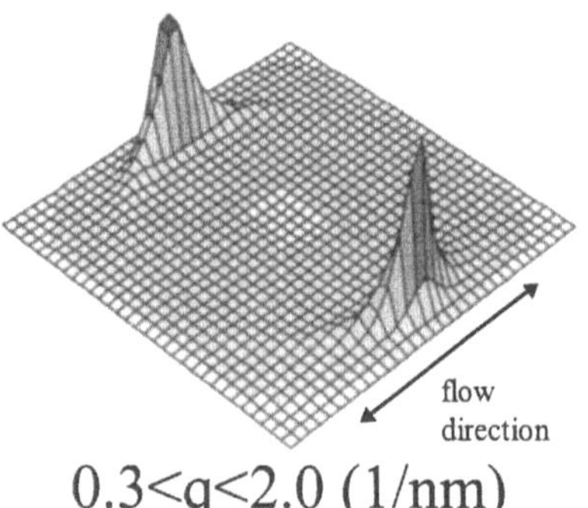

Fig. 2 Depolarized SALS patterns obtained from the hexagonal phase under shear at different shear rates

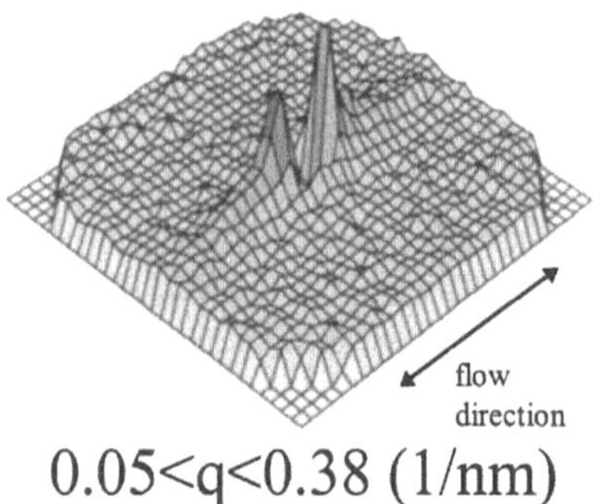

Fig. 3 SANS patterns from the hexagonal phase under shear at low shear rates and different scattering vectors

was found. The scattering vectors accessible in SALS are in the range of a few μm^{-1} and the structure responsible for the scattering patterns can also be studied by optical microscopy. A typical stripe texture was observed under crossed polarizers when the sample was slightly sheared. The stripes are aligned perpendicular to the flow direction and this gives rise to SALS pattern that is characterized by a strong scattering intensity along the flow direction. The stripe texture was discussed recently by Oswald and coworkers [10]. The director is aligned along the flow direction, but an undulation occurs that is responsible for the stripe texture. At higher shear rates, the stripe texture was destroyed and domains aligned in the flow direction were observed. This disrupted texture is responsible for the second SALS pattern shown in Fig. 2.

Obviously, the texture in the hexagonal phase is strongly affected by the shear deformation and three different mesoscopic structures of the sample can be obtained: (i) the polydomain structure in the quiescent state, (ii) a shear aligned structure characterized by the stripe texture and the SALS pattern in Fig. 2 (top) and (iii) a disrupted structure.

These three states display different rheological properties. The storage modulus was higher than the loss modulus in the polydomain sample, i.e. the elastic properties dominated. The moduli decreased when the sample

was shear oriented and storage and loss modulus were similar. Even lower moduli were observed when the stripe texture was disrupted by shearing the sample at higher shear rates [9].

So far only the influence of shear on the mesoscopic length scale was described. Since the micelles have a diameter of a few nm, it is not possible to follow the shear orientation of micelles by light scattering but neutrons can be used instead. The results obtained by small-angle neutron scattering from the hexagonal phase are shown in Fig. 3 [11]. A Bragg peak corresponding to the distance between micelles was observed for a high scattering vector. The scattering intensity was isotropically distributed on the two-dimensional multidetector from the polydomain sample at rest. However, the scattering pattern became anisotropic when the sample was sheared, as can be seen from Fig. 3. The scattering peak was only observed perpendicular to the flow direction, thus the rodlike micelles were aligned in flow direction, already at low shear rates. This observation confirms the results obtained by nuclear magnetic resonance [12]. A different scattering pattern was found at smaller scattering vectors with a higher scattering intensity along the flow direction.

The influence of shear on the micellar hexagonal phase can now be described as follows: *Microscopically* the

93

rodlike micelles are aligned along the flow direction and a well-aligned sample is obtained at low shear deformation. *Mesoscopically* this aligned sample is characterized by an undulation of the director [10], which gives rise to the stripe texture perpendicular to flow direction and the SALS pattern along the flow direction. This mesoscopic structure can be disrupted when stronger shear forces are applied.

The rheological properties of the hexagonal phase are determined by the structure on the mesoscopic length scale, i.e. by the texture. At low strain G' was larger than G'' but the moduli crossed when the sample was oriented by the shear deformation. Obviously, the grain boundaries between the domains in the polydomain sample are responsible for the dominating elastic properties. A disruption of the well-aligned sample results in a further decrease of the moduli.

The hexagonal phase can be shear oriented in simple shear flow as well as in large-amplitude oscillatory shear flow [9]. The experiment illustrated in Fig. 4 also leads to an orientation of the sample. Here storage and loss modulus at a frequency of $0.1 \, \mathrm{s}^{-1}$ are shown as a function of strain. Obviously, a reversal of the flow direction does not influence the alignment on the mesoscopic length scale. This is probably due to the undulation of the director. The director is aligned in a certain angle with respect to the flow direction. The undulation of the director means that the angle changes from "up" to "down" (zigzag) and therefore a reversal of the flow direction does not alter the average orientation of the director with respect to the direction of flow.

Lamellar phase

The lyotropic lamellar phase consists of surfactant bilayers of infinite lateral extension. The effect of shear on a lamellar phase was first studied by Roux and coworkers [13–15]. They described their results with an orientation diagram. At low and high shear rates (region I and III) the lamellae were aligned parallel to the wall, i.e. the layer normal was parallel to the direction of the velocity gradient. In region III less defects were present as compared to region I. At intermediate shear rates, multilamellar vesicles were observed (region II). The vesicles were of nearly spherical shape and were characterized by an isotropic scattering ring in small-angle light scattering.

Here we will review results obtained from two different systems involving an ionic and a nonionic surfactant. First a lamellar mesophase in the ternary system sodium dodecyl sulfate (SDS)/decanol/water will be discussed. The phase diagram was described by Quist et al. [16]. The sample that has been investigated in this study had a water mass fraction of 0.65. The molar fraction of decanol in the surfactant/cosurfactant mixture was 0.36. At this concentration of lamellar phase is observed. However, this phase is not a classical lamellar phase with homogeneous surfactant bilayers of infinite lateral extension, but is characterized by a defective structure of water holes in the surfactant lamella [17]. In the following, the defective lamellar structure will be abbreviated as L_α^H in order to distinguish it from the classical L_α lamellar phase. With this ternary system, the rheometer had to be loaded with the sample in the anisotropic phase. The viscosity was much lower as compared to the hexagonal phase and

Fig. 4 Strain dependence of the moduli of the hexagonal phase during shear orientation. G' filled symbols, G'' open symbols

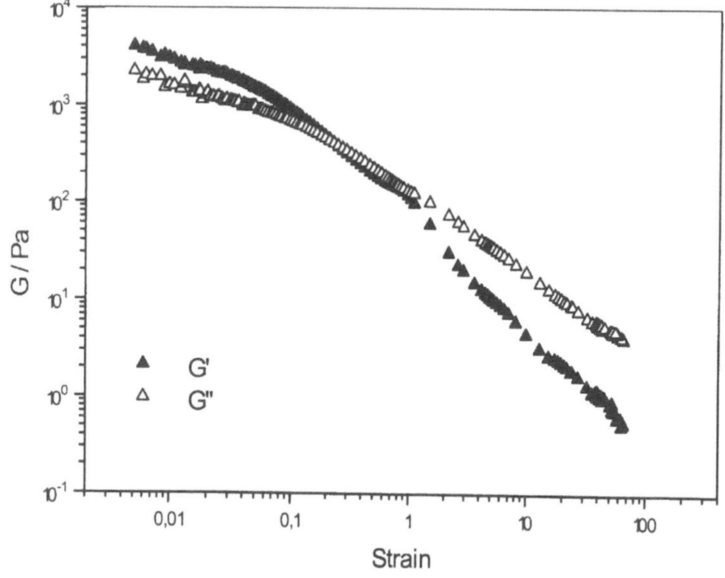

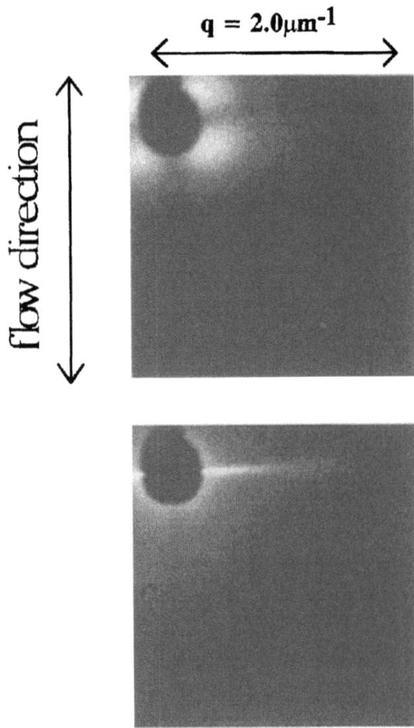

Fig. 5 Depolarized SALS patterns obtained from the defective lamellar phase under shear in the vesicle region (top) and the aligned lamellae (bottom)

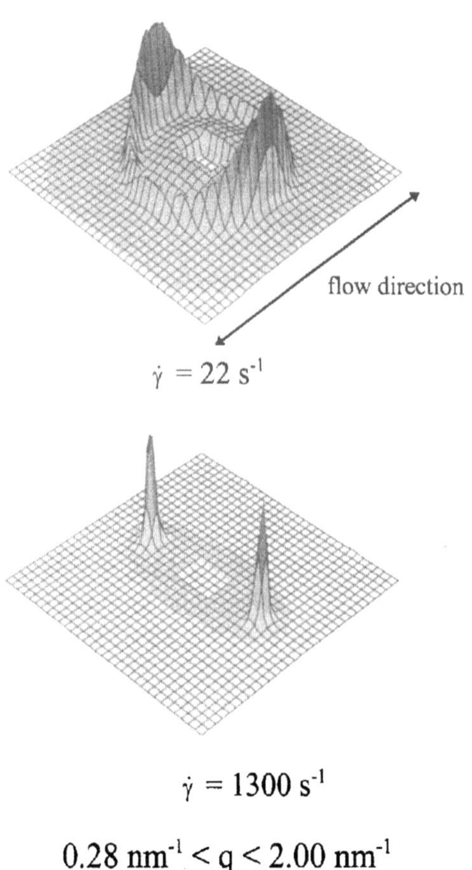

Fig. 6 SANS patterns obtained from the defective lamellar phase under shear in the vesicle region (top) and the aligned lamellae (bottom)

reproducible results could be obtained for the sample "as loaded". Nevertheless, preshearing effects could be of influence and therefore we will not describe the behavior of the unsheared sample but will only discuss the properties at higher shear rates where loading effects can be neglected.

In general the defective lamellar phase showed the same rheological properties as the classical phase studied by Roux and coworkers: three regions were found with a viscosity maximum at intermediate shear rates and a plateau at high shear rates. These two states are characterized by different scattering pattern in depolarized SALS as can be seen in Fig. 5. A four lobe pattern was observed at intermediate shear rates and a streak perpendicular to the flow direction was found at high shear rates [5].

A four-lobe scattering pattern as found in depolarized scattering is generally observed with optically anisotropic particles [18]. Optically anisotropic particles are, e.g. spherulites, but spherical or elongated vesicles could give rise to a four-lobe depolarized scattering pattern as well. This scattering pattern was recently observed also by Bergenholtz and Wagner with the AOT surfactant [19]. Light scattering is caused by fluctuations in polarizability. In the double layers of vesicles, the orientation of surfactant molecules with respect to the polarization plane of the incident light is inhomogeneous. In one part of the vesicle's

double layer, the surfactant molecules are aligned with their molecular axis along the flow direction and therefore this part of the double layer has a different polarizability in the polarization plane of the incident laser beam as compared to the bilayer in the longer axis of the vesicle. In other words, radial and tangential polarizabilities are different.

The SALS pattern at high shear rates (region III) indicates that lamellae which were stretched along the flow direction were present. With light scattering, however, it is not possible to decide in which direction the lamellae were aligned. Therefore SANS experiments have been performed that provide information on the orientation of the surfactant bilayers. Two typical scattering patterns are displayed in Fig. 6 [20]. At intermediate shear rates, the Bragg peak corresponding to the layer thickness was observed in all directions on the whole detector plane. That means that surfactant bilayers are present in all directions which is typical of a vesicular structure. The intensity of the peak, however, was not isotropic. The scattering intensity perpendicular to the flow direction was higher as compared to the intensity along the flow direction. Obviously, the vesicles were elongated during shear and this explains why the four-lobe pattern observed in SALS was slightly distorted.

At high shear rates the SANS pattern became extremely anisotropic and the Bragg peak was only detected perpendicular to the flow direction while the total scattering intensity at the peak position increased. A very good alignment of the lamellae was obtained with the layer normal being parallel to the neutral (vorticity) axis. This orientation is thus different as was observed by Roux and coworkers. The reason for this different behavior of the defective lamellar phase as compared to the classical phase is not clear but some suggestions have been discussed [20].

The second system to be reported here is a binary mixture of the surfactant C12E4 ($CH_3[CH_2]_{11}$ $[OCH_2CH_2]_4OH$) with water. The phase diagram was reported by Strey [21], the lamellar phase is stable in a broad temperature range and isotropic solution can be obtained at ca. 8 °C. Again, SALS was used to follow shear-induced structural changes on a mesoscopic length scale. Typical SALS patterns are shown in Fig. 7 [4].

Again, a four-lobe pattern typical of vesicles was observed in depolarized light scattering. The formation of vesicles also gave rise to an increase of viscosity but was obtained already at smaller shear rates as compared to the ionic system discussed above. With increasing shear rate the scattering pattern extended to larger angles, i.e. the vesicle size decreased.

A different scattering pattern was found at high shear rates. Enhanced scattering in flow direction was found in polarized light scattering. Here no scattering intensity was detected in the direction perpendicular to the flow. This characteristic scattering pattern has been observed before and was termed "*butterfly*" pattern. Butterfly scattering patterns have been observed in many different systems including semi-dilute polymer solutions, polymer networks, mixtures of molten polymers, liquid crystals and polymer liquids filled with silica spheres [22–25]. The butterfly pattern is believed to arise from shear-induced concentration and orientation fluctuations and might be a universal feature of soft two-component solids [26, 27].

SANS measurements at high scattering vectors gave similar results as with the defective lamellar phase in the vesicle region. The intensity of the Bragg peak perpendicular to the flow direction was slightly higher as compared to that along the flow direction. This pattern did not change significantly at higher shear rates when the butterfly pattern was observed in light scattering. Thus the underlying structure of the butterfly pattern was still vesicular. Obviously, the vesicle size decreased with shear rate and eventually the vesicles were too small to be detected in

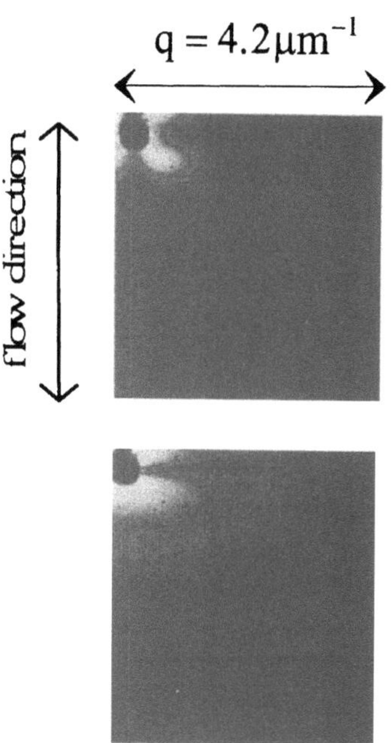

Fig. 7 SALS from the C12E4 lamellar phase under shear, top: vesicle pattern in depolarized SALS at low shear rate; bottom: butterfly pattern observed in polarized SALS at higher shear rate

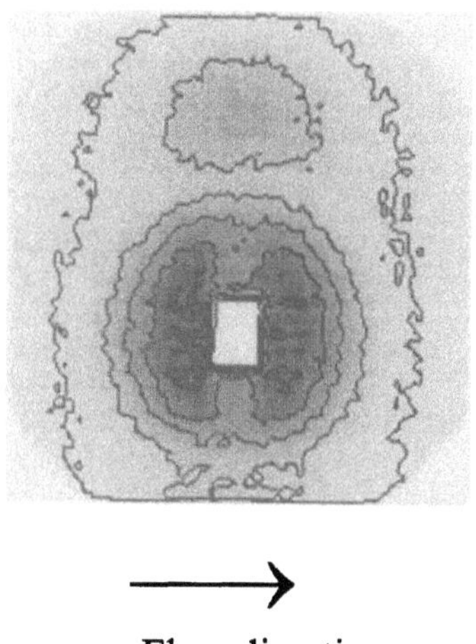

Fig. 8 Butterfly pattern observed in SANS from the C12E4 lamellar phase at high shear rate

light scattering. Instead a supravesicular structure was observed.

The formation of this specific supravesicular structure that gives rise to the butterfly pattern is strongly dependent on the shear history. A butterfly pattern was also found in neutron scattering as shown in Fig. 8 [28]. Here it was obtained when the shear rate was reduced from 1500 to 1250 s⁻¹. The butterfly pattern was accompanied by an intensity maximum perpendicular to the flow direction.

The butterfly pattern relaxed after cessation of shear and a scattering ring was observed. Apparently, an ordered vesicular suprastructure was present in the quiescent state. The shear flow induced concentration fluctuations along the flow direction giving rise to the butterfly pattern, but the ordered structure prevailed perpendicular to the direction of flow. In contrast to the ionic system, with the C12E4 water system it was not possible to destroy the vesicles and to obtain oriented lamellae at high shear rates.

Conclusions

The influence of shear on three different lyotropic mesophases, namely a hexagonal, a lamellar and a defective lamellar phase was described. All examples demonstrate that shear flow can alter the *microscopic* as well as the *mesoscopic* structure in a complex manner. Both, the shear orientation of the anisotropic micelles and the influence of mechanical deformation on the texture, in other words the defects, influence the behavior of the sample under flow but also the equilibrium properties studied by oscillatory shear within the linear viscoelastic region. In order to understand rheological properties of such anisotropic mesophases it is therefore necessary to monitor the sample structure on a broad length scale. Here scattering experiments are very helpful, since a combination of neutron, X-ray and light scattering can cover dimensions from nm to μm, but other techniques should be also used whenever possible [29].

Acknowledgment The results reported here have been obtained in cooperation with K. Berger, J. Läuger, R. Linemann, G. Schmidt, R. Weigel and J. Zipfel. The SANS experiments were performed in collaboration with P. Lindner at the ILL in Grenoble. Support by the Deutsche Forschungsgemeinschaft is gratefully acknowledged.

References

1. Tiddy GJT (1980) Phys Rep 57:1
2. Larson RG, Doi M (1991) J Rheol 35:539
3. Richtering W, Läuger J, Linemann R (1994) Langmuir 10:4374
4. Weigel R, Läuger J, Richtering W, Lindner P (1996) J Phys II France 6:529
5. Läuger J, Weigel R, Berger K, Hiltrop K, Richtering W (1996) J Colloid Interface Sci 181:521
6. Läuger J, Gronski W (1995) Rheol Acta 34:70
7. Lindner P, Oberthür RC (1984) Rev Phys Appl 19:759
8. Kratzat K, Finkelmann H (1993) Liquid Crystals 13:691
9. Linemann R, Läuger J, Schmidt G, Kratzat K, Richtering W (1995) Rheol Acta 34:440
10. Oswald P, Géminard JC, Lejcek L, Sallen L (1996) J Phys II France 6:281
11. Schmidt G, Müller S, Lindner P, Schmidt C, Richtering W, in preparation
12. Lukaschek M, Müller S, Hasenhindl A, Grabowski DA, Schmidt C (1995) Colloid Polym Sci 274:1
13. Diat O, Roux D (1993) J Phys II France 3:9
14. Roux D, Nallet F, Diat O (1993) Europhys Lett 24:51
15. Diat O, Roux D, Nallet F (1993) J Phys II France 3:1427
16. Quist PO, Fontell K, Halle B (1994) Liquid Crystals 16:235
17. Berger K, Hiltrop K (1996) Colloid Polym Sci 274
18. Samuels RJ (1971) J Polym Sci Part A-2 9:2165
19. Bergenholtz J, Wagner NJ (1996) Langmuir 12:3122
20. Zipfel J, Lindner P, Richtering W, in preparation
21. Strey R (1996) Ber Bunsenges Phys Chem 100:182
22. Bastide J, Leibler L, Prost J (1990) Macromolecules 23:1821
23. Inoue T, Moritani M, Hashimoto T, Kawai H (1971) Macromolecules 4:500
24. van Egmond JW, Fuller GG (1993) Macromolecules 26:7182
25. DeGroot JV, Macosko CW, Kume T, Hashimoto T (1994) J Colloid Interface Sci 166:404
26. Rabin Y, Bruinsma R (1992) Europhys Lett 20:79
27. Onuki A (1992) J Phys II France 2:45
28. Zipfel J, Lindner P, Richtering W, in preparation
29. Safinya CR, Sirota EB, Bruinsma RF, Jeppesen C, Plano R, Wenzel L (1993) Science 261:588

Progr Colloid Polym Sci (1997) 104:97–103
© Steinkopff Verlag 1997

W. van Megen
B.J. Ackerson

Comparison of Bragg and SALS studies of crystallization in suspensions of hard spheres

W. van Megen
Department of Applied Physics
Royal Melbourne Institute of Technology
Melbourne, Victoria, Australia

Dr. B.J. Ackerson (✉)
Department of Physics
and Center for Laser
and Photonics Research
Oklahoma State University
Stillwater, Oklahoma 74075, USA

Abstract Crystallization in nearly identical suspensions of hard spheres has been studied independently by small angle light scattering and by Bragg scattering. The results of those two measurement techniques are brought together here. Qualitatively different behavior is observed in different time domains. The reduced time and crystal size agrees for the different measurement techniques when compared at the boundary between the time domains. As a result the reduced nucleation rate density also correlates between the two techniques. Below the melting volume fraction a comparison of the characteristic length scale with either the small angle intensity or the Bragg crystal fraction indicates a constant nucleation rate density. Above the melting volume fraction an argument is given for accelerated nucleation. The theoretical classical nucleation rate varies more rapidy with volume fraction between melting and freezing than is observed experimentally.

Key words Colloids – crystallization – light scattering – hard spheres – nucleation

Introduction

The process of solidification has intrigued the scientific community ever since Fahrenheit reported his observations on the freezing of water in 1714. Yet today we do not have a thorough grasp of the microscopic mechanisms by which liquid solidifies as it is cooled. Under certain conditions suspensions of colloidal particles show transitions to ordered structures that closely resemble the freezing of molecular liquids [1]. The important difference is that the "freezing" transitions in colloids occur on experimentally accessible time scales and Bragg scattering from colloidal crystals, examples of which include opal, crystalline plant viruses and polymer particles, are visible in ordinary light. It is apparent, therefore, that colloidal suspensions present themselves as valuable experimental models by which we might gain deeper insight into crystallization process.

Crystallization in suspensions of colloidal hard spheres has been studied recently by small angle light scattering (SALS) [2, 3] and by Bragg scattering [4, 5]. SALS monitors density fluctuations, a conserved (order) parameter, while Bragg scattering monitors the nonconservative crystal order parameter. The two techniques are complementary in terms of the optimal volume fraction range for operation. Bragg scattering works better for volume fractions greater than and equal to the melting value (0.545) where large numbers of relatively small crystals are generally encountered. This ensures that a statistically significant number of Bragg reflecting crystals scatter to the detector, producing a signal proportional to the volume of crystal present in the sample. On the other hand, SALS works better for volume fractions less than the melting value, because all crystals (not just those properly oriented) contribute to the signal, the crystals are relatively larger, and the signal is proportional to the volume of crystal present times the volume of a typical crystal [2]. The SALS signal declines rapidly with increasing volume fraction, because the typical crystal size is reduced. Nearly identical samples have been studied in separate experiments by these two techniques [3–5]. These results are compared and contrasted here.

Experiment and data analyses

The colloidal "hard spheres" are uniformly sized poly-methylmethacrylate (PMMA) core particles coated with a thin (~ 10 nm) layer of poly-12-hydroxystearic acid [6]. This coating provides sufficient steric stabilization to prevent flocculation. The particles were made in two separate batches and identified as SMU20 for the SALS experiments and SMU29 for the Bragg scattering experiments. Average radii and polydispersities, determined by dynamic light scattering on dilute suspensions, are listed in Table 1. The SMU20 particles were suspended in a mixture of tetralin and decalin and the SMU29 particles in a mixture of decalin and carbon sulfide. In each case the composition of these liquid mixtures (see Table 1) was adjusted to closely match the index of refraction of the particles so that the resulting suspensions appeared transparent.

The equilibrium phase behavior of these suspensions mimics that of the hard sphere system [7]. Freezing and melting concentrations were respectively identified as the concentration where crystallization first occurred and that where the suspension became fully crystalline. Equating the observed freezing concentration with the value, $\phi_f = 0.494$, known for the freezing value of hard spheres [8], gave a multiplicative factor with which concentrations of all samples were be converted to effective hard sphere volume fractions, ϕ. The observed effective hard sphere volume fraction at melting, $\phi_m = 0.545$, agrees with that expected for perfect hard spheres [8].

Samples were prepared in a metastable fluid state by shear melting prior to measurements. Elapsed times were measured from the cessation of tumbling the sample cell. The small angle intensity, $I_S(q)$, and the intensity around the lowest order Bragg maximum, $I_B(q)$, were measured

and the respective results were analysed by the following procedures.

A given SALS spectrum was characterized by its maximum, I_{max}, and the larger wavevector, $q_{(1/2)}$, where the intensity fell to half its maximum. Scaled structure factors and wavevectors were then calculated as,

$$S_S(Q) = I_S(Q)/I_{max}, \quad Q = q/q_{(1/2)} . \tag{1}$$

The characteristic crystal length reduced by the particle radius, a, was determined from,

$$L_S = 2\pi/q_{(1/2)}a . \tag{2}$$

This expression has been confirmed by direct measurements of the crystal radius [9]. The intensities $I_B(q)$ were divided by the particle form factor and corrected for the fluid scattering. From the resulting crystal structure factors, $S_B(q)$, the following quantities were calculated. First, the total integrated area X is proportional to the fraction of crystal present. Second, the peak width, Δq, is related to a crystal length scale L_B reduced by the particle radius as

$$L_B = \frac{\pi K}{\Delta q a}, \tag{3}$$

where $K = 1.155$ is the Scherrer constant for a crystal of cubic shape. Third, assuming the crystal structure to be close-packed, the crystal volume fraction, ϕ_c, was obtained from the peak position, q_m, by

$$\phi_c = \frac{2q_m^3}{9\pi^2\sqrt{3}} . \tag{4}$$

The reader is referred to the literature [2–5] for further details of sample preparation and characterization, as well as a thorough discussion of SALS and Bragg scattering techniques.

Results

Figure 1 shows the scaled SALS intensity distribution, $S_S(Q)$, as a function of the scaled scattered wavevector, Q, for a sample near melting ($\phi = 0.549$) and parametrized by the elapsed time. Figure 2 shows the time evolution of I_{max} and $q_{(1/2)}$ for the same sample.

Figure 3 shows the Bragg scattering crystal structure factor for a sample with $\phi = 0.537$ (equilibrium coexistence region) in the vicinity of the first order maximum and parametrized by elapsed time. Figure 4 shows the crystal fraction, X, and crystal dimension, L_B, obtained from the Bragg scattering peak as a functions of elapsed time for volume fractions 0.537, below melting and 0.570, above melting. The corresponding crystal volume fractions, calculated by Eq. (4), are shown in Fig. 5.

Table 1 Showing sample designation, average particle radius, polydispersity as standard deviation to mean radius, and weight percent of the higher refractive index liquid mixed with decalin

Designation	SMU20	SMU29
Radius	215 nm	200 nm
Polydispersity	6%	5%
Solvent	46% tetralin	30% carbon disulfide

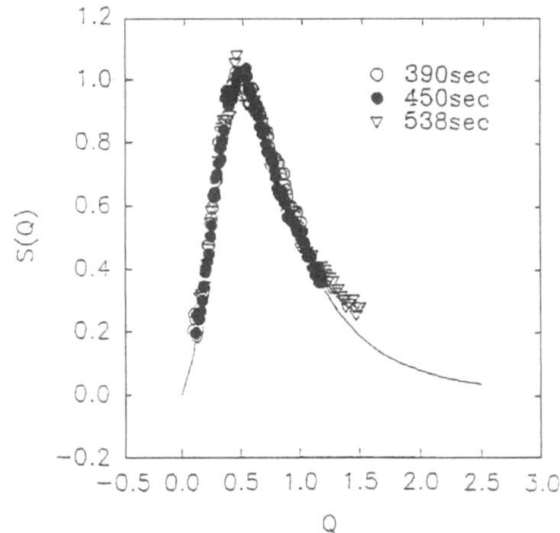

Fig. 1 Scaled SALS structure factor S_S as a function of the scaled wavevector Q and parametrized by the time in seconds after cessation of shear melting for sample with volume fraction 0.549

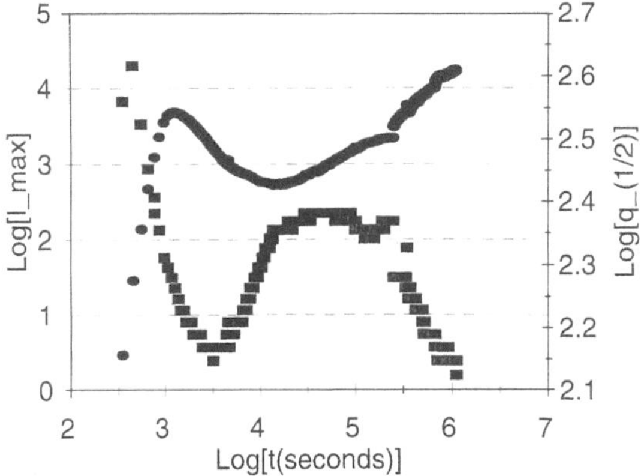

Fig. 2 SALS scaling parameters I_{max} (circles-left axis) and $q_{(1/2)}$ (squares-right axis) as a function of elapsed time for sample with volume fraction 0.549

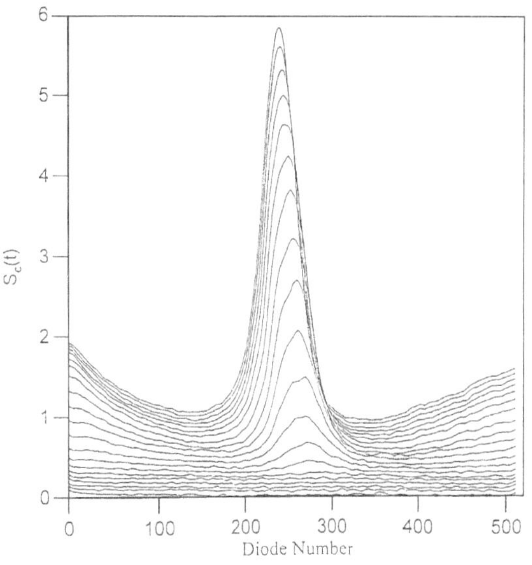

Fig. 3 Structure factor S_B for the lowest order Bragg peak, at 20 s intervals, as a function of diode channel number

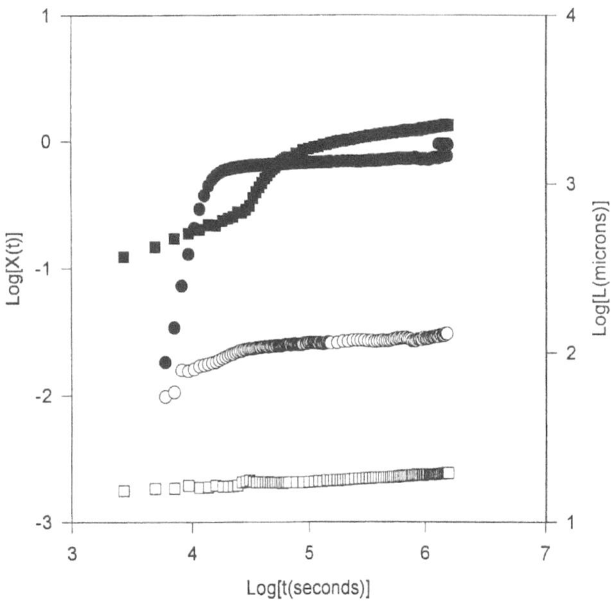

Fig. 4 The solid symbols correspond to the Bragg scattering crystal fraction X (left axis) and the open symbols to the crystal size $L = L_B a$ (right axis). Two volume fractions are shown: 0.537 (circles) and 0.570 (squares)

Following shear melting, distinct relatively rapid growth regimes are exhibited in both techniques (Figs. 2 and 4). The intensity maximum of the SALS results increases as a (quasi-) power law of elapsed time, $I_{max} \sim t^\eta$, while a measure of the position of the maximum, $q_{(1/2)} \sim t^{-\alpha}$, decreases with time. The exponents range from $\eta = 4.66 \pm 0.02$ and $\alpha = 0.75 \pm 0.02$, at one of the smallest volume fractions studied ($\phi = 0.531$), to $\eta = 7.15 \pm 0.16$ and $\alpha = 1.01 \pm 0.88$ for a sample at the melting point ($\phi = 0.545$). This initial stage of rapid increase in I_{max} and accompanying decrease in $q_{(1/2)}$ has been associated with "nucleation and growth" and is followed by the "crossover" region where the maximum intensity decreases and the position of the maximum increases.

However, these variations in the characteristic length lag the intensity variations. Finally, I_{max} increases again while $q_{(1/2)}$ decreases in the region which we refer to as "ripening". Interestingly, the dynamical scaling, illustrated in Fig. 1 for the nucleation and growth stage, is lost in the crossover regime but it is recovered during ripening.

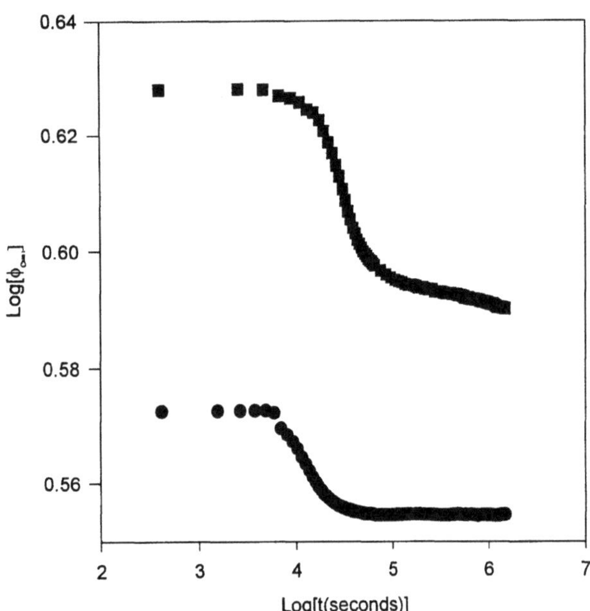

Fig. 5 Solids volume fraction ϕ_c as a function of elapsed time determined by Bragg scattering

For volume fractions up to about the melting value the area under the Bragg peak increases as $X \sim t^\mu$, during nucleation and growth, with μ increasing from 3 at volume fraction $\phi = 0.530$ to 4 at $\phi = 0.548$. The crystal characteristic length scale behavior is difficult to determine given the quality of the data except for $\phi = 0.548$ where $L_B \sim t^\beta$ with $\beta \sim 0.5$. For larger times X shows a much slower rate of growth or saturation.

Above the melting volume fraction, the integrated area, X, shows a qualitatively different behavior with an initial growth characterized approximately by power law growth with exponent ~ 0.3, which is then followed by a more rapid growth with μ ranging from 3 to 1 as the volume fraction increases from the melting value. At these elevated volume fractions the crystal length scale, L_B, evidences little change in the nucleation and growth regime. As time progresses the integrated intensity, X, again crosses over, but now in a manner that is less pronounced than at lower volume fractions, to a stage of much slower growth or saturation. Beyond this crossover L_B shows modest rates of increase. In the Bragg scattering experiments elapsed times were not followed to time scales corresponding to "ripening" observed in SALS experiments.

Discussion

For volume fractions less than the melting value the observed growth exponents may be explained with the

following model. The crystal length scale is assumed to increase as

$$L_{S,B} \sim 1/q_{(1/2)}a \sim 1/\Delta qa \sim t^\alpha . \tag{5}$$

The number of crystals is given as

$$N \sim t^\nu . \tag{6}$$

$\nu = 1$ corresponds to a constant nucleation rate. The SALS intensity maximum for coherent scattering from a collection of randomly placed crystals is given by the product of the number of crystals times the square of the volume of a crystal [2], or as stated in the introduction, the product of the total volume occupied by crystals times the single crystal volume, i.e.,

$$I_{max} \sim NL_S^6 \sim t^{\nu + 6\alpha} . \tag{7}$$

For Bragg scattering the integrated intensity indicates the total crystal volume,

$$X \sim NL_B^3 \sim t^{\nu + 3\alpha} . \tag{8}$$

In Fig. 6 we compare values of the growth exponents α obtained directly from $q_{(1/2)}$ and, assuming $\nu = 1$, from I_{max} and X (Eq. (7) and (8)). As stated earlier, in the Bragg scattering experiments L_B is poorly characterized in the coexistence region during nucleation and growth. Below the melting volume fraction approximate agreement in the values α indicates compatibility of the SALS and Bragg scattering and consistency of α obtained from I_{max} and $q_{(1/2)}$ vindicates the assumption of constant nucleation rates in the coexistence region. However, differences in the estimates of α increase for volume fractions above melting suggesting deviations from a constant nucleation rate.

We understand this behavior qualitatively in terms of classical nucleation theory. At lower volume fractions (in the equilibrium coexistence region), a situation analogous to slightly undercooled molecular liquids, the nucleation rate density predicted by classical theory is comparatively low [10]. One can then visualize nucleation and growth of crystals with large separations from one another causing only a slow reduction in the volume fraction of the metastable fluid. As a consequence, the nucleation rate remains constant until the equilibrium complement of crystals is obtained.

For volume fractions greater than the melting value, SALS indicates limited or no crystal growth (Fig. 2 and Ref. [2]) and Bragg scattering indicates no crystal growth (Fig. 4 and Refs. [4, 5]) in the nucleation and growth regime. This may happen if the nucleation rate density is so large that crystals have little time to grow before the sample is filled with crystals. For the larger metastable fluid volume fractions, classical nucleation rate density predicts an increase in nucleation rate with decreasing metastable fluid volume fraction [10]. When crystals

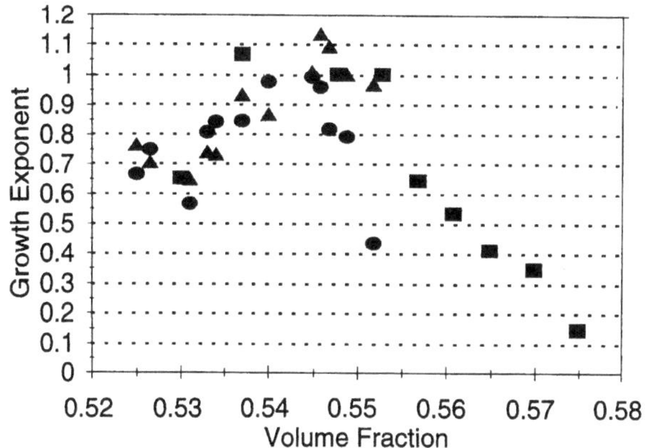

Fig. 6 Growth exponent α determined indirectly from a model for SALS $I_{\max}$ (circles), indirectly from a model for Bragg scattered intensity X (squares), and directly from $q_{(1/2)}$ (triangles) as a function of volume fraction

nucleate to form slightly denser crystals due to mechanical equilibration, the metastable fluid density is decreased and the nucleation rate increases. An accelerated nucleation rate results [4, 5]. While we have no quantitative theory to describe these observations, we speculate that this is the reason for the two stage behavior observed for X at large values of ϕ and that this behavior occurs in the absence of any significant change in the crystal size. It is also possible that relaxation of the sheared fluid competes with initial nucleation.

Following the nucleation and growth stage the crystal content, X obtained from the Bragg peak, increases much more slowly (Fig. 4), attaining values that generally exceed unity. If the total crystal content were achieved at this crossover, then naively X is expected to saturate at unity. However, the total Bragg peak intensity depends on both the crystal content and the crystal quality, more perfect crystals having a greater scattering power than less perfect ones. The shift in position of the Bragg peak indicates that the crystals are expanding in time (Fig. 5). Interestingly, this expansion is also predicted by the application of the classical theory of crystal growth to hard-sphere colloids [10]. The observed expansion may be accompanied by reduction in nonuniformities and stacking faults within the crystals, leading to the gradual increase in scattering power seen in Fig. 4.

In contrast to the monotonic behavior of the moments of the Bragg peak, SALS indicates a dramatic decrease in the characteristic length scale in the crossover region. Evidently, gravitational stresses and stresses induced by crystal expansion cause the breakup of crystals, particularly the larger crystals that grow in the coexistence region. At volume fractions in excess of the melting value the reduction in L_S in the crossover region is far less

pronounced [4], indicating that smaller crystals are less affected by these stresses. These ideas are also supported by direct observation [9]. According to Eq. (7) the observed accompanying decrease in $I_{\max}$ is expected. However, this model assumes the independent growth of randomly positioned crystals. As the sample fills with crystals, the scattering mechanism which produces the maximum intensity at finite wavevector must change from a form factor for a dense crystal surrounded by a depletion zone [13] to a correlation in position between neighboring crystals. This shift in mechanism is a possible source for the temporal separation between the maximum intensity and minimum wavevector observed in SALS. An explanation for the apparent inconsistency in the time dependence of the length scales, L_S and L_B, in the crossover region (Figs. 2 and 4) is that the crystals break up predominantly along grain boundaries leaving the Bragg reflecting planes largely intact. At the same time, the total crystal content, as measured by the area, X, under the Bragg peak remains unaffected (Fig. 4).

Next we compare the crossover times, the characteristic lengths at crossover times and the nucleation rate densities obtained by SALS and Bragg scattering. In order to eliminate differences in particle size and solvent viscosity in the two sets of measurements, time is reduced with respect to the Brownian characteristic time, $\tau_0 = D_0 t/a^2$, where D_0 is the free particle diffusion constant and a is the particle radius (Table 1). The crossover time, t_c, for SALS is defined to be the time where $I_{\max}$ reaches its first relative maximum. For Bragg scattering it is defined to be the time where X exhibits a dramatic decrease in the rate of growth. The reduced crossover times, $\tau_c = D_0 t_c/a^2$, are shown as a function of volume fraction in Fig. 7. The characteristic crystal size, $L_{B,c}$, at the crossover time and the minimum characteristic crystal size, $L_{S,c}$, in the crossover region are compared in Fig. 8.

If we assume that the equilibrium complement of crystal is realized at τ_c, then the average nucleation rate density, R_S, is determined from SALS measurements as

$$R_S = \frac{[(\phi - 0.494)/(0.545-0.494)]}{\pi L_S^3 \tau_c/6}. \tag{9}$$

This is simply the fraction of sample filled with crystal divided by the size of the average crystal and the elapsed time of the measurement. For Bragg scattering measurements a similar estimate of the nucleation rate density is constructed from the time taken to reach the maximum in the crystal density calculated from measured values of X and L_B as

$$R_B = \max[X/L_{B,c}^3]/\tau_c. \tag{10}$$

W. van Megen and B.J. Ackerson
Comparison of Bragg and SALS studies

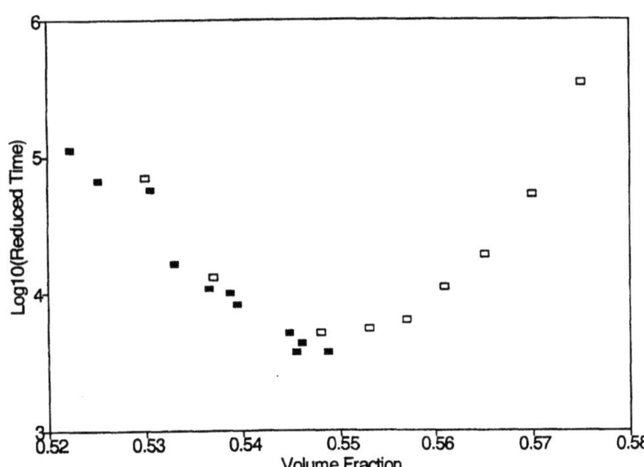

Fig. 7 Reduced time at crossover $\tau_c = D_0 t_c / a^2$ as a function of volume fraction from SALS measurements (filled squares) and Bragg measurements (open squares)

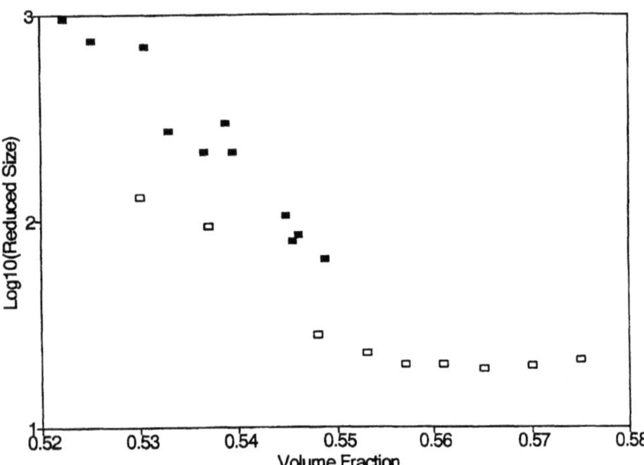

Fig. 8 Reduced sizes L_S and L_B as a function of volume fraction for both SALS measurements (filled squares) and Bragg measurements (open squares)

This estimate of the average nucleation rate density coincides with the maximum rate of crystal addition, $\max[\mathrm{d}(X/L_B^3)\mathrm{d}\tau]$ [4]. Figure 9 shows the dimensionless nucleation rate densities R_S and R_B as functions of the metastable fluid volume fraction. The results shown in Figs. 7–9 indicate general agreement for the different measurement techniques.

Also shown in Fig. 9 as a solid, dotted and dashed lines are estimates of the classical nucleation rate density for monodisperse samples with reduced surface energies $\gamma a^2 / k_B T = 0.16$ and 0.14 and a reduced surface energy 0.16 with a polydispersity 7%, respectively. Here the surface energy between liquid and crystal averaged over crystal

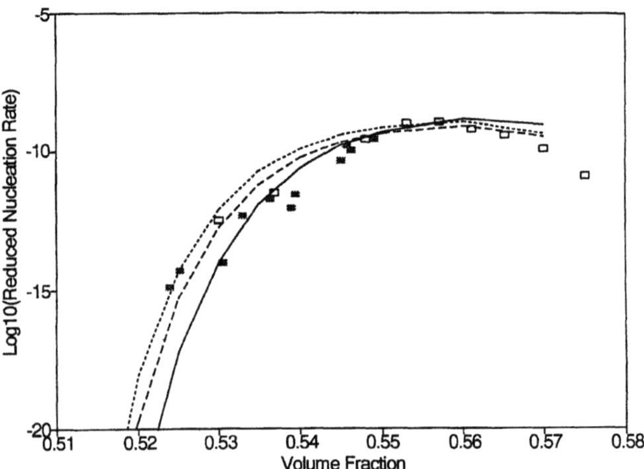

Fig. 9 Reduced nucleation rate density R_S shown as filled squares and R_B shown as open squares as a function of volume fraction. The solid line gives R_C for a surface energy 0.16 and the dotted line for 0.14. The dashed line represents a surface energy 0.16 but finite polydispersity 0.07 for the metastable fluid phase

faces is γ and $k_B T$ is the thermal energy. The polydispersity is accounted for in the equation of state for the metastable fluid only [11]. Classical theory gives the nucleation rate density as [10]

$$R_C = A\phi^{5/3} D \exp\left[-\frac{4\pi^3 \gamma^3}{27\phi^2 \Delta\mu^2} \right]. \tag{11}$$

Here the dimensionless diffusivity is taken to be $D = (1 - \phi/\phi_g)^{2.6}$ which fits to the long-time single particle diffusion coefficient measured on metastable fluids of these particles [12]. The chemical potential difference between the metastable fluid and the crystal in mechanical equilibrium with it is given by $\Delta\mu$. Significantly, the combination of SALS and Bragg scattering results are consistent with the predictions of Eq. (12) for over about 6 decades.

Conclusions

We have compared results of crystallization dynamics of suspensions of hard spherical particles by small angle and Bragg laser light scattering. These techniques are complementary in that SALS is better suited to the study of crystallization in the coexistence region, whereas Bragg scattering gives the stronger signal above the melting volume fraction. The results obtained by these techniques are compatible and, where the suspension volume fractions overlap, consistent. In particular, the two methods give concordant results for the elapsed times, from shear melting to complete crystallinity, and nucleation rate densities. In the coexistence region the results of both experiments are compatible with linear rates of nucleation and linear or marginally sublinear crystal growth. By contrast, crystal growth is virtually suppressed due to high nucleation rates, above the melting volume fraction.

Acknowledgments Support from the National Science Foundation (grant DMR 9501865) and the Australian Research Council are gratefully acknowledged.

References

1. Ackerson BJ (ed) (1990) Phase Transitions in Colloidal Suspensions, Vol 21. Gordon and Breach, New York
2. Schätzel K, Ackerson BJ (1993) Phys Rev E 48:3766
3. He Y, Ackerson BJ, van Megen W, Underwood SM, Schätzel K, Phys Rev E, accepted
4. Harland JL, Henderson SI, Underwood SM, van Megen W (1995) Phys Rev Lett 75:3572
5. Harland JL, van Megen W, to be published
6. Antl L, Goodwin JW, Hill RD, Ottewill RH, Owens SM, Papworth S, Waters JA (1986) Colloids Surfaces 17:67
7. Pusey PN, van Megen W (1986) Nature (London) 320:340; Paulin SE, Ackerson BJ (1990) Phys Rev Lett 64:2663
8. Hoover WG, Ree FH (1968) J Chem Phys 49:3609
9. He Y, Olivier B, Ackerson BJ, to be published
10. Ackerson BJ, Schätzel K (1995) Phys Rev E 52:6448
11. Salacuse JJ, Stell G (1982) J Chem Phys 77:3714
12. Russel WB (1990) Phase Trans 21:127; van Duijneveldt JS, Lekkerkerker HNW (1995) In: van Erde, Bruinsma OSL (eds) Science and Technology of Crystal Growth. Kluwer Academic, Dordrecht
13. Ackerson BJ, Schätzel K (1992) In: Garrido L (ed) Complex Fluids. Springer, Heidelberg, p 15

Progr Colloid Polym Sci (1997) 104:104–106
© Steinkopff Verlag 1997

Effective interaction potential obtained from experimental structure factors: the inverse problem

M. Quesada-Pérez
J. Callejas-Fernández
R. Hidalgo-Álvarez

M. Quesada-Pérez
Dr. J. Callejas-Fernández (✉)
R. Hidalgo-Álvarez
Biocolloid and Fluid Physics Group
Department of Applied Physics
Campus Fuentenueva s/n
18071 Granada, Spain

Abstract The structure factor $S(q)$ was measured for a colloidal dispersion of latex microspheres (40 nm of diameter) by both static and dynamic light scattering (using a Malvern Instruments spectrometer, $\lambda = 488$ nm). The structure was formed for a sample of particle number density 2×10^{13} ml^{-1}. The interaction potential was determined from the experimental $S(q)$ by means of the straightforward use of three different closures (Percus–Yevick, hipernetted chain, mean spherical approximation). Some difficulties found with this procedure (such as the cut-off effect) are discussed. However, these closures lead to very similar results. Furthermore, the results are tested and compared with the simulations based on a Yukawa potential.

Key words Structure factors – colloidal stability – polymer colloids

Introduction

During the last years, optical techniques such as dynamic light scattering (DLS) or static light scattering (SLS) have been used to get information about the interaction forces between particles forming a colloidal dispersion. Colloidal dispersions are treated as macroscopic analogs of atomic liquids, on which liquid-state theories can be applied [1]. Then, the experimentally obtained structure factor ($S(q)$) can be related to the *effective* potential of interaction between particles.

There are two ways to study the problem: The structure factor $S(q)$ can be obtained from a previously known potential (for instance, the DLVO one) applying integral-equation theories like hipernetted chain (HNC), Percus–Yevick (PY) or mean spherical approximation (MSA) [2] or applying computational methods [3]. The obtained $S(q)$ is then compared with the experimental one. This is the direct problem. Also, the effective potential of interaction can be obtained from experimental data of $S(q)$. This is the inverse problem. Recently, this problem has been solved using new theoretical approaches [5, 6].

The aim of this work is to analyze some of the difficulties which arise with the inverse problem. To test the obtained potential, we have compared the data of the interaction potential with computational simulations.

Experimental

The latex used in the experiments was prepared from butyl acrylate and styrene by emulsion polymerization with sodium dodecyl sulphate as surfactant, potassium peroxidisulphate as initiator, and NaHCO$_3$ as buffer. The mean diameter σ was measured by DLS and found to be 40 ± 3 nm. The latex was cleaned by dialysis and kept with Amberlite MB-3 mixed-ion exchange resins. The colloidal liquid was formed using a particle concentration (ρ) of 2×10^{13} ml^{-1}.

The experimental set up used is the 4700 System from Malvern Instruments with an Argon laser of 75 mW

($\lambda = 488$ nm) and a 64-channels correlator. This spectrometer measures the scattered light intensity $I(q)$ at scattering vectors q corresponding to angles from $30°$ to $140°$. After a week we observe no changes in the intensity curve. $I(q)$ was also measured for a reference dispersion ($\rho_0 = 1 \times 10^{12}$ ml^{-1}) to calculate $S(q)$ by SLS. Furthermore, $S(q)$ was determined from the effective diffusion coefficient D_{eff} (DLS). These results coincide with those obtained using SLS.

The Monte Carlo (MC) simulations were carried out with 216 particles and 6000 steps. The right performance of the programme (with these parameters) was first checked by comparing simulations with the ones found in literature [6].

Results and discussion

The radial correlation function $g(r)$ can be obtained from experimental $S(q)$ (Fig. 1) under certain assumptions. We assumed $S(q) = 1$ for values of $q \cdot \sigma$ greater than 1.28 (corresponding to $140°$) and extrapolated for $q \cdot \sigma$ lower than 0.35 (corresponding to $30°$) by fitting $S(q)$ at low q. Figure 1 also shows the *extended* $S(q)$.

The function $g(r)$ was computed by means of

$$g(r) = 1 + \frac{1}{2\pi^2\rho}\frac{1}{r}\int\limits_0^\infty (S(q) - 1)q\sin(qr)\mathrm{d}q \qquad (1)$$

(see solid line in Fig. 2). It should be noted that the main peak is close to $\rho^{-1/3}$, i.e., the mean distance between the neighbour particles. However, $g(r)$ does not tend to zero for r going to zero, and even takes negative values which

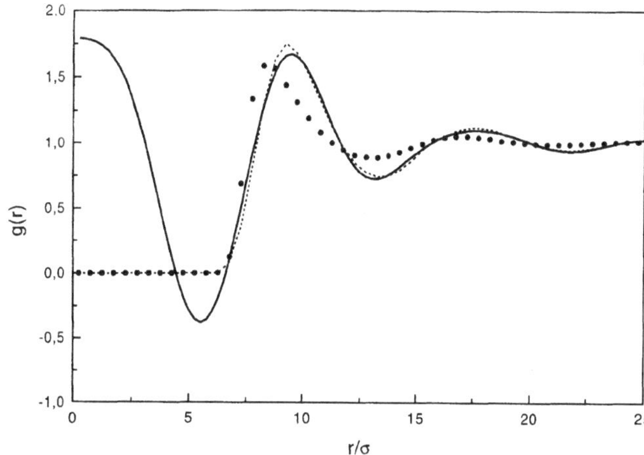

Fig. 2 Radial correlation function $g(r)$: (a) obtained from structure factor (——); (b) recovered by means of MC simulation with a potential obtained with HNC closure (------); (c) recovered by means of MC simulation with a Yukawa potential (·····)

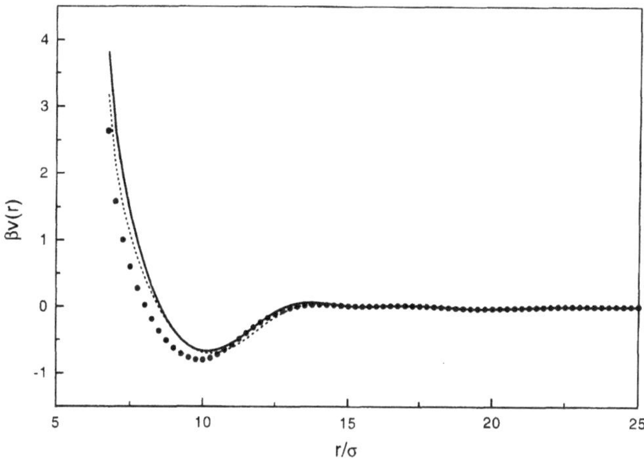

Fig. 3 Potentials obtained from $g(r)$ and $c(r)$ and the following closures: (a) HNC (——); (b) PY (- - -); (c) MSA (·····)

have no physical meaning. Additional calculations seem to point out that this behaviour does not depend on the structure factor $S(q)$ at low q. So we think it can be due to both the so-called *cut-off effect* and inaccuracies in experimental $S(q)$. The integral in (1) would not tend to zero faster than r because of these errors. This has also been pointed out in literature [4]. We put $g(r) = 0$ from $r/\sigma = 0$ to $r/\sigma = 6.75$ in order to calculate the potential. As $h(r) = g(r) - 1$, the direct radial correlation function $c(r)$ can be computed with the aid of Ornstein–Zernike relationship

$$h(r) = c(r) + \rho \int \mathrm{d}^3r' c(r')h(|r' - r|) . \qquad (2)$$

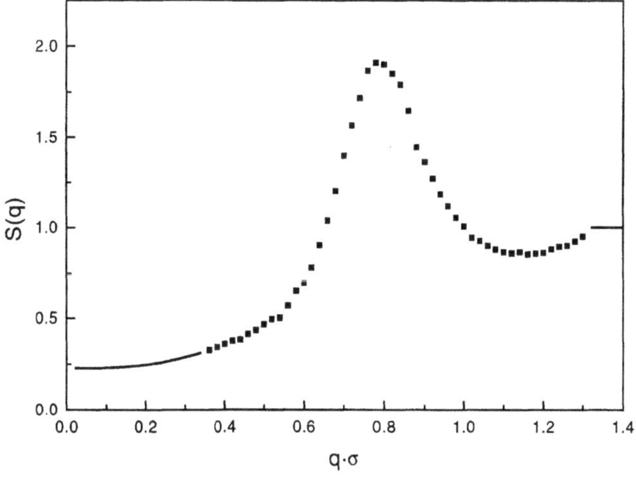

Fig. 1 Experimental structure factor (● ● ●) and *extended* structure factor (——)

Finally, three potentials $v_{\text{HNC}}(r)$, $v_{\text{PY}}(r)$ and $v_{\text{MSA}}(r)$, were obtained from $g(r)$ and $c(r)$ by means of HNC, PY and MSA approximations, resp. (shown in Fig. 3). We were not able to find out the potential for r/σ lower than 6.75 as consequence of putting $g(r) = 0$. The curves are similar and in agreement with those found by Nieuwenhuis and Vrij [4].

However, the three pair-potentials differ from the Yukawa potential due to the presence of a minimum. At present, we wonder if it could be possible to recover the function $g(r)$ from the experimental $S(q)$ by fitting the curve to a Yukawa pair-potential, or if the minimum in the functional form of $v(r)$ should be included. In order to know the role played by this minimum, we have selected $v_{\text{HNC}}(r)$ and have carried out MC simulations using both *this* previous form and the one fitted to a Yukawa expression

$$v_{\text{YK}}(r) = \frac{A}{r}\,e^{-\lambda r}\,. \tag{3}$$

As illustrated by curves b and c in Fig. 2, the correlation function $g(r)$ obtained by simulating with $v_{\text{HNC}}(r)$ and the *experimental* $g(r)$ are practically identical. However, the $g(r)$ obtained by simulating with $v_{\text{YK}}(r)$ differs from the experimental data. This *could* suggest that a Yukawa pair-potential is too simple to predict experimental $g(r)$ data in all the cases studied in this work.

It should be noted that, with this procedure, HNC and PY closures lead to very similar results. Even the MSA closure (which is expected to be a bad approximation) gives results in excellent agreement with the others. Though it has not been possible to compute the pair-potential at low r, the obtained curves show a minimum which *seems similar* to the secondary minimum of the potential with attractive part (for instance, the DLVO one). This suggests the need to work with well-characterized experimental systems using both the inverse and the direct approach.

Acknowledgments Financial support from "Comisión Interministerial de Ciencia y Tecnología", project $n°$. MAT 94-0560 is gratefully acknowledged.

References

1. Hansen JP, McDonald IR (1976) In: Theory of Simple Liquids. Academic Press, New York
2. Härtl W, Versmold H, Wittig U (1983) Mol Phys 50:815–823
3. D'Aguanno B, Méndez-Alcaraz JM, Klein R (1991) Progr Colloid Polym Sci 84:1–10
4. Nieuwenhuis EA, Vrij A (1979) J Colloid Interface Sci 72:321–341
5. Rajagopalan R (1992) Langmuir 8:2898–2906
6. Reatto L (1986) Phys Rev A 33:3451–3465

Progr Colloid Polym Sci (1997) 104:107–109
© Steinkopff Verlag 1997

R.D. Emmerzael
E.A. van der Zeeuw
G.J.M. Koper

The use of reflectometry for the study of swelling of latex particles at a silica surface

Dr. R.D. Emmerzael (✉)
E.A. van der Zeeuw · G.J.M. Koper
Leiden Institute of Chemistry
Leiden University
Gorlaeus Laboratories
P.O. Box 9502
2300 RA Leiden, The Netherlands

Abstract We show that reflectometry can be used as an accurate tool to study the swelling of latex particles adsorbed on a silica surface. The reflectivity of a layer of adsorbed latex spheres can be easily calculated using Mie theory, provided one neglects the electromagnetic interactions between particles and between particles and substrate. The validity of this simple model has been investigated thoroughly. The results can be used to advantage in the study of the swelling of polymer particles by an organic solvent. Within the validity of the assumptions underlying the simplified model we can detect changes in particle size as small as 0.5 nm.

Key words Reflectometry – polymer swelling – optical properties – latex particles

Introduction

Reflectometry is a very sensitive technique for studying the optical response of interfaces. It makes use of the property that, for a flat abrupt interface, the reflection of the electric field vector, polarized parallel to the plane of incidence (p-polarized light) vanishes at a certain, material-dependent, angle: the Brewster angle. In industry this technique is often used to roughly estimate the optical thickness of a layer of deposited material (one can, for instance, think of a polystyrene layer adsorbed on silica or a metal film on a dielectric). The analysis methods are usually equivalent to the "plane parallel plate" model. In this model, two materials with refractive indices n_1 and n_2 are separated by a homogeneous layer of refractive index n_l and of a thickness d. The criterion of homogeneity is usually not satisfied however, so that the underlying assumptions are usually very gross.

The purpose of this work is to show that, for certain systems, it is possible to extract more accurate information than the "plane parallel plate" model allows. The system that we investigated consists of polystyrene latex particles adsorbed on a silica prism. We will show that it is possible to accurately determine the size of the particles and their number density on the substrate.

The optical properties of spherical particles in free space is described by the so-called Mie theory [1]. When such spherical particles are deposited on a substrate, the problem becomes more complicated, because the electromagnetic interactions between the particles, as well as between the particles and the substrate, may be of importance. For particles that are small compared to the wavelength of light, Vlieger and Bedeaux have developed a theory which takes these interactions into account upto the dipole approximation [2]. The theory for the scattering of planar electromagnetic waves by arbitrarily sized spheres on a substrate was developed by Haarmans and Bedeaux [3]. In this theory, all the above-mentioned electromagnetic interactions are taken into account. Computation of a full reflectivity curve as measured in a reflectometry experiment may take up to 100 h, thus the exact theory is impractical for use in the data analysis. If we neglect the electromagnetic interactions between the spheres and the substrate, as well as between the spheres, the following expression for the reflectivity of an assembly of uncorrelated spheres can be derived in the far-field approximation [4]:

$$R(\theta_i) = r_p^2 + 2t_p^2 r_p \{ \mathrm{Re}(r_l)\cos\Delta + \mathrm{Im}(r_l)\sin\Delta \} + t_p^4 |r_l|^2 . \quad (1)$$

Here r_p and t_p are the Fresnel reflection and transmission coefficients (depending on the angle of incidence θ_i), Δ is the optical path difference between the light directly reflected from the substrate and the light originating from the centre of the spheres. r_l is given by [5]

$$r_l(\theta_t) = -\frac{2\phi S_p(\pi - 2\theta_t)}{k_2^2 a^2 \cos\theta_t} . \quad (2)$$

Here, ϕ is the surface coverage, k_2 is the wave vector in the ambient, S_p is the Mie-scattering function for p-polarized light [4], a is the radius of the particles, and θ_t is the angle of refraction which is related to θ_i by Snell's law. The particle parameters (radius and refractive index) are contained in S_p.

Van der Zeeuw et al. [6] compared the exact theory of Haarmans and Bedeaux with the approximate theory formulated in Eq. (1). They showed that for a system of latex particles reliable values for the particle size are obtained when using the approximate theory (deviation less than 2%), while the surface coverage is systematically underestimated (up to 15% at 10% coverage).

Since the approximate model has been proven to be sufficiently accurate for the determination of the *size* of the particles, it is therefore possible to detect changes in the particle size. This will be shown by an experiment in which polystyrene latex particles are swollen by an organic solvent dissolved in water (in our case butanone or MEK).

Experimental

A complete description of our experimental setup can be found elsewhere [4]. Here we will limit ourselves to a brief discussion of the experimental procedures. The latex particles (IDC, Portland, OR) were adsorbed onto the optically flat hypotenuse of a rectangular prism made of quartz. Two such prisms are mounted in a holder, leaving a 6 mm spacing between the two hypotenuses. The prisms were cleaned with a laboratory-use detergent (Hellmanex II; Hellma GmbH, D-7840 Mullheim) and rinsed thoroughly with deionized Millipore water, a dilute sulfuric acid solution, and again with water.

We measure the reflectivity of the p-polarized waves at a number of angles around the Brewster angle of the quartz/water interface. The light source is a 5 mV He–Ne laser, operating at a wavelength of 632.8 nm.

At the beginning of an experiment, reflectivities are measured for the pure quartz/water interface, in order to determine the instrument-dependent parameters (amplification, background intensity). After that, a dilute latex dispersion was injected into the cell. When a coverage of $\approx 6\%$ was reached, the cell was flushed with water to remove all non-adsorbed particles.

The MEK/water solution was then injected into the cell, and reflectivity curves were measured continuously for about one week. During the initial stage of the swelling, one reflectivity curve was measured in less than 2 min, in order to follow the initial rapid uptake of MEK.

Results

Examples of the obtained curves are shown in Fig. 1. During the experiment, the shape and the position of the curves change, first due to adsorption of the particles, and then due to the introduction of MEK. These changes in the curves can be related to the diameter of the particles and the surface coverage by Eq. (1).

In the analysis we assume that the solvent is homogeneously distributed in the polymer. Since the initial uptake of MEK is usually very fast [7,8], this seems a reasonable assumption. The volume fraction of MEK in the polymer $v_{1,p}$ can be calculated if we furthermore assume that all uptake of MEK results directly in an increase in size of the particles, corresponding to the volume uptaken.

The swelling as function of time, shown in Fig. 2, shows some striking features: (i) there is an initial fast uptake of solvent, resulting directly in an increase in particle size. (ii) From the logarithmic plot we see that, following this fast

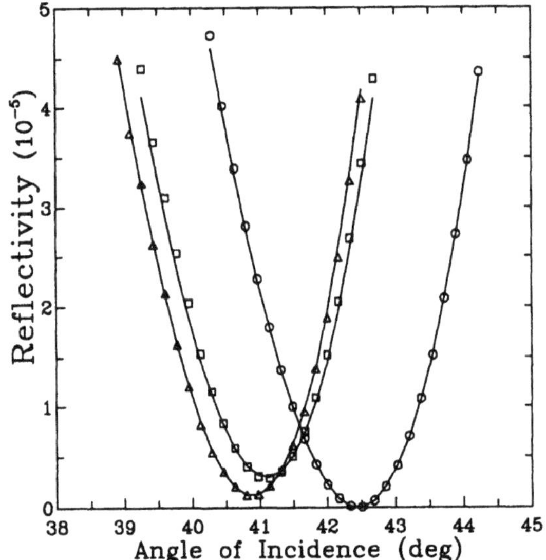

Fig. 1 Reflectivity as a function of angle: (○) reflectivity of the bare surface; (△) reflectivity just before the swelling experiment; (□) reflectivity obtained with 5 vol% MEK in water

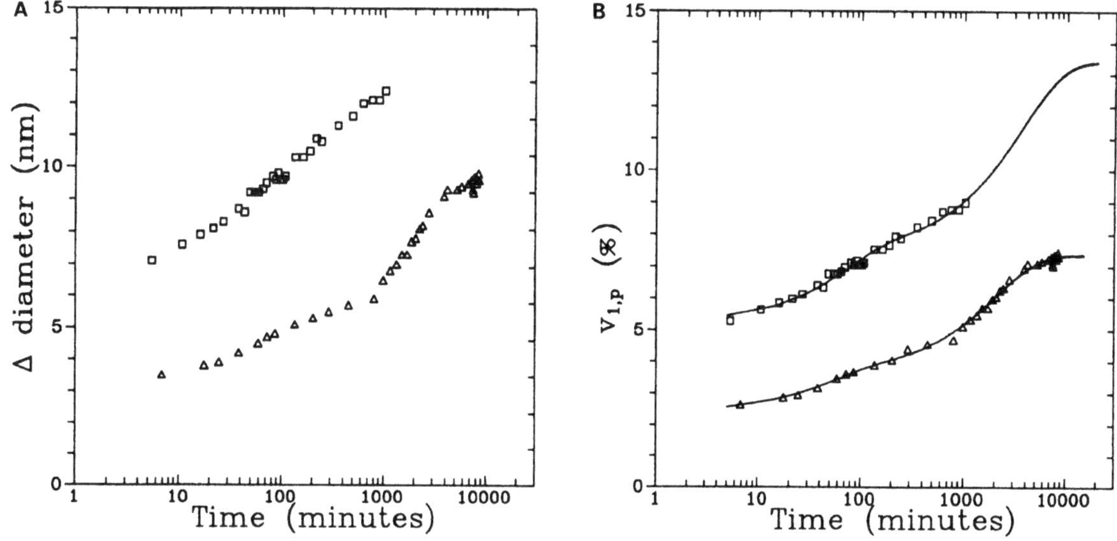

Fig. 2A Change in diameter of the particle and **B** the corresponding volume fraction MEK ($v_{1,p}$) in the polymer phase as a function of time for a 5% ($\triangle$) and a 9% ($\square$) MEK/water solution

uptake, two characteristic times seem to be involved. The physical implications of these features were discussed further in Ref. [9]. In any case, it is seriously to be doubted whether the swelling completes within an experimental time scale. If not, the standard treatment in terms of a thermodynamic model is fundamentally incorrect.

Conclusions

We have shown that reflectometry is a useful technique which can provide more information than effective parameters as "optical thickness" and effective dielectric constant. For latex spheres adsorbed on silica, it can provide information on the size of the particles and their number density at the substrate. This information can be obtained using a simple model based on Mie theory. Because of the great sensitivity of the method, the technique can be used to study the swelling of adsorbed latex particles by an organic solvent. Assuming the validity of the assumptions mentioned in the text, we conclude that it is possible to study the swelling of these polymer particles to a great accuracy (changes in diameter as small as 0.5 nm can be detected).

References

1. Kerker M (1969) The Scattering of Light and Other Electromagnetic Radiation. Academic Press, New York
2. Vlieger J, Bedeaux D (1980) Thin Solid Films 69:107
3. Haarmans MT, Bedeaux D, J Opt Soc Am (submitted)
4. Mann EK, Van der Zeeuw EA, Koper GJM, Schaaf P, Bedeaux D (1995) J Phys Chem 99:790
5. Bobbert PA, Vlieger J (1986) Physica A 137:243
6. Van der Zeeuw EA, Sagis LMC, Koper GJM, Mann EK, Haarmans MT, Bedeaux D (1996) J Chem Phys 105(4):1646
7. Berens AR, Hopfenberg HB (1978) Polymer 19:489
8. Hidi P, Napper DH, Sangster DF (1995) Macromolecules 28:6042
9. Van der Zeeuw EA, Sagis LMC, Koper GJM (1996) Macromolecules 29:801

Progr Colloid Polym Sci (1997) 104:110–112
© Steinkopff Verlag 1997

S. Will
A. Leipertz

Measurement of particle diffusion coefficients with high accuracy by dynamic light scattering

Dr. S. Will (✉) · A. Leipertz
Lehrstuhl für Technische
Thermodynamik (LTT)
Universität Erlangen-Nürnberg
Am Weichselgarten 8
91058 Erlangen, Germany

Abstract We present a novel optical scattering cell for the measurement of particle diffusion coefficients by dynamic light scattering, which may be utilised for particle size analysis and viscosity measurements. A major feature of the cell is that it allows measurements without knowledge of the refractive index of the dispersing liquid, which facilitates viscosity measurements over an extended range of temperature, where refractive indexes are often not found tabulated. This improvement is achieved by the use of a symmetrical set-up.

The light scattering system has been tested by measuring the diffusion coefficient of silica particles in water, where the results showed a total standard deviation of 0.5% including measurements at various angles of incidence. In order to prove the capability of the system for accurate viscosity measurements, n-heptane, where reliable reference values are available, has been investigated in a temperature range of 20°–80 °C. The viscosities measured showed a maximum deviation of the mean from the reference values of below 1.0%.

Key words Dynamic light scattering – n-heptane – particle diffusion – viscosity

Introduction

Dynamic light scattering (DLS) may be regarded as a successful technique both for the determination of various thermophysical properties of fluids [1, 2] and for particle size analysis [3, 4]. A useful special application of characterising fluids by DLS is the measurement of the dynamic viscosity [5, 6]. Both this technique and particle size analysis are based on the measurement of the diffusion coefficient D_p of suspended seed particles, which is related via the Stokes–Einstein relationship

$$D_p = \frac{kT}{3\pi\eta d},\qquad(1)$$

with the desired quantities viscosity η or particle diameter d, respectively, Boltzmann's constant k and temperature T.

If one illuminates the particle dispersion with laser light, the diffusion coefficient required can be obtained by observation of the fluctuations in time of the resulting particle interference pattern. The intensity fluctuations on a suitable detector may be statistically analysed by calculating the autocorrelation function $\hat{g}^{(2)}(t)$, which in the idealised case of monodisperse, freely diffusing particles is of form

$$\hat{g}^{(2)}(t) = a + b\exp(-t/\tau_c),\qquad(2)$$

where the experimental constants a, b, and the decay time τ_c may be obtained by a non-linear fit algorithm [7].

In particle size analysis, the central problem is to recover the distribution of particle sizes and thus a whole spectrum of different decay times from the experimental autocorrelation function. The problems of applying DLS to viscosity measurements are of a somewhat different nature. These are mainly connected with the finding of suitable seed particles, which must be stable in the liquids investigated and should have a narrow size distribution, and the accurate calibration of particle size. A special problem which arises with viscosity determination is the numerical relationship between the experimental decay time and the particle diffusion coefficient. In the case of a homodyne detection this is given by

$$D_p = 1/(2q^2\tau_c)\,, \tag{3}$$

with the modulus

$$q = \frac{4\pi n}{\lambda_0}\sin\frac{\Theta}{2} \tag{4}$$

of the scattering vector of the optical arrangement. Here, n denotes the refractive index of the liquid investigated, λ_0 is the laser wavelength in vacuo, and Θ is the scattering angle. Whereas in particle sizing mostly liquids with well-known refractive indices like water are used at temperatures near ambient, viscosity measurements require the use of the technique in systems with refractive indices often not found tabulated.

One approach to overcome this restriction is to estimate the refractive index on the basis of correlations derived from other fluid properties known [5]. Yet the typical uncertainty of more than 1% in the refractive index obtained in such a way alone results in an uncertainty of more than 2% in viscosity, which cannot be accepted for accurate measurements. In order to measure the fluid refractive index required for DLS measurements, several possibilities have been described in the literature [8, 9].

The solution introduced by Chu and co-workers [9] is based on a prismatic design of the sample cell and offers the advantage of an automated measurement of the refractive measurement via a position sensitive diode. On the other hand, such a solution requires additional technical effort and specialised electronic equipment. It would be more desirable to realise a set-up in which the refractive index need not to be measured at all and thus enables DLS measurements which are only based on even fewer input parameters in the evaluation formulae.

We present a new approach which is based on the principle of a symmetrical set-up and which allows exact measurements without requiring the refractive index of the dispersing liquid. The system has been used to measure the viscosity of n-heptane.

Light scattering system

Basic idea of the symmetrical geometry is to replace the term $n\sin(\Theta/2)$, which is required for the evaluation of the

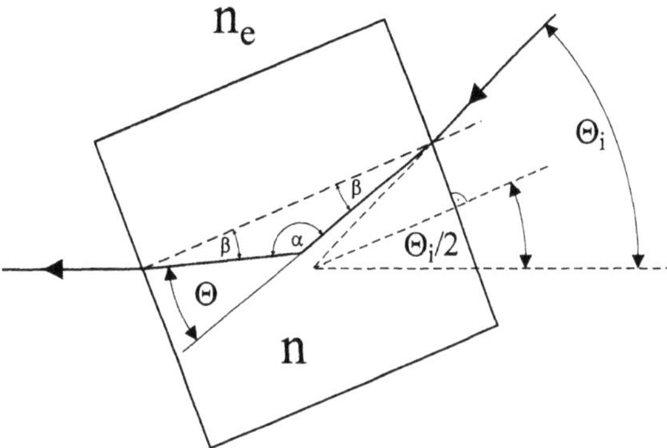

Fig. 1 Principle of the symmetrical set-up, where the cuvette is tilted from the perpendicular position by $\Theta_i/2$; here the case $n > n_e$ is illustrated

scattering vector, by quantities which are easily accessible experimentally.

With the use of a square cuvette, this can be accomplished if one tilts the cuvette from the usual position, where the exit window is perpendicular to the axis of observation, by an angle $\Theta_i/2$, where Θ_i is the angle between the incident laser beam and the axis of observation. From Snell's law and simple geometrical considerations the relation

$$n\sin\frac{\Theta}{2} = n\sin\frac{\pi - \alpha}{2} = n\sin\frac{\pi - (\pi - 2\beta)}{2}$$

$$= n\sin\beta = n_e\sin\frac{\Theta_i}{2} \tag{5}$$

follows (cf. Fig. 1). The angle of incidence can be easily measured; if the surrounding medium is air, the external refractive index may be set equal to 1, otherwise a single determination of n_e suffices, independently of the liquid to be investigated.

On the basis of the symmetrical set-up described, we have designed a thermostated scattering cell for a temperature range of approximately 280–400 K, which is made of an anodised aluminium alloy. A commercial square cuvette contains the liquid under investigation. The sample temperature is measured directly by a calibrated Pt-100-resistance probe immersed. The cuvette is fixed to an adjustable support, which can be aligned to the angle required by a rotating table. The cell contains an oil, which serves as a thermostating medium and the refractive index of which closely matches that of the glass material. The oil is temperated via an external computer-controlled water or oil circuit. Optical access is realised by several windows, which allow perpendicular incidence for angles of 30°, 60° and 90°. To take water at 20 °C as a specific example, these angles of incidence result in scattering angles of 33°, 67°

Table 1 Test results for the DLS system. For the purpose of comparison also the diffusion coefficient for a rectangular set-up is included, where the refractive index of the liquid investigated must be known; this is not necessary in the case of a symmetrical set-up

Θ_i	Geometry	D_p (10^{-12} m² s)	σ (%)
90°	Symmetrical	1.978	0.24
60°	Symmetrical	1.987	0.33
30°	Symmetrical	2.001	0.36
	Total	1.989	0.53
90°	Rectangular	1.989	0.34

and 107° with the use of the symmetrical set-up. If this range of angles should not suffice, different angles of incidence may be chosen without affecting the principle of the symmetrical set-up.

The symmetrical set-up sets the user free from measuring the refractive index of the liquid under investigation. Only that one of the index-matching fluid must be determined once. This may be accomplished by the use of commercial refractometers or by measuring the refraction at a prism immersed [10].

The cell is illuminated by a vertically polarised 10 mW He–Ne laser, which is mounted on a rotatable arm for a variation of the angle of incidence. The laser bundle is focused by two cylindrical lenses with different focal lengths. The detection system simply consists of two apertures at a distance of 0.5 m and a photomultiplier tube, connected to a single-board correlator.

The light scattering system has been tested by measuring the diffusion coefficient of commercial native silica particles in water, where the results showed a total standard deviation of 0.5% including measurements at various angles of incidence (see Table 1).

Viscosity of *n*-heptane

In order to prove the capability of the system for accurate viscosity measurements, a liquid with a well-documented viscosity has to be chosen. For a measurement of absolute viscosity values, it is of central importance to calibrate the particle sizes in liquids of similar chemical properties. For this reason we have selected various *n*-alcanes, where reliable reference values both for the calibration at 298 K and an extended range of temperature have been published [11]. Commercially produced silica particles with a surface modification by stearyl groups were used as a seed. The particle size was calibrated to give 222.8 nm $\pm$ 0.6% for *n*-hexane and 222.4 nm $\pm$ 1.1% for *n*-octane.

On basis of this calibration the viscosity of *n*-heptane was then determined in a range of 293–373 K. The total standard deviation of the experimental data was about 0.9% throughout the whole range of temperatures, including measurements at various scattering angles, particle concentrations and a temperature cycle, where the maximum deviation of the mean from the reference values amounted to below 1.0%. The results show that it is possible to perform reliable measurements of particle diffusion coefficients and absolute viscosity values by DLS over an extended range of temperature and that the novel scattering system is a valuable tool for this aim, as measurements may be performed without a priori knowledge of any other data of the liquid under investigation.

References

1. Kraft K, Will S, Leipertz A (1995) Measurement 14:135–145
2. Shaumeyer JN, Gammon RW, Sengers JV (1991) In: Wakeham WA, Nagashima A, Sengers JV (eds) Measurement of the Transport Properties of Fluids. Blackwell Scientific, Oxford, pp 197–213
3. De Jaeger N, Demeyere H, Finsy R, Sneyders R, Vanderdeelen R, van der Meeren P, van Laethem M (1991) Part Part Syst Char 8:179–186
4. Dahneke B (ed) (1984) Measurement of Suspended Particles by Quasi-Elastic Light Scattering. Wiley, New York
5. Williams DF, Byers CH (1987) J Chem Eng Data 32:2534–2538
6. Will S, Leipertz A (1995) Int J Thermophys 16:433–443
7. Will S, Leipertz A (1993) Appl Opt 32:3813–3821
8. Straub J (1965) PhD thesis, Technische Universität München
9. Chu B, Xu R, Maeda T, Dhadwal HS (1988) Rev Sci Instrum 59:716–724
10. Will S, Leipertz A (1996) Rev Sci Instrum 67:3164–3169
11. Dymond JH, Øye HA (1994) J Phys Chem Ref Data 23:41

Progr Colloid Polym Sci (1997) 104:113–116
© Steinkopff Verlag 1997

A.W. Willemse
J.C.M. Marijnissen
A.L. van Wuyckhuyse
R. Roos
H.G. Merkus
B. Scarlett

Photon correlation spectroscopy for analysis of low concentration submicrometer samples

Dr. A.W. Willemse (✉)
J. C.M. Marijnissen
A.L. van Wuyckhuyse · R. Roos
H.G. Merkus · B. Scarlett
Delft University of Technology
Faculty of Chemical Engineering
Particle Technology Group
Julianalaan 136
2628 BL Delft, The Netherlands

Abstract Photon Correlation Spectroscopy is a technique to rapidly measure particle size in the submicrometer region. The use of PCS is, however, limited by concentration. The upper limit is due to multiple scattering of the incident light, the lower limit is determined by the fact that fluctuations of the number of particles in the measuring zone have a significant influence on the apparent diffusion coefficient. In this article a new signal processing method is described which differentiates this influence. With this system the lower limit is no longer limited to about 100 particles in the measuring volume corresponding to a concentration of 10^9 particles/cm^3. The limitation is now the intensity of the scattered light which becomes too weak at a concentration of about 50 particles/cm^3. As a consequence of this work, a revision to the basic theory of PCS may be necessary. Moreover, the new processing method enables also the measurement of particle concentration in the sample.

Key words Light scattering – particle-size analysis – photon correlation spectroscopy

Introduction

For measuring particle-size in the submicrometer range, photon correlation spectroscopy (PCS) is a well-suited technique. PCS, also referred to as quasi-elastic light scattering (QELS) or dynamic light scattering (DLS) has become a standard technique for measuring the diffusion coefficient and hence the particle size in liquid dispersions of macromolecules and submicrometer particles. This is achieved by measuring the fluctuations of scattered laser light from an ensemble of particles which exhibit Brownian motion. The term "quasi-elastic" in QELS refers to the fact that the frequency of the scattered light experiences a slight Dopplershift due to the particle motion. The PCS measuring technique is used not only in research but also in the industrial environment. The technique is however limited by concentration. Very low particle concen-

trations are normally necessary which leads to extreme dilution of process streams. This is to prevent multiple scattering of the incident light by the particles in the sample before it reaches the detector. However, a very dilute sample also cannot be measured. The intensity fluctuations measured then arise not only from the Brownian motion, but also from the changing number of particles in the scattering volume.

Number fluctuations

There are two contributions to the autocorrelation function. One is a Gaussian contribution, including the diffusion coefficient, the other is a non-Gaussian term. If the number of particles is large, then $N^2 \gg N$ ($N > 100$) and no significant influence of the non-Gaussian term is observed. When the number is small, $N < 100$, the

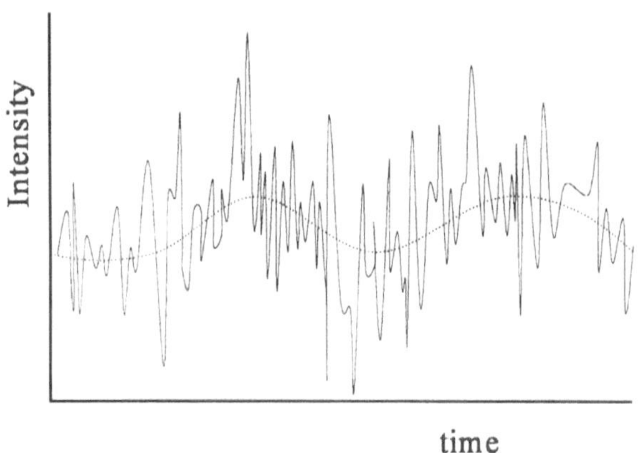

Fig. 1 Raw data signal from a low concentration PCS sample

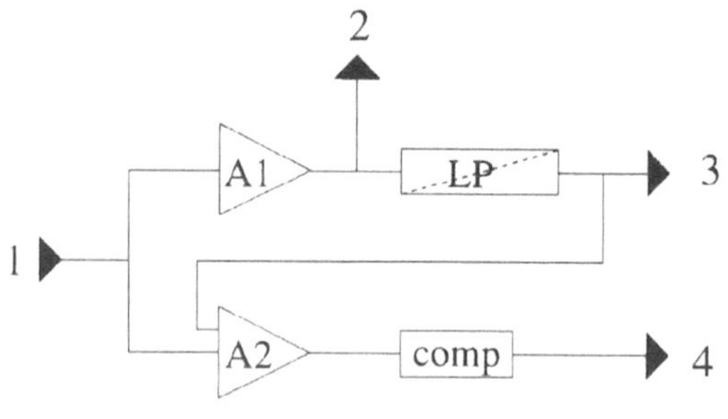

1 = PCS signal from PMT in   A1   =   1st amplifier
2 = Amplified signal out   A2   =   2nd amplifier
3 = LF signal out   LP   =   low pass filter
4 = HF signal out (TTL)   comp   =   comparator

Fig. 2 Electronic filter for the low concentration PCS setup

fluctuations measured arise not only from the Brownian motion but also from the changing number of particles in the scattering volume. This results in an apparent decrease in the diffusion coefficient and thus an apparent increase in the particle diameter. This phenomenon is described in detail by Weber [1], who explains that the autocorrelation function consists of three parts. They are a DC-component, a part consisting of the product of an exponential and a Gaussian decay which is caused by the diffusional and convective motions and lastly a non-Gaussian decay.

In traditional PCS measurements these number fluctuations have been considered to determine the lower concentration limit of the technique. A typical PCS measuring volume is about $10^{-6}\,\mathrm{cm}^3$ [2], which implies a lower detection limit of 10^8 particles/cm^3. Several authors have noticed the effect of number fluctuations and have described the phenomena. Figure 1 gives a typical representation of what is observed as the raw signal of a PCS measurement when the concentration is low.

Compared to a normal PCS signal, this figure shows that there is superposed a second signal of lower frequency. A normal PCS signal, with $N > 100$, would have a constant average value, but with $N < 100$, there is a clear slower fluctuation. The key postulate of this paper is that the particle fluctuations and the number fluctuations each have their own frequency domain which can be differentiated.

Experimental setup

Since the particle fluctuations and number fluctuations each have their own frequency domain they can be differentiated by applying a low-pass filter. A suitable filter is shown schematically in Fig. 2. This filter was build and used to separate the low and high frequencies of the signal.

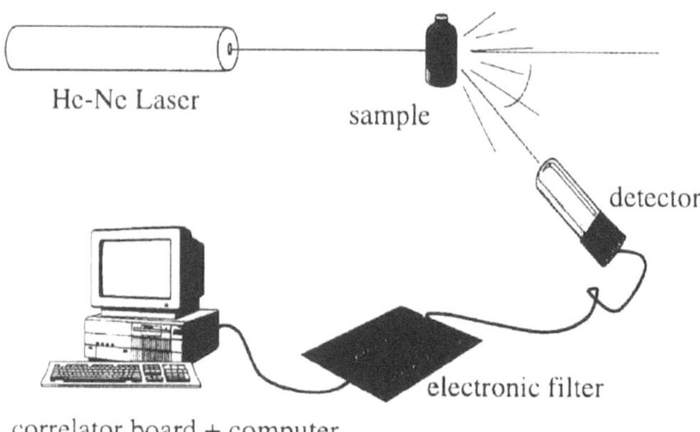

Fig. 3 Low concentration PCS setup

The cut-off frequency of the filter could be adjusted to match the experimental. In order to prevent electronic noise and to prevent disturbances from other sources, the filter was mounted in a shielded metal cage. The frequency of the particle Brownian motion was estimated from

$$\psi = D q^2 . \tag{1}$$

Here ψ is the expected characteristic frequency of the detected signal, D is the diffusion coefficient and q is the modulus of the scattering vector. The term ψ can be used to estimate the optimum cut-off frequency of the filter.

This filter was coupled to a standard PCS goniometer setup, as shown in Fig. 3 with which it was possible to measure both the particle size and particle concentration simultaneously. The correlator board used was a Brookhaven BI-8000AT, the goniometer setup a Brookhaven SM200. The signal was detected by a Hamamatsu R1635

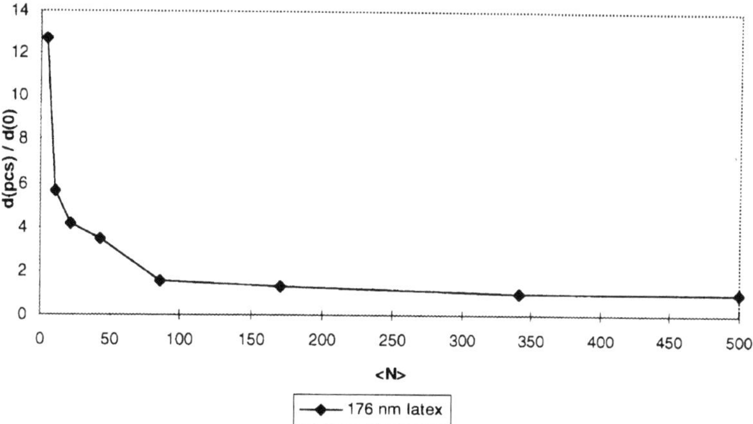

Fig. 4 Results of PCS measurements using a conventional PCS setup. The sample used was a 176 nm latex

photo multiplier tube. The photo multiplier received its signal through a standard lens-and-pinhole setup. The laser used was a Lexel 95 argon-ion laser, operating at 514.5 nm.

Measurements were made on a BASF latex of 176 nm diameter with a nominal stock concentration of 50% vol/vol solids and on a Duke Scientific latex of 501 nm diameter with a nominal stock concentration of 1% vol/vol solids. The samples were prepared by diluting the stock latex with double distilled water. This water was filtered, prior to use, through a 0.22 μm. Millipore Sterivex filter to minimise contamination of the samples.

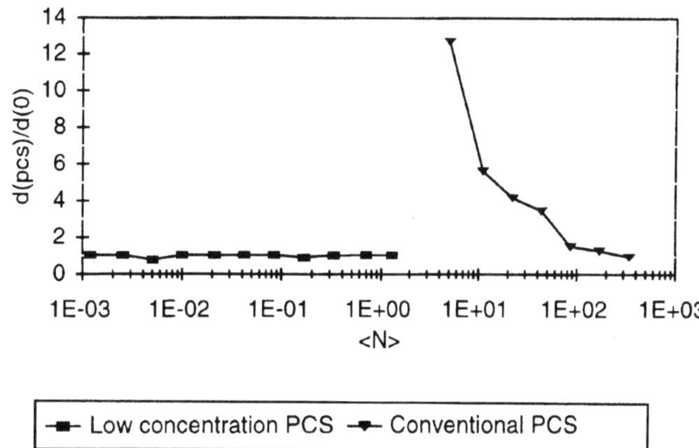

Fig. 5 Results obtained with the low concentration and the conventional PCS setup. The sample used was a 176 nm latex

Results

Conventional PCS

Initially, measurements were made with a conventional PCS setup without the electronic filter. With this setup, a Helium–Neon laser was used, operating at a wavelength of 632.8 nm. A range of dilution of the 176 nm latex was made and measured. The results are given in Fig. 4. This figure shows the apparent particle diameter as a fraction of the calibrated diameter plotted against concentration. The concentration is expressed as number of particles in the scattering zone, N. It is seen that at concentrations of about $N = 350$ the apparent particle diameter increases with decreasing N.

Figure 4 shows, in fact, that it is not possible to reliably measure the particle size with less than 350 particles in the measuring volume. It should also be noted that for $N < 80$, the results of the measurements were no longer reproducible. The values in Fig. 4 for $N < 80$ are simply an average value of a large spread and should not be interpreted as an absolute value.

PCS with new electronic filter

The same samples were now measured using the setup shown in Fig. 3 which incorporates the electronic filter. Because of its high output power, a water-cooled Lexel 95 Ar-Ion laser operating at 514.5 nm was used. The laser power was adjustable up to 1800 mW. The electronic filter was coupled between the output of the photomultiplier and the input of the correlator board. The angle of observation was 75° for the measurements with 176 nm latex and was 90° with the 501 nm latex. The pinhole was initially set at 200 μm but was increased with decreasing concentration to 3 mm in order to obtain a reasonable countrate. The sample temperature was maintained at 298 K. The measurements with the 176 nm latex are shown in Fig. 5. It is seen that this setup is well able to measure in the regime which includes number fluctuations ($N < 100$). These measurements were repeated with the 501 nm latex and the results are shown in Fig. 6. In both figures the results obtained are compared with the

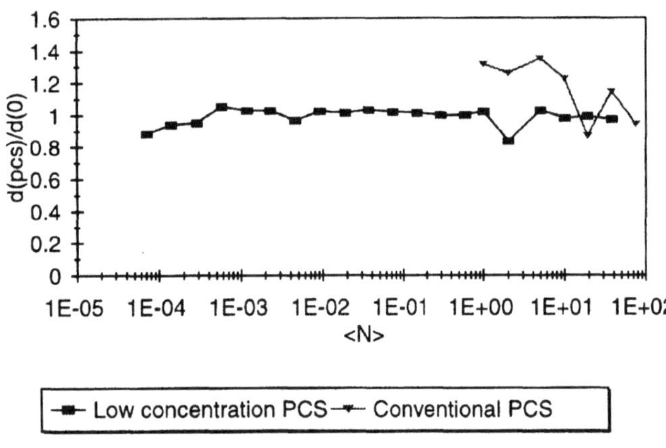

Fig. 6 Results obtained with the low concentration and the conventional PCS setup. The sample used was a 501 nm latex

conventional PCS setup. No measurements were made in the conventional concentration regime using the filter. It is seen that for particles of 501 nm compared to conventional PCS, the lower concentration limit is reduced by a factor of 1×10^7.

Conclusion

An extension to conventional PCS has been made by reducing the lower concentration detection limit. As the new system only works on the raw signal from the photo multiplier, standard PCS equipment can be used. The standard software can also be used without any modifications.

References

1. Weber R, Rambau R, Schweiger G, Lucas K (1993) Analysis of a flowing aerosol by correlation spectroscopy: Concentration, aperture, velocity and particle size effects. J Aerosol Sci 24:485

2. Drunen van M (1995) Measurement and modeling of cluster formation, PhD thesis, Delft University of Technology

Progr Colloid Polym Sci (1997) 104:117–120
© Steinkopff Verlag 1997

E. Overbeck
C. Sinn
T. Palberg

Approaching the limits of multiple scattering decorrelation: 3D light-scattering apparatus utilising semiconductor lasers

E. Overbeck · Dr. C. Sinn (✉) · T. Palberg
Institut für Physik
der Johannes-Gutenberg-Universität
Staudingerweg 7
55099 Mainz, Germany

Abstract Light scattering as a function of scattering angle can be regarded as a standard method to investigate the dynamics of dilute colloidal suspensions. Concentrated suspensions, which are of interest if interactions between the particles are to be investigated, usually show strong multiple scattering. Decorrelation of multiple scattered light, which isolates single scattering events at the expense of a reduced signal-to-noise ratio, has been proven to work using the two-colour cross-correlation scheme.

In this contribution we demonstrate for the first time the suppression of multiple scattering of a concentrated colloidal suspension at different scattering angles using the 3D cross-correlation technique. Our set-up is designed to extend both measurements towards smaller q values and the scattering intensity of the samples under study beyond the limits of existing apparatus. The latter feature will enable us to approach the photon-diffusion regime as far as possible using decorrelation methods.

Key words Dynamic light scattering – multiple scattering – cross-correlation – colloidal physics

Introduction

Since long there have been theoretical publications to account for the effects of multiple scattering [1–5]. On the other hand, different experimental methods were suggested [6, 7] and proved to work [8–10]. Most recently, the two-colour scheme demonstrated by Drewel et al. is commercially available and its suitability for the investigation of different aspects of colloidal physics has been demonstrated [11]. However, the complexity of this set-up hinders a broad application of multiple scattering decorrelation techniques. We therefore describe in this contribution the 3D set-up, which is by far more easy and stable compared to the two-colour set-up, but offers comparable experimental possibilities.

Experimental details

The experimental set-up is shown in Fig. 1. For a detailed discussion of the "Evolution" of light scattering we refer to [12].

The near IR wavelength ($\lambda = 790$ nm) of the semiconductor lasers guarantees a small scattering cross section and the availability of a comparably low q. The laserdiodes operate under single-mode condition which assures a coherence length of about 1.5 mm, which is large compared to the scattering volume dimensions of about 250 μm. A temperature stabilisation circuit prevents the diodes from mode hopping due to temperature changes. Both lasers share the same wavelength to within $\Delta\lambda = 0.05$ nm, which is necessary for a large intercept in cross-correlation mode [7].

Fig. 1 Experimental set-up of the 3D cross-correlation scheme in side view and top view, respectively

side view for $\theta = 0$:

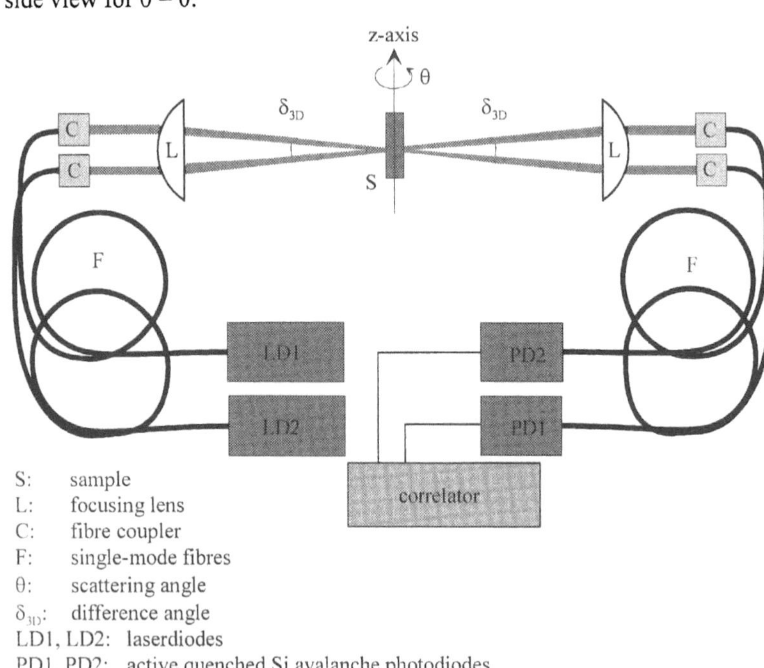

S: sample
L: focusing lens
C: fibre coupler
F: single-mode fibres
θ: scattering angle
δ_{3D}:   difference angle
LD1, LD2:   laserdiodes
PD1, PD2:   active quenched Si avalanche photodiodes

top view for arbitrary θ:

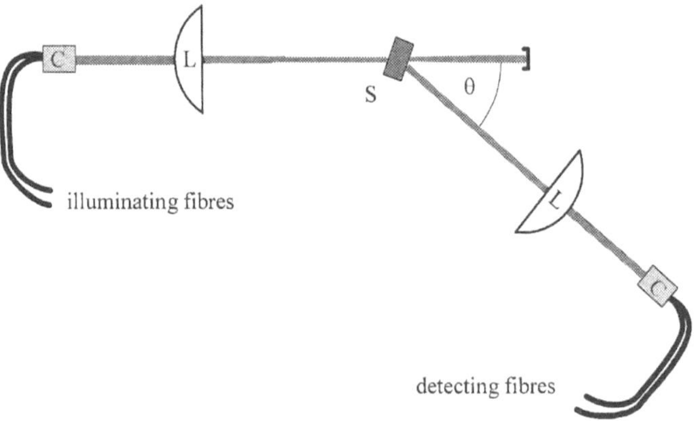

The use of photomultipliers as single photon counters is impossible because of their low responsivity at the wavelength chosen. Therefore, we use Si avalanche photodiodes as detectors. The active quenching electronics is home-build and quenches the bias voltage periodically ($v = 12$ MHz) below breakdown. This leads to an estimated dead-time of the detectors of approximately 40 ns.

Fibre optic components (fibre couplers, single-mode fibres on the detection sides, polarisation maintaining fibres on the transmission sides) as well as the common large lens have been chosen for a maximum compactness and easy alignment of the set-up. On the transmission side, the fibres generate a diffraction limited Gaussian beam out of the transversally multi-mode structure of the laser diode radiation field. On the detection side they enable the maximum intercept possible to be obtained [13].

While the incident beams are polarised (25 dB) perpendicular to the plane of symmetry of the experiment, the scattered light is detected unpolarised. Because the scattering plane in general is inclined with respect to the plane of symmetry, polarisers in the detection optics must be used with care (details will be discussed in a forthcoming publication). In the present investigation by far most of the light is scattered polarised so that the usual V_V geometry may be assumed for the interpretation of the scattering data.

Fig. 2 A) Intensity correlation functions for two polystyrene latex samples ($R = 57.5$ nm) of different turbidity, measured under a scattering angle of $\theta = 36°$ ($q = 6.37\ \mu\mathrm{m}^{-1}$); open symbols: $T = 0.69$, closed symbols: $T = 0.02$; circles: auto-correlation mode, squares: cross-correlation mode. B) The same as above, measured under a scattering angle of $\theta = 90°$ ($q = 14.64\ \mu\mathrm{m}^{-1}$). For details see text

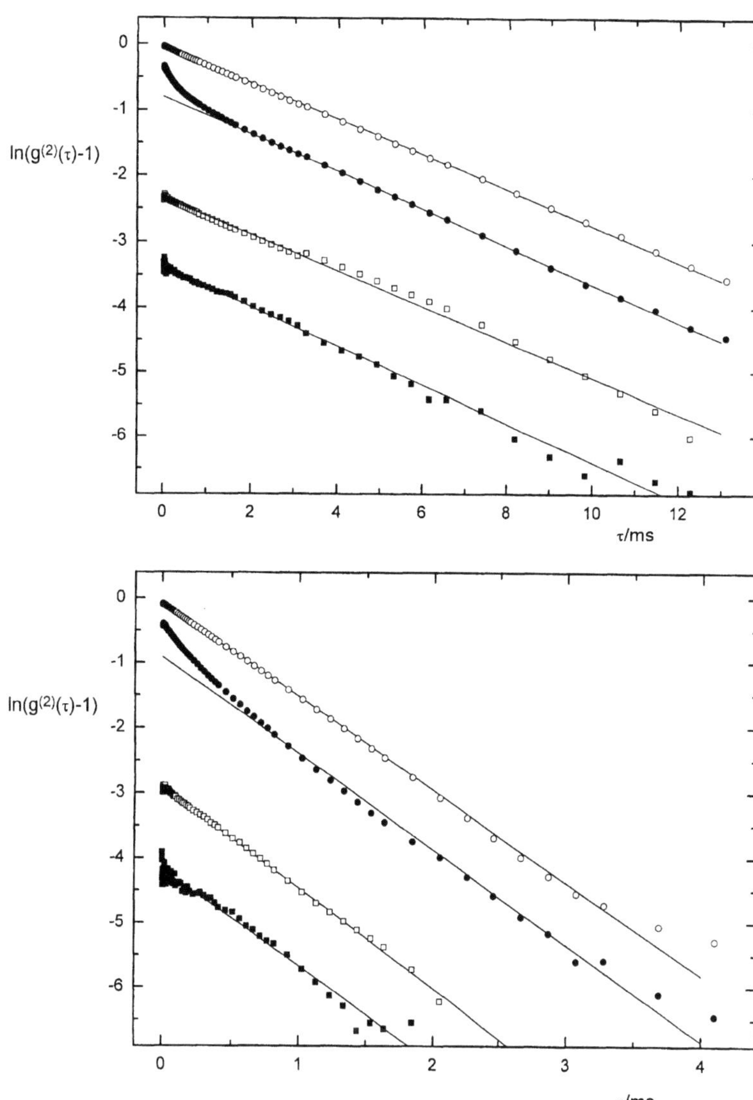

Results and discussion

Figure 2 show intensity correlation functions determined from polystyrene latex sphere ($R = 57.5$ nm) samples with two different concentrations. The transmission $T = I/I_0$ of the samples was measured to be $T = 0.69$ and $T = 0.02$, respectively. Figure 2A shows the results for a scattering angle of $\theta = 36°$, Fig. 2B those obtained for a scattering angle of $\theta = 90°$ ($q = 6.37\ \mu\mathrm{m}^{-1}$ and $q = 14.64\ \mu\mathrm{m}^{-1}$, respectively).

In auto-correlation mode, the turbid sample shows a strong distortion at small lag times due to multiple scattering. This distortion can be completely eliminated by cross-correlating the scattered intensities. Fitting either the long-time tail or all data points, respectively, to an exponential decay, the sphere's radius is obtained from all measurements to be $R = 61.6$ nm ($\pm\ 3\%$), which is in reasonable agreement with the value supplied by the manufacturer (obtained by TEM).

The intercept decreases from auto-correlation mode to cross-correlation mode because of the decorrelation of multiple scattering, which contributes to the background only. However, even under optimal conditions only an intercept of 0.13 has been measured (the theoretical one being approximately 0.40). This unsatisfactory value clearly indicates a necessary improvement of the alignment of the apparatus.

Conclusion

It could be demonstrated for the first time that the proposed innovative cross-correlation scheme, namely the 3D technique, which allows the decorrelation of multiple scattering events in a large angular range, is working as expected. The suitability of multiple scattering decorrelation for the investigation of the physics of strongly scattering colloidal dispersions already has been demonstrated with the two-colour set-up, which is a comparable complicated apparatus. The set-up introduced in this contribution is by far less complex, which might offer the possibility of a broader application of multiple scattering decorrelation techniques.

References

1. Sorensen CM, Mockler RC, O'Sullivan WJ (1976) Phys Rev A 14:1520–1532
2. Sorensen CM, Mockler RC, O'Sullivan WJ (1978) Phys Rev A 17:2030–2035
3. Bøe A, Lohne O (1978) Phys Rev A 17:2023–2029
4. Böheim J, Hess W, Klein R (1979) Z Physik B 32:237–243
5. Dhont JKG, de Kruif CG (1983) J Chem Phys 79:1658–1663
6. Phillies GDJ (1981) J Chem Phys 74:260–262
7. Schätzel K (1991) J Mod Opt 38:1845–1865
8. Phillies GDJ (1981) Phys Rev A 24:1939–1943
9. Mos HJ, Pathamanoharan C, Dhont JKG, de Kruif CG (1986) J Chem Phys 84:45–49
10. Drewel M, Ahrens J, Podschus U (1989) J Opt Soc Amer 7:206–210
11. Segrè PN, van Megen W, Pusey PN, Schätzel K, Peters W (1995) J Mod Opt 42:1929–1952
12. Overbeck E, Sinn Chr, Palberg T, Schätzel K (1997) Colloid Surf A, accepted for publication
13. Ricka J (1993) Appl Opt 32:2860–2875

Progr Colloid Polym Sci (1997) 104:121–125
© Steinkopff Verlag 1997

L.B. Aberle
S. Wiegand
W. Schröer
W. Staude

Suppression of multiple scattered light by photon cross-correlation in a 3D experiment

Presented at the International Workshop on Optical Methods and the Physics of Colloidal Dispersions (Mainz 30.9-1.10, 1996) in Memory of Klaus Schätzel

L.B. Aberle · S. Wiegand
W. Schröer (✉)
Institut für Anorganische
und Physikalische Chemie
Fachbereich 2 Biologie-Chemie
Universität Bremen
Leobener Straße NWII
28359 Bremen, Germany

W. Staude
Institut für Experimental-Physik
Universität Bremen
Leobener Straße NWII
28359 Bremen, Germany

Abstract In strongly scattering media, the presence of multiple scattered light prevents the straightforward interpretation of photon-auto correlation functions in terms of single scattering processes. In order to suppress the influence of multiple scattering Schätzel suggested a so-called 3-D cross-correlation technique. This technique operates by cross-correlating the intensities of the scattered light of two coherent laser beams illuminating the same scattering volume and so defining two scattering geometries. The cross-correlation function is identical to the auto-correlation function from single scattering if the scattering vector **q** is chosen to be identical for both scattering geometries. Based on this idea an experimental set-up has been developed, which appears to be a fairly simple modification of a conventional light scattering experiment. Test measurements with solutions of standard latex particles with a diameter of 109 nm at various concentrations show, that contributions due to multiple scattering are well suppressed even in the range of strong multiple scattering.

Key words Dynamic light scattering – multiple scattering – cross-correlation

Introduction

Photon correlation spectroscopy of scattered laser light is an important tool for characterising fluids comprising particles of mesoscopic size [1] as colloidal suspensions, emulsions or fluids in the vicinity of phase transitions. The evaluation of the scattering data relies commonly on the assumption of singly scattered light for which the interpretation of the correlation functions (CF) in terms of microscopic properties is straightforward [2]. However, such conventional light scattering theory applies only to weakly scattering samples while, in general, the scattered light is a sum of contributions due to single, double and multiple scattering. The presence of multiple scattering in strongly scattering samples is recognized as a major limitation for the application of dynamic light scattering in many cases of practical interest.

A comparably simple model is available only for the limit of very strong multiple scattering where single scattering is negligible. In this model the propagation of the light in the medium is approximately described as a succession of random scattering events in the medium, resulting in a diffusion process of photons [3–5]. Considering sequences of scattering processes, important details as the wave-vector dependence of the dynamic structure factor are lost due to the averaging procedure. Consequently, this method, termed "diffusing wave spectroscopy", cannot yield all informations available by analysing auto-correlation functions (ACF) of single scattering.

Theories which assess double and multiple scattering [6–9] are very involved and far too complicated to be applied in routine spectroscopy. Therefore, as reviewed by Schätzel [10], experimental methods suppressing the multiple scattering contributions from the measured photon correlation data are of vital importance for many applications of light scattering spectroscopy. A suitable set-up was first proposed and demonstrated by Phillies [11] which, however, is restricted to a 90°-scattering geometry. The development of a method allowing the separation of the single scattering to the ACF for arbitrary scattering vectors is one of the major achievements of Schätzel and his collaborators [12, 13].

The common principle of the three suggested methods [10] is the following. Two coherent laser beams are focused into a sample and build two different scattering geometries with a common scattering volume. In the two geometries the light scattered with identical scattering vectors reflects the same microscopic process and is, therefore, correlated. This condition for correlation is satisfied by single scattering events in the common scattering volume, but not by the majority of multiple scattering processes, which contribute to the ACF and, for example, involve particles outside the volume where the two laser beams intersect. Therefore, the cross-correlation function (CCF) of the intensity in the two scattering geometries is identical with the desired single scattering ACF. The method acts as a filter for multiple scattering contributions which obscure the ACF in the conventional light scattering experiment.

Schätzel [10] distinguishes three different methods for selecting the light of the two scattering geometries: "two-colour coding", "time coding" and "3-D coding". The method of "two-colour coding" [12, 13] is now applied in various light scattering laboratories [14]. Here the cross correlation of the light of two laser beams of different frequency is investigated. The "colour coding" allows for an elegant separation of the two scattering geometries by means of interference filters.

Schätzel also proposed an experiment applying the method of "time coding" [10]. He considers a modification of the set-up used by Phillies [11] which, however, is not restricted to a 90°-scattering geometry. As in the original experiment, the set-up consists of a pair of counter propagating laser beams of the same colour with detectors on either side. However, rapid synchronous switching of the incident and the scattered beams ensures that either of the two detectors receives scattered light only of one of the lasers so that only light of a definite scattering vector is received. Without the application of time coding the detectors would receive signals corresponding to two different scattering vectors [15]. The technique of cross correlation of "time coded" scattering has proved to be useful in a quite different context. It was applied by Staude and Schmidt [16] in investigations of turbulent flow.

In this contribution we report the realisation of a "3-D coding" cross-correlation experiment for 90°-scattering geometry which is shown in Fig. 1. The set-up appears as a fairly simple modification of a conventional light scattering experiment in the vertical to the scattering plane. We report measurements on solutions of latex particles with 109.0 ± 2.7 nm diameter in the range of $c = 0.1\text{--}3.5\%$ mass fraction (mf). By comparing the ACF and the CCF it is demonstrated that the CCF yields indeed the desired ACF of singly scattered light.

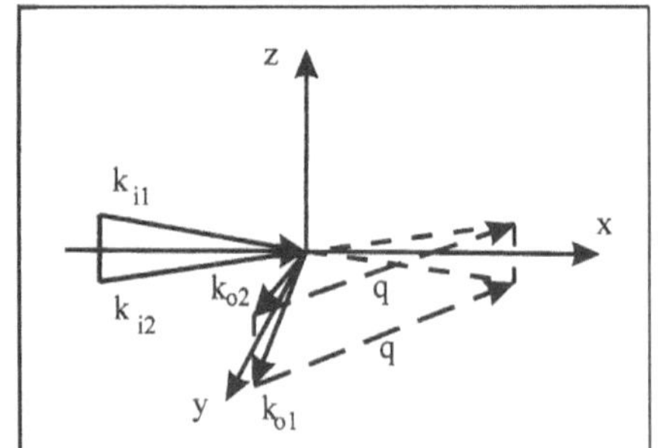

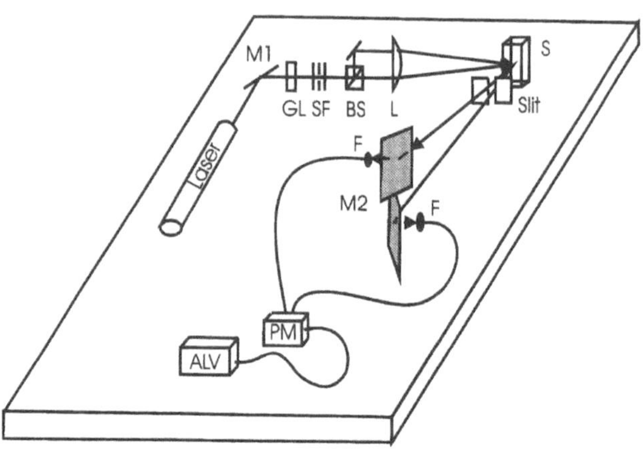

Fig. 1 **A** Scattering geometry: The wave vectors $\mathbf{k}_{i1}$ and $\mathbf{k}_{i2}$ denote of the incoming beam while $\mathbf{k}_{o1}$ and $\mathbf{k}_{o2}$ indicate the direction of the scattered light. The scattering vector $\mathbf{q}$ is identical in the two scattering planes. **B** Set-up of the "3-D coding" experiment. A detailed description is given in the text

Theoretical background

The ACF of the scattered intensity I may be separated into a contribution of uncorrelated light and a sum of contributions due to single, double and higher multiple scattering processes which are correlated. Assuming the validity of the Siegert relation [2] the correlated parts are represented by squares of CFs of the electric field. Correlations between different scattering processes, for example, between single and double scattering do not contribute in this approximation. Introducing the normalized field CFs $c_i(t)$ with $c_i(0) = 1$ and the factors β_i for the relative amplitudes, the ACF is given by

$$\langle I(0)I(t)\rangle = \langle I\rangle^2 \cdot (1 + \sum \beta_i^2 c_i(t)^2) \,. \tag{1}$$

Denoting the contributions to the scattered intensity $\langle I\rangle$ due to single scattering by $\langle I_1\rangle$, due to double scattering by $\langle I_2\rangle$, etc., the amplitude factors β_i satisfy the inequality $\beta_i \leq \langle I_i\rangle/\langle I\rangle$. For simplicity of notation the dependence of $c_i(t)$ and of the other terms in Eq. (1) from the scattering vector $\mathbf{q}$ (see below) is not written out explicitly. The experimental field CF represents the square root of the sum in Eq. (1) and may formally be written as product of an overall amplitude β and $c(t)$, a normalized CF. The amplitude factor β_1 may come near to the theoretical limit $\beta_i = 1$ when single scattering is observable only and the set-up is optimized with respect to, for example, speckle size, polarisation effects, etc. This can quite easily be achieved in experiments using monomode fibres [17]. If multiple scattering is present the overall amplitude β is restricted by the inequality

$$\beta \leq \sqrt{\sum \frac{\langle I_i\rangle^2}{\langle I\rangle^2}} < 1 \,. \tag{2}$$

In many cases the time dependency of the single-particle CF is well described by an mono exponential function $c_1(t) = \exp[-\Gamma_1 \cdot t]$ involving a decay coefficient Γ_1. If the diffusion model applies, we have $\Gamma_1 = q^2 D$, where D is the diffusion coefficient. In order to determine the particle sizes in colloidal solutions the Stokes–Einstein theory is frequently applied. According to this theory $D = kT/6\pi\eta a$ where a, η, k, T denote the radius of the particles, the shear viscosity, Boltzmann's constant, and the temperature, respectively.

The $c_i(t)$ of multiple scattering processes usually decay faster than $c_1(t)$. Even if exponential decay for the various $c_i(t)$ with a characteristic decay constants Γ_i may be assumed, $c(t)$ becomes nonexponential. The initial decay of $c(t)$ is then determined by

$$\bar{\Gamma} = \frac{\sum \beta_i^2 \Gamma_i}{\sum \beta_i^2} \,. \tag{3}$$

Erroneous interpretation of $\bar{\Gamma}$ as $q^2 D$ yields estimates of the particle size which are too small. The purpose of the cross-correlation experiments discussed here is to eliminate this error.

Figure 1A shows the scattering geometry of our set-up. In principle, the scattering geometry differs only in the coordinate system from that discussed in Schätzel's review [10]. Two coherent, parallel and vertically aligned laser beams are focused into the sample and constitute the incoming wave vectors $\mathbf{k}_{i1}$ and $\mathbf{k}_{i2}$. The scattered light is detected in the directions $\mathbf{k}_{o1}$ and $\mathbf{k}_{o2}$. The wave-vector pairs $(\mathbf{k}_{i1}, \mathbf{k}_{o1})$ and $(\mathbf{k}_{i2}, \mathbf{k}_{o2})$ form two scattering planes

which can be obtained by rotating the horizontal plane about the common scattering vector $\mathbf{q}$ given by $\mathbf{q} = \mathbf{k}_{o1} - \mathbf{k}_{i1} = \mathbf{k}_{o2} - \mathbf{k}_{i2}$. The angles between the wave vectors of the incident beams $(\mathbf{k}_{i1}, \mathbf{k}_{i2})$ are equal to that between the wave vectors $(\mathbf{k}_{o1}, \mathbf{k}_{o2})$ of the scattered beams. The plane constituted by the corresponding bisectors is horizontal and the scattering geometries labelled 1 and 2 are symmetric in respect to this plane. The intensity I^1 received by detector 1 is the sum of two parts $I^{11}(\mathbf{q}, t)$ and $I^{12}(\mathbf{q} + \Delta\mathbf{q}_z, t)$ which represent the intensity scattered of the incident beams with the wave vectors $\mathbf{k}_{i1}$ and $\mathbf{k}_{i2}$ into the direction $\mathbf{k}_{o1}$, where $\mathbf{k}_{o1} - \mathbf{k}_{i2} = \mathbf{q} + \Delta\mathbf{q}_z$. Analogously, the intensity I^2 received by detector 2 is the sum of two parts $I^{22}(\mathbf{q}, t)$ and $I^{21}(\mathbf{q} - \Delta\mathbf{q}_z, t)$ which represent the intensity scattered of the incident beams with the wave vectors $\mathbf{k}_{i1}$ and $\mathbf{k}_{i2}$ into the direction $\mathbf{k}_{o2}$. The terms I^{11} and I^{22} represent the same microscopic process and are correlated. All other terms give no contribution to the cross-correlation function but enhance the background provided $\Delta\mathbf{q}_z$ is sufficiently large. Assuming that on average all contributions $\langle I^{11}\rangle$, $\langle I^{21}\rangle$, etc. to the intensity are equal to the intensity $\langle I\rangle$ of just one of the scattering geometries, the intensity is finally

$$\langle I^1(0)I^2(t)\rangle = 3 * \langle I\rangle^2 + \langle I^{11}(\mathbf{q},0) * I^{22}(\mathbf{q},t)\rangle$$
$$\cong 4 * \langle I\rangle^2(1 + \beta_c^2 c_1(\mathbf{q},t)^2) \,. \tag{4}$$

The maximal value for the amplitude β_c of the cross-correlation function, which may be achieved by this method is $\beta_c \leq 1/2$.

Apparatus

The set-up

The set-up is built for the measurement of cross- and auto-correlation functions in the same experiment and designed to allow for convenient alignment. The apparatus is largely simplified by taking advantage of the properties of mono mode fibers [17] with a gradient-index (GRIN) lens integrated functioning as a collector. The monomode fiber with this integrated optic accepts only almost parallel light within a small spatial angle (0.3°). In conventional light scattering such filtering requires a fairly large, sophisticated optical system. The fiber system needs no further optical system and therefore allows for a compact design of the experiment.

A sketch of the experiment is given in Fig. 1(b). The vertical polarized beam of a 5 mW Helium-Neon laser is spatially filtered (SF), split into two parallel beams by a mirror/beam splitter system (BS) and focused by the lens (L) with the focal length $f = 0.16$ m into the sample (S). In order to avoid spherical aberration the vertical distance z between the two parallel beams ($z = 1.1$ cm) is chosen small enough so that both beams can be considered as paraxial. The diameter of the approximately cylindrical shaped scattering volume is determined by the diameter of the focused beams (0.1 mm) and the width of the slit (5 mm).

In the present experiment the detection arm is positioned perpendicular to the incoming beams. For each of the given scattering planes the scattered light is directed via a mirror (M2) to a monomode fiber (F). The mirrors allow for convenient placement of the mounts of the fibers (F) which are too large to be placed directly into the beams of the scattered light. The two mirrors (M2) change the direction of the scattered light by 90°. They are tilted in such a way that in both scattering planes the light to the same scattering vector is directed towards the fibers. The fibers (F) lead the scattered light to the double photon multiplier (PM). Finally, the output of the detector (Single-Dual ALV/SO-SIPD) is transferred to a PC with the correlator card (ALV-5000) and processed here.

Alignment

The set-up appears to be fairly straightforward and simple. However, the conditions for yielding correlated signals in different scattering geometries – scattering from the same scattering volume and identical scattering vectors – cannot be satisfied easily. Therefore, the alignment turns out to be quite delicate and it seems appropriate to draw attention to some details which are crucial for a successful experiment. The important steps are to ensure that (a) the incident beams have a large cross-section, (b) the fibers receive light from the same scattering volume and (c) the scattering vectors of the light detected in the two scattering geometries are identical.

(a) A large cross-section of the incident beams requires the two beams obtained by the beam splitter–mirror system to be perfectly parallel to each other. By checking the alignment of the beams with the help of a plumb-line the alignment of the beams becomes independent of orientation and flatness of the desk.

(b) In order to test that the light, finally received in the two fiber optics, results from scattering processes in the same scattering volume, we place a pin hole in the diagonal of the quadratic sample at the position of the scattering volume. We check that both incident beams pass the pinhole. The path of the scattered light is probed by taking advantage of the property of the fibers to allow for the transmission of light in both directions and considering the reversed path. Proper alignment is achieved when all

four beams, the two incident and the two reversed 'scattered' beams, pass the pinhole.

(c) The two mirrors M2 select the components with the appropriate wave vectors $\mathbf{k}_{o1}$ and $\mathbf{k}_{o2}$ from the scattered light and direct it into the fiber optics. The alignment of the mirrors (M2) is achieved with the help of a mirror placed in the diagonal of the square sample. This mirror mimics the scattering of the sample. The wave vectors of the beams reflected by this mirror correspond exactly to the wave vectors $\mathbf{k}_{o1}$ and $\mathbf{k}_{o2}$ in the scattering experiment. The tilt of the mirrors M2 must be chosen so that the wave vectors of the reflected beams are now parallel to the bisector of the angle ($\mathbf{k}_{i1}$, $\mathbf{k}_{i2}$) which is exactly the direction accepted by the fibers.

Results and discussion

In order to demonstrate the functioning of the set-up the ACF and the CCF of the scattered intensity are measured for solutions of latex particles with a diameter of (109 ± 2.7) nm in the 90°-scattering geometry. Quadratic samples with 1 cm thickness are used. The concentrations c investigated ranging from 0.01% to 0.35% mf include very transparent and highly turbid samples. The turbidity $\tau = \ln(I_i/I_t)$ of the samples increases linearly from 0.14 to 4.2 cm^{-1} with the mass fractions of the samples. The intensity of the incoming beam I_i and the intensity of the transmitted beam I_t are determined using a power meter (Field Master, Coherent Radiation).

Apart from minor statistical deviations the same ACFs are found for the two scattering geometries ($\mathbf{k}_{i1}$, $\mathbf{k}_{o1}$) and ($\mathbf{k}_{i2}$, $\mathbf{k}_{o2}$). The amplitudes of the ACFs are close to the theoretical limit of $\beta = 1$ which is typical for a set-up involving a monomode fiber [17]. Only 10–15 min are required for measuring an ACF yielding a correlation time with an accuracy near 6%. The decay of the ACFs become faster with increasing concentration of the scatterers. Deviations from a single exponential are noticeable at higher concentrations. The apparent diffusion coefficient, determined from the initial slope of the ACF, increases in a non-linear manner with concentration. This is shown in Fig. 2. While the apparent diffusion coefficient D shows a strong dependence of the concentration of the scatterers the amplitudes β are affected very little. A minor decrease of the amplitudes from $\beta = 0.97$ at $c = 0.2\%$ mf to $\beta \approx 0.87$ at the highest concentration of $c = 3.5\%$ mf is noticeable. All this observations are in accordance of what is expected for the strong scattering samples where multiple scattering is present.

The measurements of the CCF turn out to be much more difficult. The amplitudes of the CCF are quite small. They decrease from 0.18 at mf of $c = 0.01\%$ to 0.06 at

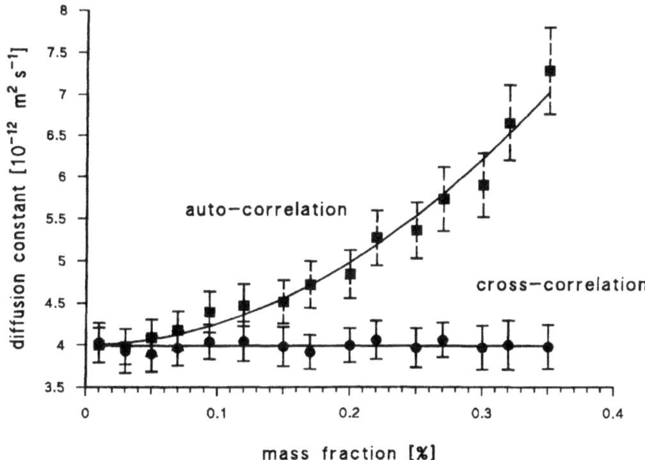

Fig. 2 Apparent (■) and the "true" (●) Diffusion coefficient of solutions of latex particles at 21 °C (diameter 109 nm) determined from the auto- and cross-correlation function, respectively

$c = 0.35\%$ mf. Therefore, measuring times of 1–20 h and more are required here to get a CCF of similar statistical quality as the ACFs. The amplitudes are well below the theoretical limit of $\beta_c = 0.5$. The decrease with concentration reflects the increasing contributions of multiple scattering, which, as intended, do not contribute to the CCF. The fairly small amplitude in the single scattering region at low concentrations is probably caused by light scattered in regions which are illuminated by only one of the incident beams and consequently cannot contribute to the CCF.

However, the set-up proves to work as a very good filter, removing the contributions due to multiple scattering. We find, that the CCFs at all concentrations can be described by mono exponential functions. Furthermore, the decay constants do not vary with concentration and are in agreement with the value determined from the ACF at the lowest concentration. This shows that the CCF is identical to the ACF of single scattered light. Assuming diffusion like dynamics we calculate the diffusion coefficients, shown in Fig. 2.

The diffusion constants D determined from the CCFs are $(1.39 \pm 0.05) * 10^{-12}\,\mathrm{m^2\,s^{-1}}$ at 21 °C. Using the Stokes–Einstein relation with the shear viscosity $\eta_{H_2O}^{21} = 0,9779 \times 10^{-2}\,\mathrm{g/cm\,s}$ [18] leads to a particle diameter of (110 ± 3) nm, which is in good agreement with the specified size of $(109 \pm 2,7)$ nm of the latex particles.

Outlook

We have demonstrated for 90°-scattering geometry that by "3-D coding" contributions of multiple light scattering to the intensity CF can very effectively be eliminated. In principle, the set-up can be applied for other scattering angles as well. Comparing the amplitudes of the CFs, at present the "two colour" technique reaching amplitudes of 0.8 [14] is superior to our 3-D method. Improved alignment is expected to allow for amplitudes near the theoretical limit of $\beta_c = 1/2$ while additional application of the "time coding" method may even yield $\beta_c \approx 1$. The most important advantage of the "3-D coding" method is founded in the use of only one frequency: the alignment is not affected by the dispersion of the refractive index which, if compared to the "colour coding" technique, may simplify the work if measurements in different solvents and at varying temperatures are considered.

Acknowledgments We are very much indebted to M. Kleemeier and J. Sebald for valuable discussions and advice. The technical assistance of C. Rybarsch–Steinke and the support of the mechanical workshops of the departments of Physics and Chemistry is highly appreciated. L.B.A. acknowledges the PhD grant given by the University of Bremen for this project. Support of the Deutsche Forschungsgemeinschaft and the Fonds der Chemischen Industrie is gratefully acknowledged.

References

1. Brown Wyn (ed) (1993) Dynamic Light Scattering. Clarondon Press, Oxford
2. Berne BJ, Pecora R (1976) Dynamic Light Scattering. Wiley, New York
3. Maret G, Wolf PE (1987) Z Phys B 65:409–416
4. Pine DJ, Weitz DA, Chaikin PM, Herbolzheimer E (1988) Phys Rev Lett 60:1134–1137
5. Weitz DA, Pine DJ (1993) In: Brown Wyn Ed, Dynamic Light Scattering. Clarondon Press, Oxford, pp 652–720
6. Boots HMJ, Bedeaux D, Mazur P (1975) Physica 79a:397–419
7. Lakoza EL, Chalyi AV (1983) Sov Phys Usp 26:573–590
8. Fererel AF, Bhattacharyee J (1979) Phys Rev A19:348–369
9. Shanks JG, Sengers JV (1988) Phys Rev A 38:885–896
10. Schätzel K (1991) J Mod Optics 38:1849–1865
11. Phillies GDJ (1981) J Chem Phys 74:260–62
12. Schätzel K, Drewel K, Ahrens K (1990) J Phys.: Condens Matter 2:SA393–398
13. Drewel M, Ahrens J, Podschus U (1990) J Opt Soc Am A 7:206–210
14. Segrè PN, van Megen W, Pusey PN, Schätzel K, Peters W (1995) J Mod Opt 42:1929–1952
15. Dhont JKG, De Kruif CG (1983) J Chem Phys 79:1658–1663
16. Schmidt U (1990) PhD Thesis, Bremen
17. Ricka J (1993) Applied Optics 32:2860–2875
18. Weast RC ed (1978) CRC Handbook of Chemistry and Physics, 58th tir CRC Press, Cleveland

Progr Colloid Polym Sci (1997) 104:126–128
© Steinkopff Verlag 1997

R. Lenke
G. Maret

Coherent backscattering of light in multiple scattering media

Dr. R. Lenke (✉) · G. Maret
Institut Charles Sadron (CRM-EAHP)
6 rue Boussingault
67083 Strasbourg Cedex, France

Abstract We will give a short overview of the phenomenon of coherent backscattering of light, essentially with respect to its possible application for the characterization of multiple scattering samples. We will present some basic experiments and their evaluation in comparison to Monte-Carlo simulations.

Key words Coherent backscattering – multiple light scattering – Monte-Carlo simulation

Coherent backscattering (CB) of waves in multiple scattering (ms-) media has been well known for electrons [1] and light [2, 3] for some time. During this period it has become more and more clear that CB is a fundamental effect in nature. It is closely related to the fact that light paths are reversible, i.e., "if I can see you, you can see me". More precisely, it is closely related to the theorem of reciprocity [4], which states that the scattering matrix of the reversed path is the transposed matrix of the direct path. In exact backscattering direction a light path and its reversed path have exactly the same length and are thus always interfering constructively. Off exact backscattering direction this coherent backscattering enhancement by – theoretically – a factor of two [5], disappears and the scattered intensity decreases to the "normal", non-coherent intensity. The angular width of this so-called CB-cone is inversely proportional to the optical density of the medium, i.e., the transport mean free path l^* which is the characteristic length of the random walk of the light in the medium. In most cases, the width of the cone is smaller than about $1°$. CB is present in any ms-medium. It is a very "stable" effect, i.e., it is not destroyed by the movement of the scatterers, nor by absorption, nor by a short coherence length of the incident light (only the shape of the cone might change but not its maximum value); in fact, CB can even be seen with sun light (Fig. 1). The only effect to our knowledge which violates the theorem of reciprocity and thus destroys

CB is Faraday Rotation [6, 7]. We observed this effect for the first time experimentally [8]. Of course, CB is also destroyed by non-elastic scattering, for example, by relativisticly moving scatterers [6].

In a standard CB experiment the sample is illuminated with an expanded plan-parallel laser beam. In order to observe the light which is scattered in the backward direction, one uses a semi-transparent mirror. After this mirror the light is focused onto a CCD camera, for example, which is placed in the focal plane of the lens in order to obtain an angularly resolved image of the scattered light. According to the theorem of reciprocity a CB enhancement of two is only obtained if incident and detected polarization states are the same. Therefore, a polarizer and analyser are used in the experimental setup, normally. In order to destroy the speckle pattern the sample needs to be moved during the measurement if it is not liquid. Figures 2 and 3 show the experimental results for a sample of poly-styrene beads in water. However, it is very difficult to observe an enhancement factor of two experimentally. In the case of linear polarization (same linear polarization state for both incident and detected light) this is mostly due to both single scattering and direct reflections which do not contribute to CB since CB is a ms-effect. Another reason, apart from experimental problems, is the limited coherence area of the incident light and a non-constant intensity over this area. The area should be much larger

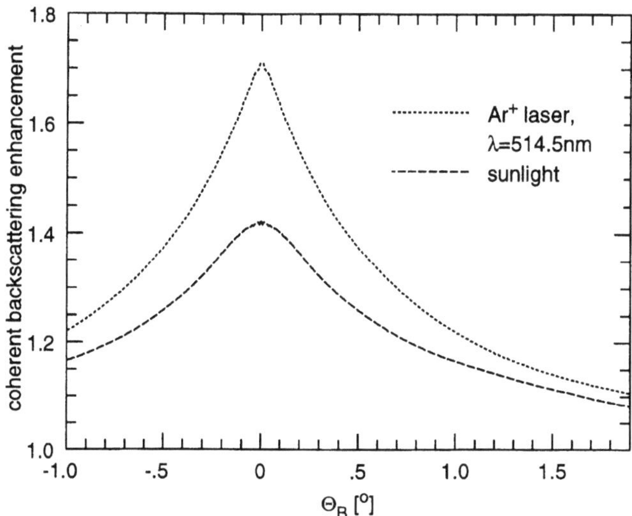

Fig. 1 CB-cone, i.e., intensity as a function of the backscattering angle θ_B, of a powder of BaSO$_4$. Light was linearly polarized. The coherence area of sun light is about $(20 \, \mu m)^2$. As l^* of this sample is smaller, i.e., in the order of a few micron, CB is visible. This coherence area corresponds to the divergence of sunlight of $\pm 16'$; the cone measured is a convolution of sources which are distributed over the same angle range

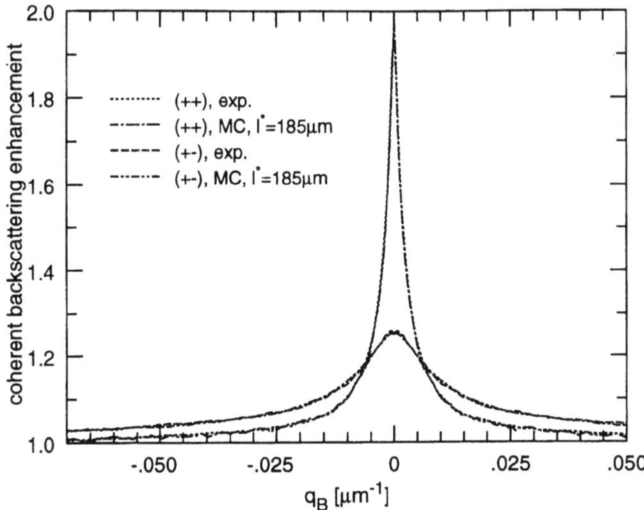

Fig. 2 CB-cone, i.e., intensity as a function of (back-) scattering vector $q_B \approx 2\pi\theta_B/\lambda$, of a sample of polystyrene beads in water. Volume fraction 2.9%, diameter ≈ 100 nm, Same circular incident and detected polarization state $(++)$. A MC-simulation for Rayleigh scatterers gives $l^* = 185 \, \mu m$. Due to the short paths, a small cone is also visible for opposite circular polarization states $(+-)$. This behavior is characteristic for Rayleigh scatterers. The experimental curves have been rescaled in order to obtain the theoretical CB enhancement values at $q - 0$. The values measured at $q - 0$ were 1.69 $(++)$ and 1.21 $(+-)$

than the average distance between the beginning and final point of the light paths in the sample. If this is not the case, however, the shape of the CB-cone can be obtained by convolution.

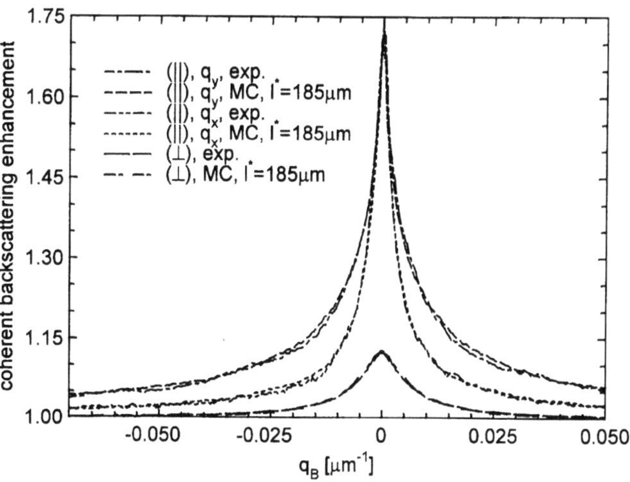

Fig. 3 Same measurements as in Fig. 2, but for parallel ($\parallel$) and perpendicular ($\perp$) linearly polarized incident and detected light. Due to the anisotropic scattering matrix of Rayleigh scatterers the cone becomes anisotropic as well, i.e., $I(q_x) \neq I(q_y)$ (polarization of incident light parallel to the x-axis). In fact, the cone is also slightly anisotropic in the case ($\perp$), similar to a four-leafed clover. The intensity of the experimental curves was rescaled. The values measured for $q = 0$ were 1.61 ($\parallel$) and 1.11 ($\perp$)

Yet, the incident wave need not to be parallel. In fact, to some extend CB acts like a phase conjugating mirror: by using a spherical incident wave of a point source the maximum value of CB will occur precisely at the place of the source. Generally, one can show that the scaling factor for the images of the camera in a CB experiment is proportional to $F/(1 + (F - D)/\pm L)$, where F is the focal length of the lens, D the distance between sample and lens and $\pm L$ the distance between the sample and the divergent $(+)$ or convergent $(-)$ point source of light. However, one has to verify (or to take into consideration) that the path length distribution of the light paths in the sample does not differ too largely from the parallel case and that l^* is relatively small. Consequently, light is considered to be parallel if $(F - D)/L \ll 1$. We verified this scaling behavior experimentally. Concerning applications, parallel incident light has the advantage that the position of the sample is not important whereas non-parallel light has the advantage that the scaling factor can be varied by changing the position of the sample.

Evaluation of CB by analytic calculations is quite difficult. This is due to the fact that in backscattering direction about 50% of the light had only been scattered less than 10 times. Thus diffusion approximations do not apply very well. Therefore, we compare our experimental results to Monte-Carlo (MC) simulations. As one can see in Figs. 2 and 3 the simulations fit very well for every single state of polarization.

Generally, CB is explained by the interference between a direct light path and its reverse. However, one can understand CB also as a contribution of the ensemble of scatterers to the scattering matrices of the (independent) scatterers. In fact, there are predicitions that CB is not only present outside of a ms-medium, but also inside the medium where it may change the diffusion constant of light. This is the so-called "weak localization" of light. In very strongly scattering media even a "strong localization" had been predicted, i.e., light might be trapped by the random potential of a ms-sample [9].

References

1. Bergmann G (1984) Phys Rep 107:1–58
2. van Albada MP, Lagendijk A (1985) Phys Rev Lett 55:2692–2695
3. Wolf PE, Maret G (1985) Phys Rev Lett 55:2696–2699
4. Saxon DS (1955) Phys Rev 100:1771–1775
5. First precise measurements of theoretical factor of two: Wiersma DS, van Albada MP, Lagendijk A (1995) Rev Scientific Instr 66:5473–5476
6. Golubentsev AA (1984) Sov Phys JETP 59:26–32
7. MacKintosh FC, John S (1988) Phys Rev B 37:1884–1897
8. Lenke R, Maret G (1994) Proc OSA Topical Meeting Adv Optical Imaging and Photon Migration 21:16–19
9. Anderson PW (1985) Phil Mag B 52:505–509

Progr Colloid Polym Sci (1997) 104:129–131
© Steinkopff Verlag 1997

W. Schärtl
C. Graf
M. Schmidt

Polyorganosiloxane-microgels as probes for forced Rayleigh scattering

Dr. W. Schärtl (✉) · C. Graf · M. Schmidt
Institut für Physikalische Chemie
Johannes-Gutenberg-Universität Mainz
Jakob-Welder-Weg 11
55099 Mainz, Germany

Abstract We describe the synthesis of microgel spheres of 10 nm radius which are suitable as probes to study diffusion by forced Rayleigh scattering (FRS), a holographic grating technique. Those particles are obtained by a copolycondensation in microemulsion. The main advantage of organosiloxanes compared to purely organic monomers as styrene or methacrylate is the simple chemical functionalization of the particles. A rich choice of silane monomers which may be copolycondensated with the standard monomer trimethoxymethylsilane are commercially available. One of those, chlorobenzyltrimethoxysilane, is used as a coupling agent to attach the photoreactive dye orthonitrostilbene (ONS) to the microgel spheres. All samples are characterized by light scattering, GPC and electron microscopy. In addition, we determined the content of photoreactive dye chemically attached to the spherical particles by UV/visible absorptionspectroscopy. Our nano-particles are redispersible in organic solvents such as toluene or THF up to weight fractions as high as 50 wt%. To prove applicability of the particles as FRS probes we show some preliminary results obtained from FRS measurements of highly concentrated toluene solutions (particle concentration 45–50 wt%).

Key words Polyorganosiloxane – microgels – colloids – forced Rayleigh scattering

Introduction

Colloidal particles are, due to their characteristic time- and length-scales, ideal model systems to study dynamics and phase behavior of condensed matter [1]. One class of these model systems are polystyrene-microgels, i.e., crosslinked nano-spheres made from styrene by polymerization in microemulsion [2].

Recently, crosslinked polyorganosiloxane particles which may be redispersed in organic solvents have been synthesized by polycondensation in microemulsion [3]. The major advantage of these new materials as compared to polystyrene microgels is the easy chemical modification.

Due to a variety of commercially available methoxysilanes, which may be used as comonomers in the polycondensation reaction, it is possible to introduce functional groups such as Si–H, Si–S–H, Si–NH$_2$ or Si–CH$_2$–Cl into the spherical particles.

Here, we are going to describe a method to obtain polyorganosiloxane microgels which are useful as a probe for Forced Rayleigh Scattering experiments [4], a holographic grating technique based on a photoreaction. We found a way to chemically incorporate a photoreactive dye, orthonitrostilbene (ONS) [5], into polyorganosiloxane spheres. We should note that dye-labeled silica particles useful for fluorescence recovery after photobleaching have been prepared previously by van Blaaderen et al. [6].

Experimental

In order to attach the photochromic label ONS to the spheres, we applied the same kind of reaction as developed for labeling of polystyrene [7], i.e., an esterification of the ONS cesium salt [8] with a chloro-benzyl functional group. Two different types of chlorobenzyl-functionalized microgels were synthesized by adding chloro-benzyl trimethoxy silane (2%/10%) to the monomer trimethoxymethylsilane during the polycondensation reaction. The principle of the synthesis is sketched in Fig. 1.

To verify applicability of our product for FRS measurements, we prepared highly concentrated solutions of ONS labeled spheres in toluene of 45–50 wt%. The

Fig. 1 Reaction scheme of synthesis of ONS-labeled polyorgano-siloxane microgels

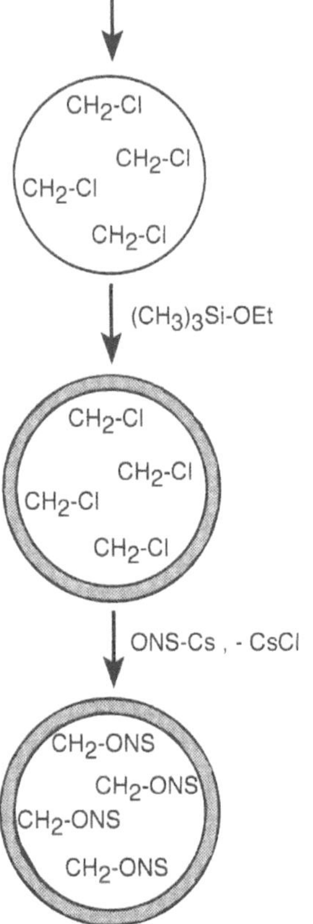

FRS setup which has been used is described in detail elsewhere [9].

Results and discussion

Molecular masses and radii given in Table 1 have been determined by GPC, static (SLS) and dynamic (DLS) light scattering, dye content by GPC used as UV/visible-absorption spectrometer. Note here the decrease in molecular mass, M_w, from non-labeled to labeled spheres, which is not yet understood. Samples have also been characterized by transmission electron microscopy (TEM) (Fig. 2).

Table 1 Characterization of non-labeled and labeled polyorgano-siloxane-microgels

		ONS-23	ONS-24
Non-labeled spheres	R_h (DLS)	9.7 nm	10.5 nm
	R_g (SLS)	< 10 nm	< 10 nm
	ρ (= R_g/R_h)	< 1.03	< 0.95
	M_w (SLS)	2.7×10^6 g/mol	2.8×10^6 g/mol
	M_w/M_n (GPC)	1.05	1.06
Labeled spheres	R_h (DLS)	10.3 nm	10.4 nm
	R_g (SLS)	< 10 nm	< 10 nm
	ρ (= R_g/R_h)	< 0.98	< 0.97
	M_w (SLS)	2.1×10^6 g/mol	2.6×10^6 g/mol
	M_w/M_n (GPC)	1.06	1.06
	Labels/sphere (UV/VIS-GPC)	87	12

Fig. 2 Transmission electron micrograph of sample ONS-23 (Table 1)

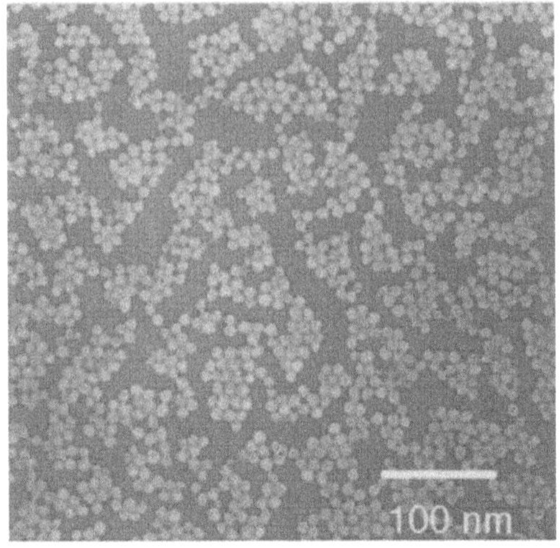

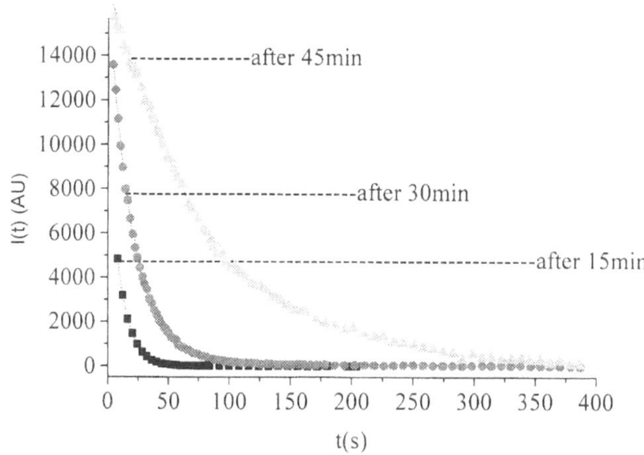

Fig. 3 FRS signals from sample ONS-23 (Table 1) in toluene solution (47 wt%), bleaching angle 4.5°, bleaching pulse 5/100 s at 100 mW. Signals are obtained at different times after sample preparation (see labels). See text for discussion

In Fig. 3, we present first FRS measurements of the sample ONS-23 (Table 1), obtained from a 47 wt% (as freshly prepared) solution in toluene. Since the sample cell which consists of 2 glass disks of 1.5 cm diameter separated by a Teflon ring of 100 μm thickness is not sealed perfectly, solvent may evaporate from the solution, thereby increasing the effective particle concentration. We believe at present that this is the reason for the slowing of particle dynamics with time after preparation, as can be seen from the slower decay of the signals after 30 and 45 min compared to the signal after 15 min (Fig. 1). It should be noted that the signal intensity is as large as 10 000 counts compared to a background scattering (i.e., without holographic grating) of 10 counts. The new system therefore seems to be excellently suited as a probe for FRS studies.

References

1. Pusey PN (1991) In: Levesque D, Hansen JP, Zinn-Justin J (eds) Liquids, Freezing and the Glass Transition, Les Houches Sessions LI. Elsevier, Amsterdam
2. Bremser W, Antonietti M, Schmidt M (1990) Macromolecules 23:3796
3. Baumann F, Schmidt M, Deubzer B, Dauth J, Geck M (1990) Macromolecules 23:3796
4. Kogelnik H (1969) Bell System Tech J 48:2909
5. Splitter JS, Calvin M (1955) J Org Chem 20:1086
6. van Blaaderen A, Peetermans J, Maret G, Dhont JKG (1992) J Chem Phys 96:4591
7. Antonietti M, Sillescu H (1985) Macromolecules 18:1162
8. Pfeiffer P (1915) Chem Ber 48:1777
9. Sillescu H, Ehlich D (1989) In: Fouassier JP, Rabek JF (eds) Application of Lasers in Polymer Science and Technology. CRC press, Boca Raton Fl
10. Schaertl W, Tsutsumi K, Kimishima K, Hashimoto T (1996) Macromolecules 29:5297

Progr Colloid Polym Sci (1997) 104:132–134
© Steinkopff Verlag 1997

Collective and self diffusion of PS microgels in solution as observed by thermal diffusion forced Rayleigh scattering

W. Köhler
R. Schäfer
E. Bartsch
S. Stölken

Dr. W. Köhler (✉) · R. Schäfer
Max-Planck-Institut für Polymerforschung
Postfach 31 48
55021 Mainz, Germany

E. Bartsch · S. Stölken
Institut für Physikalische Chemie
Universität Mainz
Jakob-Welder-Weg 15
55099 Mainz, Germany

Abstract Thermal diffusion in solutions of polystyrene micro-network spheres (microgels) in toluene has been studied by the holographic scattering technique of thermal diffusion forced Rayleigh scattering (TDFRS) and by photon correlation spectroscopy (PCS). Size distributions of microgels of different crosslink ratios are obtained from TDFRS measurements on dilute solutions at very low q-values around $4000\,\mathrm{cm}^{-1}$. At low concentrations a single diffusive mode is observed and the diffusion coefficient increases with concentration. It is attributed to the collective diffusion of the microgels and the solvent. At high concentrations an additional slow mode appears whose diffusion coefficient decreases with increasing concentration. Both diffusive modes are observed with PCS and TDFRS. Contrary to PCS, heterodyne TDFRS-measurements reveal a negative amplitude of the slow mode. We attribute the slow mode to self-diffusion of the microgels, made visible by the polydispersity of their size distribution. It is discussed in terms of a fast coupled thermal diffusion with subsequent decoupling of the individual microgels and relaxation into a new Soret equilibrium by self-diffusion of the microgels.

Key words Microgels – thermal diffusion – self diffusion – forced Rayleigh scattering

Thermal diffusion of polystyrene micronetwork spheres (microgels) in toluene has been studied by the holographic scattering technique of thermal diffusion forced Rayleigh scattering (TDFRS) [1, 2] and by photon correlation spectroscopy (PCS). In TDFRS a holographic temperature grating is written into a solution. The Ludwig–Soret effect leads to a coupling between temperature and concentration and gives rise to a superimposed concentration grating and this can be read by Bragg diffraction of a readout laser beam. The time-dependent diffraction efficiency of the concentration grating in a heterodyne TDFRS experiment is obtained from a solution of an extension of Fick's second law of diffusion [3, 4]:

$$\frac{\partial c}{\partial t} = D\Delta c + D_T c[1 - c]\Delta T \,, \tag{1}$$

$D = (\tau q^2)^{-1}$ is the diffusion coefficient, D_T the thermal diffusion coefficient, and c the concentration in weight fractions. The spatial temperature distribution $T(\mathbf{r}, t)$ is obtained from the heat equation with the absorbed laser intensity as the source term.

Dilute solutions

In case of polydisperse samples at infinite dilution, both the heterodyne TDFRS decay function and the PCS field autocorrelation function can be expressed as a sum over all contributions from the individual molar masses M_k with diffusion times τ_k, however with different statistical weights p_k:

$$\sum_k p_k e^{-t/\tau_k}, \quad p_k = \frac{c_k M_k^\alpha}{\sum_k c_k M_k^\alpha} . \tag{2}$$

The exponent α depends on the type of experiment performed:

$$\alpha = \begin{cases} 0 & \text{TDFRS } (\tau_p \ll \tau_k), \\ b & \text{TDFRS } (\tau_p \gg \tau_k), \\ 1 & \text{PCS} . \end{cases} \tag{3}$$

τ_p is the duration of the TDFRS excitation pulse and b the scaling exponent, which relates the diffusion coefficient to the molar mass: $D \propto M^{-b}$. $b = \frac{1}{2}$ for random coils under θ-conditions and $b = \frac{1}{3}$ for solid spheres. In the ideal case of short exposure TDFRS, concentration proportional contributions to the signal are obtained, independent of molar mass [2].

Size distributions of microgels with crosslink ratios of 1:10, 1:20, and 1:50 [5,6] have been determined from TDFRS measurements on dilute solutions at q-values of approximately $4000\,\text{cm}^{-1}$, using CONTIN [7] for the Laplace inversion. The low scattering angles around $2°$ ensure that the experiments are performed within the hydrodynamic limit, where pure center of mass diffusion is observed and internal modes can be neglected. In Fig. 1

Fig. 1 Normalized heterodyne diffraction efficiencies for a concentration series of 1:20 crosslinked PS microgels in toluene. The insert shows the rate distribution for the lowest concentration (c in weight fractions)

the heterodyne diffraction efficiencies are shown for a concentration series of 1:20 crosslinked microgels together with the rate distribution for the lowest concentration, from which a weight average hydrodynamic radius $\langle R_\text{h}\rangle_c = kT(6\pi\eta D)^{-1} = 19\,\text{nm}$ and a distribution width of $\langle \Delta R_\text{h}^2\rangle_c/\langle R_\text{h}\rangle_c^2 = 0.28$ are derived. With the solid sphere value of $b = \frac{1}{3}$, this translates to a molar mass polydispersity of $M_\text{w}/M_\text{n} = 1.7$.

Semidilute and concentrated solutions

A single diffusive mode is observed at low concentrations. Its diffusion coefficient increases with concentration, and it is attributed to the collective diffusion of the microgels in the good solvent. At higher concentrations an additional slow mode appears, whose diffusion coefficient decreases with increasing concentration. Both diffusive modes are observed with PCS and TDFRS (Fig. 2). Contrary to PCS, however, heterodyne TDFRS-measurements reveal a negative amplitude of the slow mode. We attribute the slow mode to mutual diffusion of the large and small microgel particles with respect to each other. Due to their narrow size distribution, this mode is essentially identical to self-diffusion of the microgels.

To understand the effect, three time domains must be distinguished. For very short times, the solvent and the microgels diffuse with respect to each other. The corresponding collective diffusion coefficient D_c

Fig. 2 Fast and slow mode (arrow) in semidilute solutions of 1:10 crosslinked microgels in toluene ($c = 21.7\%$) as observed by PCS and TDFRS. For the lower concentration ($c = 2.6\%$) there is only one fast mode. The PCS measurement has been measured at $q = 75\,000\,\text{cm}^{-1}$ and time shifted according to $\tau \propto q^{-2}$ to the scattering vector $q = 4000\,\text{cm}^{-1}$ of the TDFRS measurement

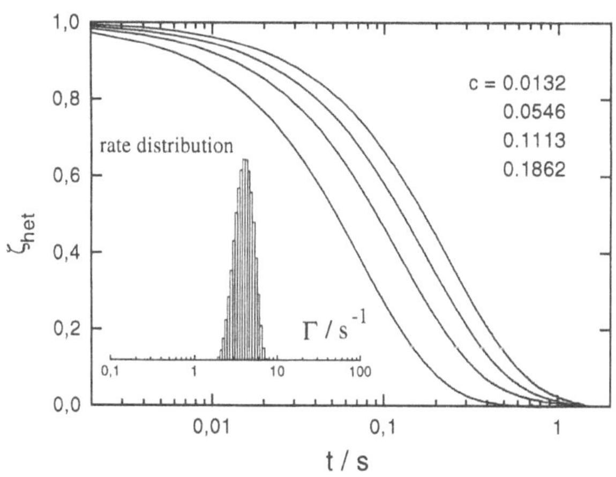

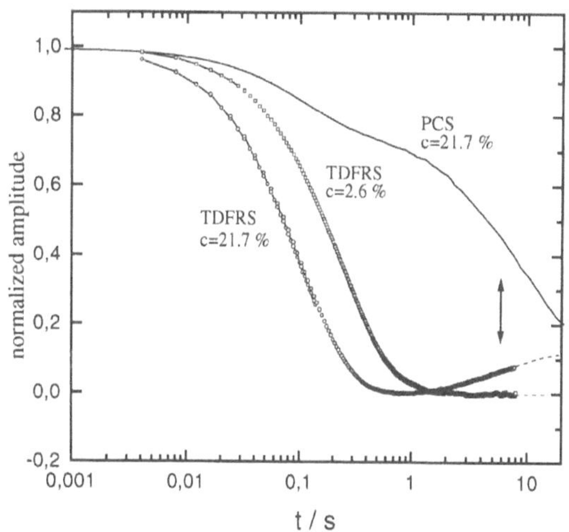

increases linearly with increasing concentration according to

$$D_c = (q^2 \tau_c)^{-1} = D_0(1 + k_D c + \cdots) . \tag{4}$$

For intermediate times, $\tau_s > t > \tau_c$, the microgels are decoupled from the solvent, but the individual microgel particles are still coupled with respect to each other and cannot move independently. For long times, $\tau_s < t$, the microgels start to decouple, and every species (size) adjusts to its respective Soret equilibrium, which is defined by a cancellation of the thermal diffusion and translational diffusion mass flows within the periodic temperature distribution of the holographic grating. The negative amplitude of the slow mode is caused by the decay of the concentration amplitude of the small particles after decoupling, which overcompensates the further increase of the concentration amplitude of the large particles. Experiments to the Ludwig–Soret effect in semidilute and concentrated polymer solutions or colloidal dispersions are hardly available, and the following argument is based on dilute solution data. Nevertheless, it gives at least a qualitative explanation.

At the Soret equilibrium, the achievable concentration change is determined by the Soret coefficient $S_T = D_T/D$:

$$\frac{\nabla c}{c} = S_T \nabla T . \tag{5}$$

It is well established that the thermal diffusion coefficient D_T is independent of both molar mass and particle morphology [8–10], and the same value of D_T has been measured in our laboratory for the PS microgels and for linear PS of different molar masses. Since D depends inversely on the effective hydrodynamic diameter, lower S_T values are expected for the smaller microgel particles. As a consequence, the Soret equilibrium modulation depth of the concentration grating of the smaller microgel particles will be lower than the one of the larger particles. Details of the respective amplitudes depend on the exact particle size distribution and require a correct mathematical treatment of thermal diffusion in semidilute polymer solutions and colloidal dispersions.

The three time domains of the evolution of the concentration grating are sketched in Fig. 3, where a spatial

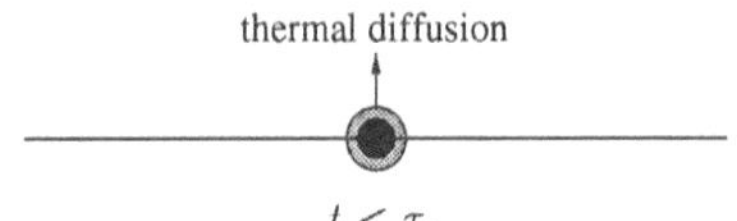

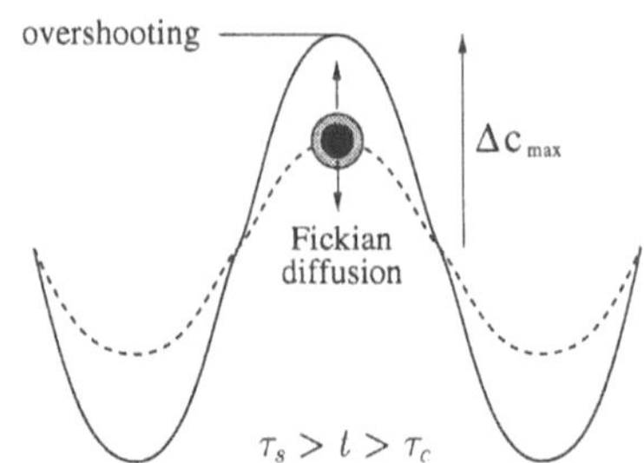

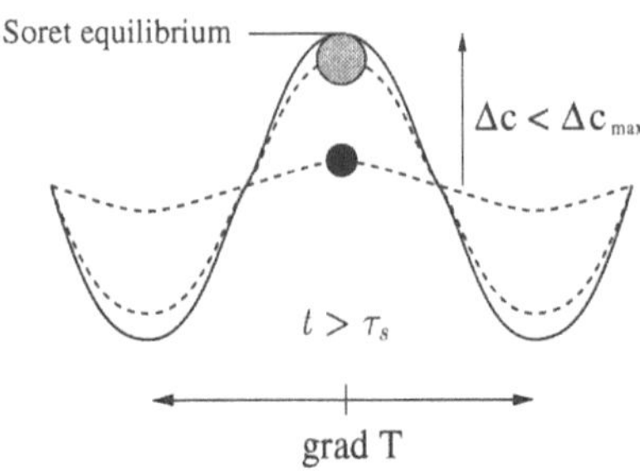

Fig. 3 Evolution of the fast (coupled) and slow (decoupled) thermal diffusion mode within the holographic grating

coordinate in the direction of the grating vector is plotted along the x-axis and the local microgel concentration along the y-axis. The dashed lines represent the concentration grating of the respective species, the solid line the total concentration grating, which overshoots at the intermediate time scale.

Summarizing, it has been shown that TDFRS can advantageously be employed for the characterization of microgels without restriction to a certain concentration regime. Though there are some similarities to PCS, the information content can be very different, as can be seen from the even statistical weights in dilute solutions and from the negative amplitude of the slow mode in the semidilute system. A detailed understanding for the semidilute case is still missing.

References

1. Köhler W (1993) J Chem Phys 98:660
2. Rossmanith P, Köhler W (1996) Macromolecules 29:3203
3. Tyrrell HJV (1961) Diffusion and Heat Flow in Liquids. Butterworth, London
4. Kolodner P, Williams H, Moe C (1988) J Chem Phys 88:6512
5. Stölken S, Bartsch E, Sillescu H, Lindner P (1995) Progr Colloid Polymer Sci 98:155
6. Renth F, Bartsch E, Kasper A, Kirsch S, Stölken S, Sillescu H, Köhler W, Schäfer R (1996) Prog Colloid Polym Sci 100:127
7. Provencher SW (1982) Computer Phys Commun 27:213
8. Schimpf ME, Giddings JC (1990) J Polym Sci Polym Phys B28:2673
9. Schimpf ME, Giddings JC (1987) Macromolecules 20:1561
10. Köhler W, Rosenauer C, Rossmanith P (1995) Int J Thermophys 16:11

Progr Colloid Polym Sci (1997) 104:135–137
© Steinkopff Verlag 1997

E. Ghenne
F. Dumont
C. Buess-Herman

Properties of TiO₂ hydrosols synthesized by hydrolysis of titanium tetraethoxide

E. Ghenne · F. Dumont (✉)
C. Buess-Herman
Université Libre de Bruxelles
Faculté des Sciences CP 255
Service de Chimie Analytique
2 Bld. du Triomphe
1050 Bruxelles, Belgium

Abstract The TiO_2 hydrosols studied in this work were synthesized by hydrolysis of titanium tetraethoxide according to Zukoski's method. Critical coagulation concentrations (CCC) were measured for pH values ranging from 2 to 12 in the presence of various monovalent ions. The observed ionic adsorption sequences are explained in terms of an extension of Gurney concept of ion–ion interactions. The influence of the volume fraction of the solid is also reported and explained by the coagulation–repeptization theory.

Key words Stability – TiO_2 – hydrosol – titanium tetraethoxide – hydrolysis

Introduction

Most ceramics are formed as powder compacts and made dense by sintering. Powder used for fabrication of high performance ceramics will only sinter to dense materials if the particle size is small enough (5–0.05 μm). In this size range colloidal forces will often be more important than the force of gravity. Therefore, the stability of particles in liquids is an important parameter in the processing of high-performance ceramics. Moreover, metal alkoxide precipitation reactions in liquids are of particular interest in the development of new ceramic materials. It is therefore important to study the mechanisms controlling the stability of hydrosols obtained by hydrolysis and condensation of metal alkoxide.

Presented here is a study of the stability of TiO_2 hydrosols obtained by hydrolysis of $Ti(OC_2H_5)_4$ as a function of two parameters: the nature of the monovalent ions and the volume fraction of the solid in solution.

Experimental and results

Sample preparation

TiO_2 hydrosols were obtained by the hydrolysis of titanium tetraethoxide method due to Zukoski [1] adapted to our experimental conditions. Synthesis is performed at 10 °C under nitrogen, using glassware dried overnight at 150 °C before use. Anhydrous ethanol is divided into two equal portions. Ethoxide and water are dissolved into separate portions of ethanol. A small quantity of hydrochloric acid is added in the water–ethanol portion. The acid is used to avoid the agglomeration by an increase of the surface charge of the particles at low pH. After filtration through 0.45 μm pore size filters, both solutions are mixed under stirring, giving a solution of 4×10^{-4} M HCl, 0.45 M H_2O and 5×10^{-2} M $Ti(OC_2H_5)_4$. The resulting TiO_2 alcosols are purified by centrifugation (5000 g) and ultrasonic dispersion cycles, replacing the supernatent by bidistilled water prior to each redispersion step. Finally, the TiO_2 particles are ultrasonically dispersed in aqueous solutions of $HClO_4$ 10^{-4} M or KOH 10^{-3} M in order to study the effect of monovalent cations and anions on the stability of the hydrosols at pH values higher and lower than the point of zero charge (pzc) (pH $= 5.2$ [2]). A digestion of the hydrosols is also carried out at 100 °C for 5 days until no change in the turbidity spectrum is detected. Transmission electron micrographs of the hydrosols show that the particles are spheres formed by the agglomeration of small entities (5–10 nm) and are 103 nm in diameter with a relative standard deviation of 22%.

Coagulation values

The stability of hydrosols can be characterized by an experimental value known as the critical coagulation concentration (CCC) corresponding to the transition concentration between the domains of slow and fast coagulation. This parameter was measured by an apparatus built in our laboratory. This equipment measures the variation of the scattered light at a fixed angle, resulting from the coagulation of the hydrosol, as a function of the electrolyte concentration. It must be pointed out that all these measurements were carried out at a volume fraction of TiO$_2$ in the range of 10^{-4}. Figures 1 and 2 present the CCC-pH relations obtained with different anions and cations, respectively, below and above the pzc. Ionic adsorption sequences at the TiO$_2$–water interface can be easily deduced from these coagulation experiments since a low CCC value indicates a large specific adsorption. The results lead to the following sequences:

In acidic media:

$$IO_3^- \gg BrO_3^- > Cl^- \simeq NO_3^- \simeq ClO_3^- \simeq ClO_4^-.$$

In alkaline media:

$$Li^+ > Na^+ > K^+.$$

Influence of the volume fraction of TiO$_2$ on the coagulation mechanism

Figures 3 and 4 show several experimental coagulation curves obtained at different volume fractions of the solid. The coagulating ions used in these experiments are BrO$_3^-$ and K$^+$. The first part of the experimental curves is characterized by a decrease of the signal resulting from the

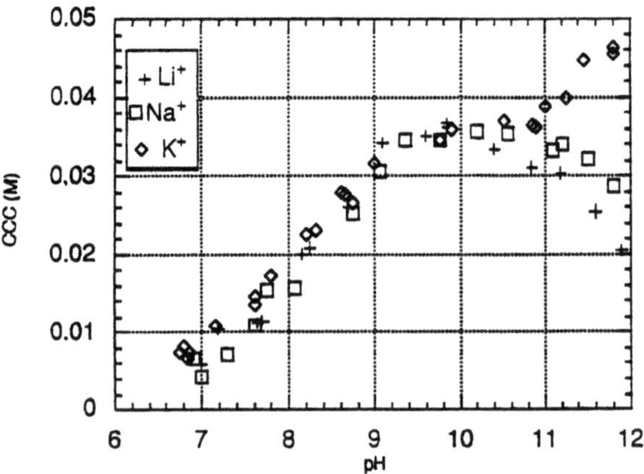

Fig. 2 Critical coagulation concentrations of the cations measured above the pzc

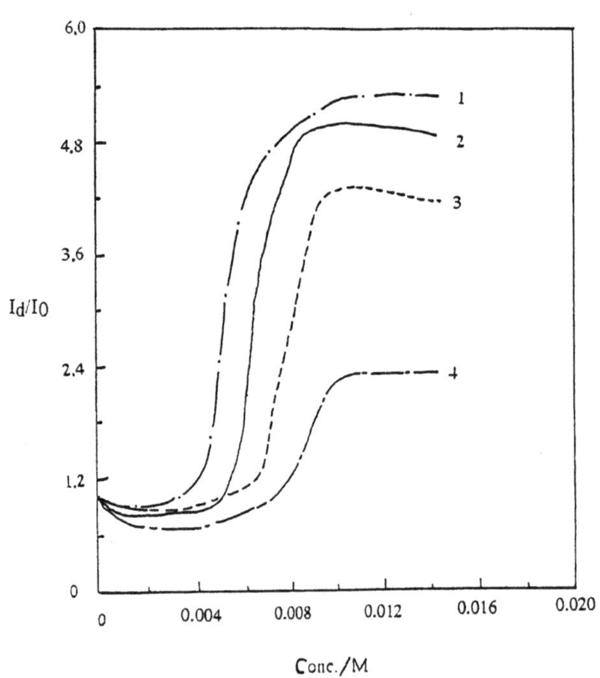

Fig. 3 Influence of the volume fraction of the solid on the coagulation process, curves obtained with BrO$_3^-$ at pH 4.2: (1) $\phi = 7.56 \times 10^{-4}$; (2) $\phi = 3.78 \times 10^{-4}$; (3) $\phi = 1.51 \times 10^{-4}$ and (4) $\phi = 7.6 \times 10^{-5}$

dilution of the hydrosols due to the addition of electrolyte. By increasing the electrolyte concentration an abrupt increase of the signal is detected, this phenomenon corresponds to the coagulation of the system. We observe that the relative increase of the signal is more important for the highest volume fraction system. Furthermore, we also observe that the slope of the curve corresponding to the coagulation process is greater for the most concentrated

Fig. 1 Critical coagulation concentrations of the anions measured below the pzc

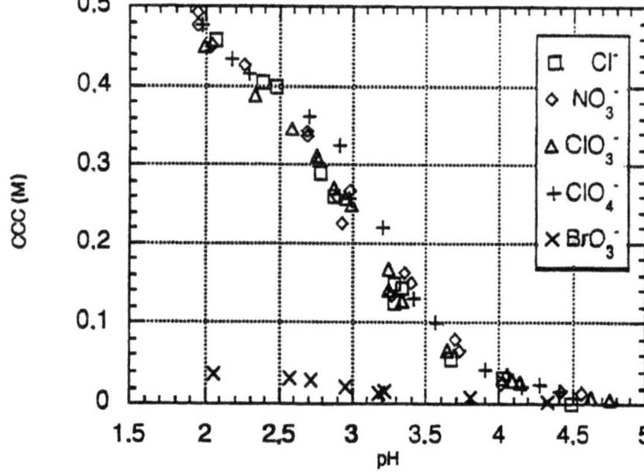

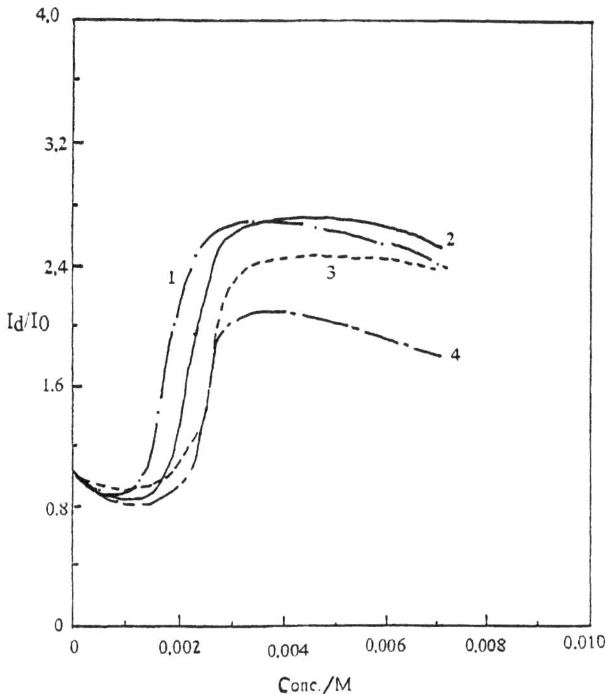

Fig. 4 Influence of the volume fraction of the solid (ϕ) on the coagulation process, curves obtained with K^+ at pH 6.8: (1) $\phi = 5.58 \times 10^{-4}$; (2) $\phi = 2.79 \times 10^{-4}$; (3) $\phi = 1.12 \times 10^{-4}$ and (4) $\phi = 5.6 \times 10^{-5}$

system. Finally, a bending of the curves due to the fact that the system no longer coagulates or that it goes out of Rayleigh's domain is observed. The influence of the volume fraction of TiO_2 on the coagulation mechanism of the hydrosols is found to be more significant for pH values lower than the pzc.

Discussion

The ionic adsorption sequences

The effect of various ions on the stability of the hydrosols is deduced from the measurement of the critical coagulation concentrations. The cationic adsorption sequence at the negatively charged surface (pH > pzc) is found to be $Li^+ > Na^+ > K^+$. On the positively charged surface (pH < pzc), the adsorption of the anions decreases according to the sequence $IO_3^- \gg BrO_3^- > Cl^- \simeq NO_3^- \simeq ClO_3^- \simeq ClO_4^-$. These sequences can be explained in terms of the respective actions of the ion and of the interface on the solvent structure according to the theory of Gierst [3] which is a generalization of the Gurney [4] concept of ion–ion interactions in solution. This theory can be summarized as follows: the structure-maker ions will be more adsorbed at a structure-maker surface than the structure-breaker ones and inversely. In agreement with data published previously [5–7], these observations confirm that oxides with a pzc greater than 4 are structure-makers and, inversely, those with a pzc lower than 4 are structure breakers.

Influence of the volume fraction of TiO_2 on the coagulation mechanism

We have observed that the volume fraction of the solid (ϕ) has an influence on the coagulation mechanism of the hydrosols. These observations are well supported by the Smoluchowski coagulation theory modified by Dumont [8] in order to take into account the existence of a coagulation–repeptization equilibrium. The influence of the volume fraction in alkaline media is weaker than in acidic media. This difference could be explained in terms of the influence of an adsorbed layer of water molecules on the interparticle interaction forces. At the negative interface, this layer is less structured so that during the collision the distance between the two particles can become very small. Consequently, the attraction forces are higher so that only a very weak repeptization takes place. At the positive interface, this water layer being more structured, the distance between two particles remains higher and the repeptization of the particles is favoured due to the lower attraction forces.

Acknowledgment One of us (E.G.) acknowledges the FRIA Council for financial support during this work.

References

1. Look J-L, Bogush GH, Zukoski CF (1990) Faraday Discuss Chem Soc 90:345–357
2. Barringer EA (1983) PhD Thesis MIT
3. Gierst L, Vandenberghen L, Nicolas E, Fraboni A (1966) J Electrochem Soc 113:1025
4. Gurney RW (1953) In: Ionic Process in Solution. Dover, New York
5. Healy TW, Herring AP, Fuerstenau DW (1966) J Colloid Interface Sci 21:435
6. Dumont F (1973) PhD Thesis, ULB
7. Dumont F, Warlus J, Watillon A (1990) J Colloid Interface Sci 138:543
8. Dumont F, Van Tan Dand Watillon A (1976) J Colloid Interface Sci 55:1678

Progr Colloid Polym Sci (1997) 104:138–140
© Steinkopff Verlag 1997

M. Tirado-Miranda
A. Schmitt
J. Callejas-Fernández
A. Fernández-Barbero

Experimental study of fractal aggregation by static and dynamic light scattering

M. Tirado-Miranda (✉) · A. Schmitt
J. Callejas-Fernández
Grupo de Física de Fluidos y Biocoloides
Departamento de Física Aplicada
Universidad de Granada
Campus de Fuentenueva
18071 Granada, Spain

A. Fernández-Barbero
Grupo de Física de Fluidos Complejos
Departamento de Física Aplicada
Universidad de Almería
Cañada de San Urbano s/n
04120 Almería, Spain

Abstract The aggregation kinetics and cluster morphology of polystyrene particles in rapid colloidal aggregation are studied. Experimental results for three aggregating systems with different amount of surfactant adsorbed on the particles are presented. The homogeneity parameter λ of the Van Dongen model was determined from the time evolution of the mean hydrodynamic radius, measured by dynamic light scattering. This parameter allowed the aggregation mechanism to be identified. The cluster fractal dimension was measured by static light scattering.

Key words Colloidal aggregation – fractal structure – static light scattering – dynamic scaling – dynamic light scattering

Introduction

In the last decade, interest in the aggregation of colloidal particles has grown considerably. In particular, the kinetics of aggregation processes and the internal cluster structure has attracted a great deal of attention. Two universal regimes have been found for colloidal aggregation: diffusion-limited cluster aggregation (DLCA) and reaction-limited cluster aggregation (RLCA). The structure of the aggregates is usually described considering that clusters are fractals [1, 2]. Aggregation kinetics can be described by Smoluchowski's coagulation equation [3]. Most coagulation kernels used in the literature are homogeneous functions of i and j, at least for large i and j. Van Dongen and Ernst [4] introduced a classification scheme for homogeneous kernels, based on the relative probabilities of large clusters sticking to large clusters, and small clusters sticking to large clusters. This model shall be used in this work to characterize the aggregation mechanism. In practice, it is difficult to ensure DLCA conditions due to the remaining interaction between the particles [5] and thus, deviations from pure diffusion aggregation is found. So,

cluster restructuring is sometimes used for explaining these behaviours [6, 7].

Light scattering techniques have been successfully applied for monitoring the cluster-size distribution. In this work dynamic light scattering (DLS) was employed to determine the mean hydrodynamic radius of the aggregates. The λ parameter was estimated from the long-time scaling behaviour of the mean radius. Static light scattering (SLS) allowed the internal structure of aggregates to be determined by measuring the cluster fractal dimension.

In this paper, we present results on aggregation of monodisperse polystyrene microspheres induced at high electrolyte concentration. Different behaviours were found depending on the amount of surfactant adsorbed to the surface of the particles.

Materials

The colloidal samples consisted of three different sulphate polystyrene latexes. Table 1 summarize their main characteristics. Size and polydispersity were determined by

Table 1 Characteristics of the particles

Latexes	Diameter [nm]	Polydispersity index	Surfactant adhered
AS5	241 ± 20	1.017	None
SS180	185 ± 2	1.006	Intermediate
SS40	40 ± 1	1.062	High

transmission electron microscopy and dynamic light scattering (DLS). Latexes were cleaned by serum replacement. The main difference between the systems was the amount of surfactant (SDS) adsorbed to the particles surface. The samples were prepared with water purified by inverse osmosis and sonicated for 15 min in order to break any initial clusters up. Aggregation was started by mixing equal amount of colloidal suspension and KBr solution with Y-shaped mixing device. The final electrolyte concentration was always well above the CCC (1.01 M KBr). The temperature was stabilized at $(25.0 \pm 0.1)\,°C$.

Methods

Static light scattering was used to assess the internal structure of aggregates by measuring the cluster fractal dimension, d_f. Dynamic light scattering allowed to monitor the time evolution of the mean hydrodynamic radius. Both techniques present the advantage that they do not perturb the system during aggregation.

The cluster fractal dimension was determined from the angular dependence of the scattered light, $I(aq) \sim S(aq) \sim (aq)^{-d_f}$ which holds for $a\langle R_h \rangle^{-1} \leq aq \leq a\langle R_h(t=0) \rangle^{-1}$ [8]. $S(aq)$ is the static structure factor for fractal aggregates, q is the scattering vector and a is the particle size.

The time evolution of the mean hydrodynamic radius was described by the dynamic scaling asymptotic prediction, $\langle R_h \rangle \sim t^{1/[d_f(1-\lambda)]}$ [8], valid for large aggregation time. The exponent of this power law contains the Van Dongen and Ernst's [4] homogeneity parameter, λ, which controls the long-time behaviour of the cluster-size distribution and governs the overall rate of aggregation. This parameter describes the tendency of a big cluster to join another big cluster. λ takes the value 0 for DLCA and 1 in the case of RLCA. The experimental setup used was a Malvern 4700M (UK), operating at a wavelength of $\lambda = 488$ nm. The scattering vector range was $2.98 \times 10^6 \ \mathrm{m}^{-1} \leq q \leq 3.23 \times 10^7 \ \mathrm{m}^{-1}$.

Results and discussion

In this paper we present results on two fundamental questions in cluster growth: the kinetics and its relation to the

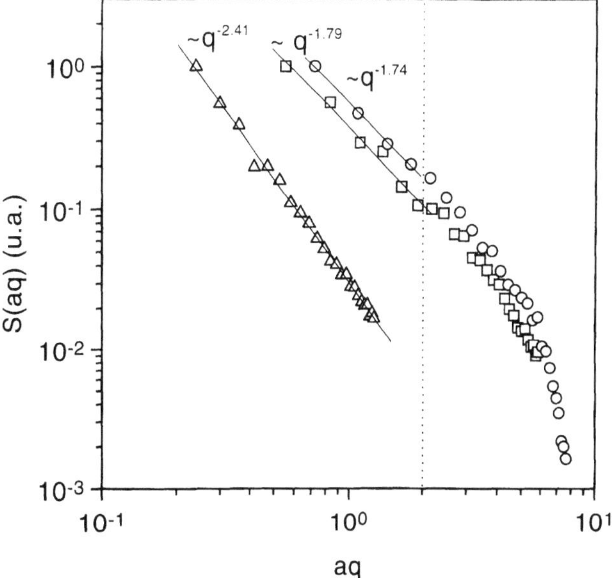

Fig. 1 Structure factor as a function of aq. The curves correspond to AS5 (○), SS180 (□) and SS40 (△)

Table 2 Characteristic aggregation kinetics and fractal structure

Latexes	d_f	λ
AS5	1.74 ± 0.06	0.03 ± 0.05
SS180	1.79 ± 0.12	0.05 ± 0.08
SS40	2.41 ± 0.03	0.06 ± 0.04

structure of aggregates. The homogeneity parameter λ and cluster fractal dimension were measured at high salt concentration for samples with different amount of surfactant on the particle surface.

Figure 1 plots the structure factor $S(aq)$ for the three systems. A decreasing power law is observed up to the limit for which this prediction is valid (vertical dashed line). For higher aq values, the scaling of $S(aq)$ is not suitable for describing our experimental data. Table 2 shows the fractal dimensions obtained directly from the exponent.

Figure 2 plots the mean hydrodynamic radius as a function of time for the three samples. An asymptotic increasing power law is observed in every case for long aggregation times. This result is in good agreement with the dynamic scaling prediction (see methods). From the time exponent and the fractal dimension, measured by SLS, the parameter λ was determined (Table 2).

For every case the homogeneity parameter λ is close to 0 and thus, the aggregation is totally controlled by the diffusion of the particles. This means that probability of two big cluster joining up is the same that between two

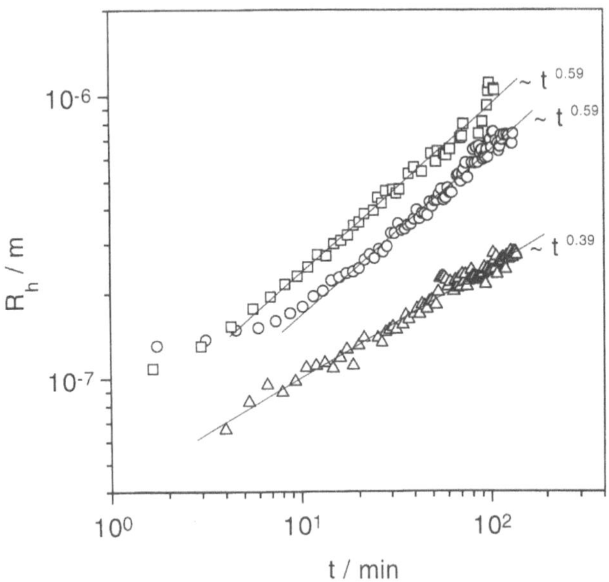

Fig. 2 The mean hydrodynamic radius of the aggregates as a function of time for the particles AS5 (○), SS180 (□) and SS40 (△)

other two samples. This is somewhat surprising because the value $\lambda \approx 0$ indicates that the process is controlled by diffusion, but the morphology found is usually related to clusters growing under repulsion between the particles [14]. On the one hand, at high salt concentration the repulsive electrostatic interaction between the particles falls off at long range. On the other hand, the extra charge due to SDS molecules increase the repulsive interaction at small distances. So, the clusters would approximate themself by diffusion and at short distances a restructuring process could occur [6, 7, 15, 16].

Conclusions

The cluster structure and aggregation kinetics of latex particles were studied for systems with different SDS surface coverage. Static and dynamic light scattering was employed to determine the fractal dimension and the homogeneity parameter λ, respectively. The value of λ indicated that the aggregations were controlled by diffusion of the clusters. Nevertheless, the fractal dimension increased as the interparticle repulsion energies grew, due to amount of SDS adsorbed. A restructuring mechanism is suggested in this paper to explain this experimental behaviour. Macromolecules adsorbed on the particles have influence only on the structure of the aggregates but not on the kinetics of growth.

Acknowledgments This work was supported by CICYT (Spain), Project MAT 94-0560. A.S. is grateful for the fellowship granted by the Gottlieb Daimler- und Karl Benz-Stiftung.

small clusters. This value is in line with other DLCA experiments [5, 9–11].

The fractal dimension is around 1.8 for the latex without surfactant (AS5). When the amount of surfactant increases (SS180) no difference in the fractal dimension is observed. The value 1.8 is generally accepted for diffusion aggregation in three dimensions [12, 13] which agrees perfectly with the value $\lambda \approx 0$. For highly covered particles (SS40), the fractal dimension was found to be 2.4, which means that the clusters were more compact than for the

References

1. Vicsek T (ed) (1992) In: Fractal Growth Phenomena. Word Scientific, Singapore
2. Mandelbrot BB (ed) (1982) In: The Fractal Geometry of Nature. Freeman WH, New York
3. Smoluchowski MV (1917) Z Phys Chem 92:129
4. Van Dongen PGJ, Ernst MH (1985) Phys Rev Lett 54:1396–1399
5. Fernández-Barbero A, Cabrerizo-Vílchez M, Martínez-García R, Hidalgo-Álvarez R (1996) Phys Rev E 53 (5):4981–4989
6. Zhu PW, Napper DH (1995) Colloids Surfaces A: Physiochem Eng Aspects 98:93–106

7. Liu J, Shih WY, Sarikaya M, Aksay IA (1990) Phys Rev A 4:3206–3213
8. Lin HY, Lindsay HM, Weitz DA, Ball RC, Klein R, Meakin P (1990) Phys Rev A 41:2005–2020
9. Asnaghi D, Carpineti M, Giglio M, Sozzi M (1992) Phys Rev A 45: 1018–1023
10. Lin HY, Lindsay HM, Weitz DA, Ball RC, Klein R, Meakin P (1989) Nature 339:360 and (1990) J Phys Condens Matter 2:3093
11. Fernández-Barbero A, Schmitt A, Cabrerizo-Vílchez M, Martínez-García R (1996) Physica A 230:53–74

12. Carpineti M, Ferri F, Giglio M, Paganini E, Perini U (1990) Phys Rev A 42:7347–7354
13. Bolle G, Cametti C, Codastefano P, Tartaglia P (1987) Phys Rev A 35:837–841
14. Martin JE, Wilcoxon JP, Schaefer D, Odinek J (1990) Phys Rev A 41:4379–4391
15. Aubert C, Cannel DS (1986) Phys Rev Lett 56:738–1986
16. Dimon P, Sinha SK, Weitz DA, Safinya CR, Smith GS, Varady WA, Lindsay HM (1986) Phys Rev Lett 57:595–598

Progr Colloid Polym Sci (1997) 104:141–143
© Steinkopff Verlag 1997

A. Fernández-Barbero
A. Schmitt
M.A. Cabrerizo-Vílchez
R. Martínez-García

Single cluster light scattering and photon correlation spectroscopy: Two powerful techniques for monitoring cluster aggregation

Dr. A. Fernández-Barbero (✉)
Grupo de Física de Fluidos Complejos
Departamento de Física Aplicada
Universidad de Almería
04120 Almería, Spain

M.A. Cabrerizo-Vílchez
R. Martínez-García · A. Schmitt
Grupo de Física de Fluidos y Biocoloides
Departamento de Física Aplicada
Universidad de Granada
18071 Granada, Spain

Abstract The aggregation kinetics of monodisperse polystyrene microspheres was studied in processes induced at high salt concentration. Measurements were taken using two alternative techniques: single cluster light scattering and photon correlation spectroscopy. We present results at different pH in order to study the influence of the remaining interaction between the particles on the aggregation rate.

Key words Mesoscopic systems – colloidal aggregation – single particle light scattering – photon correlation spectroscopy – light scattering

Introduction

The aggregation of colloidal particles is a good model for describing the growth of dynamic structures. Two universal mechanisms, have been found for colloidal aggregation: diffusion-limited cluster aggregation (DLCA) and reaction-limited cluster aggregation (RLCA) [1]. The time evolution of the cluster-size distribution is usually described by Smoluchowski's coagulation equation [2], and the structure of the aggregates is characterized by a fractal dimension, d_f [3].

Single cluster light scattering (SCLS) allows detailed cluster-size distributions to be measured by directly counting clusters. Nevertheless, the forces involved in cluster separation can break the aggregates up under extreme experimental conditions. As an alternative technique, photon correlation spectroscopy (PCS) presents the advantage that the aggregating systems are not altered during measurements.

For aggregation processes induced at a high salt concentration, the repulsive electrostatic interaction between the particles falls off at long range, leaving a large remaining interaction at small distances. The remaining interaction depends on particle surface charge density and so,

may be controlled by modifying the surface groups degree of ionization by changing the pH of the solvent. We investigated how the remaining interaction affects the rate of aggregation. Measurements were taken using both light scattering techniques, offering the possibility of covering a wider range of particle concentration and comparing results. Important differences were found in experiments carried out in glass or plastic containers. X-ray fluorescence spectroscopy was employed to check the chemical composition of the samples.

Single cluster light scattering

The measurements presented in this paper were carried out with a single particle optical instrument built in our laboratory [4]. In this technique, single clusters, insulated by hydrodynamic focusing of a colloidal dispersion, are forced to flow across a focused laser beam. A measurement of the cluster-size distribution is taken by analyzing the light intensity scattered by single clusters at low angle.

A simpler way to determine the Smoluchowski rate constant k_s, is to measure the time evolution of the monomers during the aggregation. The inverse square root of the monomer concentration N_1, shows the following linear function of time [1]:

$$\frac{1}{\sqrt{N_1(t)}} = \frac{1}{\sqrt{N_0}}(1 + N_0 k_s t) . \tag{1}$$

It is then possible to obtain the initial monomer concentration N_0 and the rate constant k_s from the intersection and the slope, respectively.

Photon correlation spectroscopy

Our aim is to measure the Smoluchowski rate constant. Following Herrington et al. [5], we fitted the theoretical field autocorrelation function for polydisperse systems

$$g^{\text{field}}(\tau) = \frac{\sum_{j=1}^{j_{max}} N_j(t/t_{\text{agg}}) \langle P_j(\theta) \rangle \exp(-\Gamma_j \tau)}{\sum_{j=1}^{j_{max}} N_j(t/t_{\text{agg}}) \langle P_j(\theta) \rangle}$$

to the experimental one, with the fitting parameter being the normalized time $T = t/t_{\text{agg}} = N_0 k_s t$. In this formula $\langle P_j(\theta) \rangle$ is the average form factor for a j-fold cluster, given by the Rayleigh–Gans–Debye (RGD) approximation. N_j is the cluster-size distribution (particle number density of each j-fold cluster). $\Gamma_j = q^2 D_j$ where q is the scattering wave vector and D_j is the diffusion coefficient of the j-fold cluster. For each time t, we obtained a normalized time T. So, the rate constants k_s are calculated from the slope, once the initial monomer concentration N_0 is known. The photon correlation instrument used to measure the intensity autocorrelation function was a Malvern 4700 (U.K.). For more details see [6].

Experimental system

The experimental systems were three sulphate monodisperse polystyrene lattices. Table 1 shows the principal characteristics. The size was determined by transmission electron microscopy and the C.C.C. was estimated directly by dilution of latex particles in solutions of different KCl concentration. Buffers were used to set the pH: acetate at pH 3, 4, 5, phosphate at pH 6 and 7, and borate at pH 9 and 10. All chemicals used were of A.R. quality and twice-distilled water was purified using Millipore equipment. Electrophoretic mobilities of the particles, μ_e, were measured as a function of the pH using a Zeta-Sizer IIc (Malvern Instruments, U.K.) (Fig. 1). $\mu_e < 0$ confirms the negative sign of the surface charge. The mobility remains constant between pH 5 and 9. At low pH (less than 5), the mobility drops significantly due to the decrease in surface charge density. The latex particles in the sulphate-

Table 1 Summary of the lattices employed

Sample	Source	Radius [nm]	C.C.C. [mM]
RP-300	Rhône-Poulenc	150 ± 11	≈ 110
AS8	Univ. Granada	290 ± 14	≈ 100
SK-580	Sekisui	292 ± 19	≈ 110

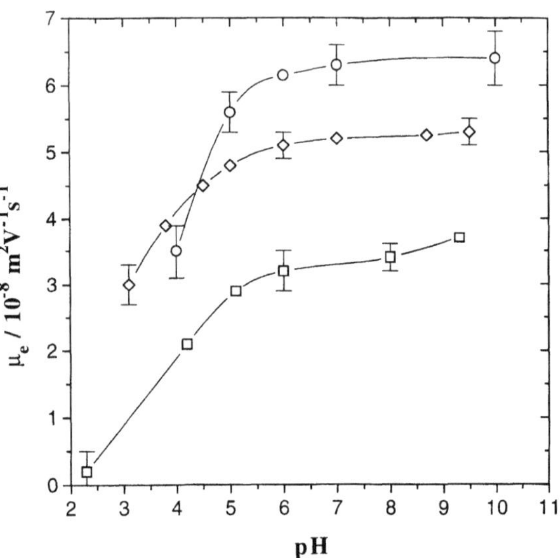

Fig. 1 Electrophoretic mobilities of the particles as a function of pH: (□) SK-580, (◇) RP-300, (○) AS8

form adsorb enough protons to eliminate the surface charge.

Results and discussion

We measured the Smoluchowski rate constant at high salt concentration and for different pH of the solvent. The fact that electrophoretic mobility drops significantly with pH (Fig. 1) demonstrates the decrease in surface charge density and the reduction of the remaining potential.

Figure 2 plots the rate constant, k_s, as a function of the pH for lattices AS8 (SCLS), RP-300 (PCS) and SK-580 (SCLS). The experiments were carried out well above the C.C.C. at 0.5 M of KCl in polypropylene containers. The rate constant remains pH-independent between pH 2.0 and 9.5, and thus, the remaining potential does not affect the aggregation rate. The experimental value for k_s is about half the value predicted for Brownian aggregation (continuous line) and may be explained by taking into account the viscous interaction (dashed line) [6, 7]. The aggregation is totally controlled by the diffusion of the

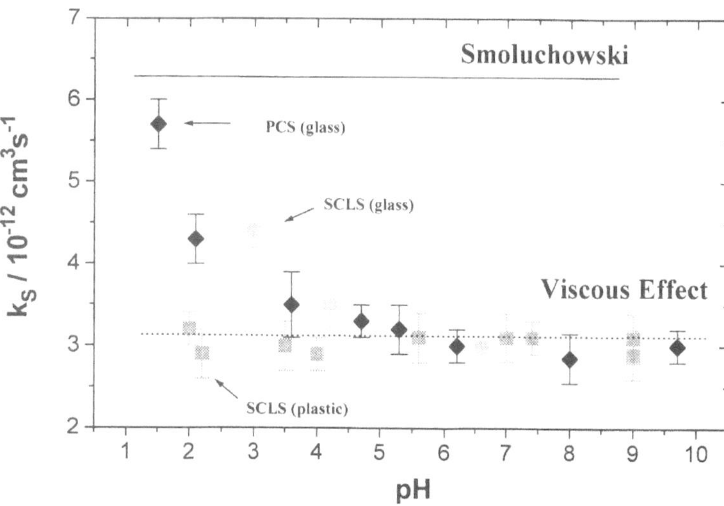

Fig. 2 Rate constant k_s measured by SCLS and PCS, as a function of the pH. The continuous line is the theoretical Smoluchoski's value and the dashed line represents the rate constant corrected by viscous interaction

particles. Nevertheless, this is only valid for the early aggregation stages. At longer times the remaining potential plays an important role and alters the aggregation mechanism [8]. The same experiments were carried out in pyrex glass containers. It may be observed (Fig. 2) that the rate constant remains pH-independent for pH higher than 4.5 and coincides with the value measured for experiments performed in polypropylene containers. At lower pH, however, the aggregation rate increases for decreasing pH. This is somewhat surprising since this behaviour was not observed for plastic containers. Two identical samples were prepared with HCl and stored in pyrex glass and polypropylene containers. X-ray fluorescene spectroscopy was employed to determine the chemical composition of the samples after several days. The only difference found was a slight trace of silicon for the sample stored in the glass container. A possible specific adsorption of silicon ions on the particle surface could give rise to a reduction in the remaining potential. Nevertheless, it does not explain the drastic increase of the aggregation rate at low pH since, at higher pH, the aggregation processes were already found

to be diffusion controlled. Therefore, an additional aggregation mechanism must be present in order to explain the differences.

Conclusions

The rate constants were measured at different pH to determine how the remaining interaction between the particles affects the aggregation kinetics. The processes were observed to be controlled by diffusion of the clusters and are pH independent. The rate constants for the samples aggregated in glass tubes at low pH significantly exceeded the limit set by diffusion. The findings for silicon did not explain the difference in behaviour and thus an additional aggregation mechanism should be considered.

Acknowledgments A.S. is grateful for the fellowship granted by the Gottlieb Daimler- und Karl Benz-Stiftung. This work was supported by CICYT (Spain), Project MAT 96-1035-C03-02.

References

1. Family F, Landau DP (eds) (1984) Kinetics of Aggregation and Gelation. North-Holland, Amsterdam
2. Von Smoluchowski M (1917) Z Phys Chem 92:129
3. Vicsek T (1992) Fractal Growth Phenomena. Word Scientific, Singapore
4. Fernández-Barbero A, Schmitt A, Cabrerizo-Vílchez M, Martínez-García R (1996) Physica A 230:53
5. Herrington TM, Midmore BR (1989) J Chem Soc Faraday Trans 85:3529
6. Fernández-Barbero A, Schmitt A, Cabrerizo-Vílchez M, Martínez-García R, Hidalgo-Alvarez R, Phys Rev E, submitted
7. Spielman LA (1970) J Colloid Interface Sci 33:562
8. Fernández-Barbero A, Cabrerizo-Vílchez M, Martínez-García R, Hidalgo-Alvarez R (1996) Phys Rev E 53:4981

Progr Colloid Polym Sci (1997) 104:144–147
© Steinkopff Verlag 1997

Experimental evidence regarding the influence of surface charge on the bridging flocculation mechanism

A. Schmitt
A. Fernández-Barbero
M. Cabrerizo-Vílchez
R. Hidalgo-Álvarez

A. Schmitt (✉)
M. Cabrerizo-Vílchez · R. Hidalgo-Álvarez
Grupo de Física de Fluidos y Biocoloides
Departamento de Física Aplicada
Universidad de Granada
Campus de Fuentenueva
18071 Granada, Spain

A. Fernández-Barbero
Grupo de Física de Fluidos Complejos
Departamento de Física Aplicada
Universidad de Almería
Cañada de San Urbano s/n
04120 Almería, Spain

Abstract In this work, the influence of particle surface charge density on the mechanism of bridging flocculation is studied. Different amount of bovine serum albumin (BSA) molecules were adsorbed onto the surface of two almost identical systems of polystyrene particles which differ only in their surface charge density. Flocculation was induced by adding a small amount of electrolyte to a dilute suspension. Single particle light scattering was used to monitor the flocculation processes. It was found that steric stabilisation does not prevent aggregation in all cases and at least some weak flocculation occurs. Nevertheless, it inhibits complete flocculation of the sample. The initial rate constants are obtained and it is shown that the constant kernel solution for Smoluchowski's system of rate equations cannot describe the flocculation processes. No clear evidence for bridging flocculation was found for the particles with low surface charge. For the higher charged particles, however, a pronounced maximum for the initial flocculation rate was measured at intermediate surface coverage. This finding gives clear evidence for bridging flocculation and that the particle surface charge indicater is a very important parameter for the formation of protein bridges between the particles.

Key words Colloidal aggregation – bridging flocculation – single particle detection – surface charge – protein adsorption

Introduction

It is well known that macromolecules which irreversibly adsorb onto the surface of colloidal particles strongly affect the stability of the colloid [1–3]. At high surface coverage, steric effects impede flocculation and lead to increased colloidal stability, whereas at lower surface coverage, *bridges* of macromolecules may form between the particles and, thus, give rise to bridging flocculation. The concept of *bridging* was already introduced by Ruehrwein and Ward in 1952 [4]. Nevertheless, it is still not completely clear how the different experimental parameters affect the mechanisms of bridging.

In this work, we focus on the influence of the particle surface charge. Therefore, two almost identical systems of colloidal particles which differ only in their surface charge density were prepared. Different amount of macromolecules were adsorbed onto the particle surface. Flocculation was then induced by adding a small amount of electrolyte to a dilute suspension of particles. The flocculation processes were monitored by single-particle light scattering. This technique allowed a detailed cluster-size distribution to be determined directly. Not like for other techniques, no more or less justified assumptions on the detailed cluster-size distribution have to be made for proper interpretation of experimental results.

Theory

The kinetics of aggregation processes in colloidal suspensions may be described by the time evolution of the number $N_n(t)$ of different species of n-fold aggregates. In this meanfield approach, one only distinguishes between clusters formed by different number n of primary particles (n-mer) and neglects the possible influence of shape and orientation of the clusters. For dilute colloidal systems, where only binary collisions have to be taken into account, von Smoluchowski proposed the following system of rate equations [5]:

$$\frac{dN_n}{dt} = \frac{1}{2} \sum_{i+j=n} k_{ij} N_i N_j - N_n \sum_{k=1}^{\infty} k_{nk} N_k \ . \tag{1}$$

The term on the left-hand side represents the rate at which the overall number of n-mers changes. The term on the right-hand side stands for rate at which n-mers are formed from the aggregation of smaller i-mers and j-mers and the rate at which they disappear when they collide with a k-mer and form a $(k + n)$-mer. The parameters k_{ij} represent the rate at which i-mers bind to j-mers. They may be interpreted in terms of a sticking probability for two clusters diffusing towards one another.

Considering constant kernel, i.e., $k_{ij} = k_{11}$, and monomeric initial conditions ($N_1(t = 0) = N_0$ and $N_{n>1}(t = 0) = 0$) one obtains for the inverse square root of the monomer concentration N_1 the following linear relationship in time [6]:

$$\frac{1}{\sqrt{N_1(t)}} = \frac{1}{\sqrt{N_0}}\left(1 + \frac{N_0 k_{11}}{2} t\right) . \tag{2}$$

The slope of this curve is directly proportional to the rate constant k_{11} which may be calculated once the initial number of particles N_0 is known.

For particles, covered with different amounts of adsorbing macromolecules, bridging flocculation takes place only when an covered part of the surface of one particle collides with the uncovered part of another particle. In this case, the rate constant k_{11} should dependent on the degree

of surface coverage θ, i.e., the fractional coverage of the particle surface by adsorbed molecules. LaMer assumed the rate of bridging flocculation to be proportional to the number free sites on one particle and the number of occupied sites on the other. He postulated [7]

$$k_{11} \sim \theta(1 - \theta) \ . \tag{3}$$

This relationship implies that maximum flocculation should occur at half surface coverage ($\theta_{\max} = \frac{1}{2}$) and no flocculation at all for uncovered and totally covered particles, respectively. This simplified approach to bridging flocculation, however, cannot be entirely true since it assumes that bridging flocculation is the only mechanism of aggregation. It does not take into account that many disperse systems are significantly unstable in the absence of adsorbed molecules (coagulation) as well as when particles are completely covered (weak flocculation) [8]. More detailed models were proposed for example by Hogg [9], Ash and Clayfield [10], Moudgil [11] and Molski [8]. Nevertheless, all of them show the common feature that bridging flocculation is the most efficient at certain intermediate surface coverage.

Materials and methods

Two monodisperse suspensions of spherical polystyrene particles, AS2 and AS8, were prepared by standard emulsifier-free emulsion polymerization [12]. The particles were cleaned by serum replacement and ion exchange over a mixed bed. Particle size and polydispersity index were measured by transmission electron microscopy (TEM). Conductimetric titration was employed to determine the surface charge density. All relevant data of the two samples are summarized in Table 1.

Bovine serum albumin (BSA, Pentex) was chosen as a model for adsorbing macromolecules. The only treatment prior to the adsorption measurements was a cleaning step by dialysis. Different amounts of BSA were then added to a fixed quantity of buffered colloidal suspension. The pH of the suspension was fixed near to the isoelectric point of BSA at pH 5 in order to facilitate the adsorption of the added protein onto the particle surface. After about two hours of incubation, the samples were centrifuged and the amount of non-adsorbed protein was measured

Table 1 Characteristics of the polystyrene particles

Sample	Diameter [nm]	Polydispersity index	Surface charge groups	Surface charge density [μC/cm^2]
AS2	600 ± 17	1.002	Sulphate	-7.6 ± 0.4
AS8	580 ± 27	1.005	Sulphate	-2.4 ± 0.4

photometrically. The adsorption isotherms for both samples show the high affinity of BSA. It adsorbs until a final plateau is reached which indicates the adsorption of a complete monolayer of protein. This gave us the possibility to obtain particles with a well known degree of surface coverage by simply controlling the amount of added protein. The covered particles were redispersed and stored at pH 9 in order to avoid spontaneous aggregation.

Flocculation was induced by adding a small amount of electrolyte to a monodisperse sample with a well known degree of surface coverage. Mixing was done by injecting equal amounts of sample and buffered electrolyte into a Y-shaped mixing cell. The initial particle concentration in the reaction vessel was 1.0×10^8 cm^{-3} and the electrolyte concentration 0.1 M KCl. The pH was set to pH 9 by a borate buffer of low ionic strength and the temperature was stabilized at $(21 \pm 1)\,°C$. Immediately after mixing, the timer was started.

During the flocculation process, a small part of the sample was continuously extracted from the reaction vessel and injected into a single particle light scattering instrument. This instrument allows the number of different species of aggregates up to heptamers to be measured as a function of time. Its principle of operation is based on the hydrodynamic focusing of the suspension. The particles are forced to flow one by one across a focused laser beam. As the particles pass, they scatter a pulse of light which is detected at small scattering angle. Since the light pulse intensity is directly related to the size of the scatterer, a detailed size distribution can be obtained by counting the number of pulses as a function of pulse height. More

detailed descriptions of the experimental set-up are given in ref. [13, 14].

The initial flocculation rate constants were obtained from the time evolution of the number of monomeric particles. According to Eq. (2), a straight line was fitted to the onset of the inverse square root of the number of monomeric particles. The initial rate constants were calculated from the slope of the fitted curve using the known initial particle concentration N_0.

Results and discussion

A series of aggregation experiments was carried out for two systems of particles, AS2 and AS8. As was described earlier, both systems have very similar characteristics. The only significant difference lies in their surface charge (see Table 1).

Figure 1 contains the results obtained for particles AS2. It shows the time evolution of the number of monomers with different degree of surface coverage. All curves exhibit a steep decline at the beginning of the aggregation process and flatten for longer aggregation times. The initial drop is observed even for the totally covered particles, which should be sterically stabilized and should not aggregate at all. This makes clear that steric stabilisation does not prevent aggregation in all cases and at least some weak flocculation may occur [8]. The flattening of the curves at longer aggregation times indicates that the aggregation process is slowing down. A complete standstill, however, is observed only for particles with high degree of

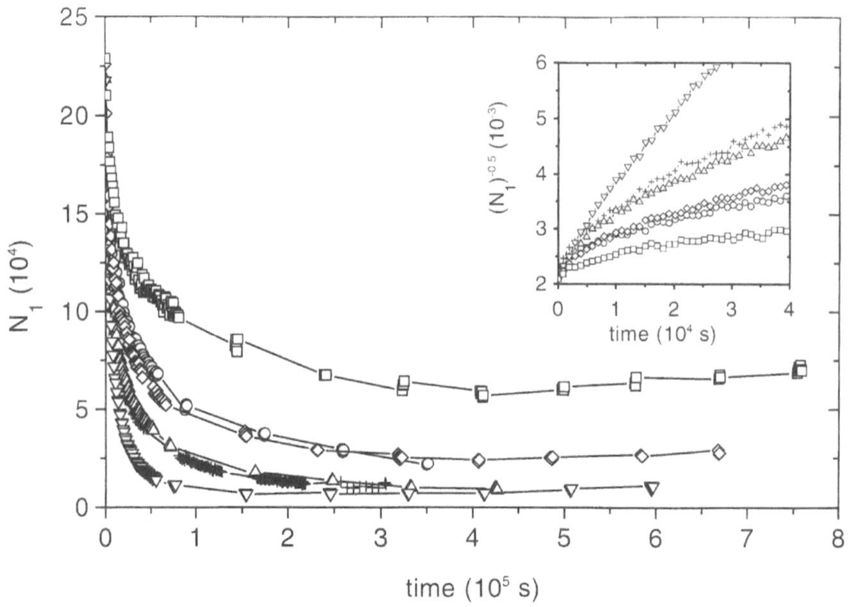

Fig. 1 Number of monomeric particles N_1 as a function of time t. The curves correspond to particles AS2 with (○) 0%, (△) 25%, (▽) 50%, (◇) 75% and (□) 100% of their surface covered by BSA molecules. Curve (+) shows the behaviour for the untreated particles. The smaller figure in the upper right corner represents the initial time dependence of the inverse square root of the number of monomeric particles $(1/\sqrt{N_1})$ during early stages of aggregation

surface coverage. This shows that steric stabilisation allows some weak aggregation but inhibits complete flocculation of the sample. This finding would have been impossible to observe with other techniques like spectroscopy or nephelometry which only give some average information about the cluster-size distribution. The fact that the curves for the untreated particles and the particles with no protein on its surface (0% coverage) do not exactly superimpose may be explained by some alteration of the particle surface during the adsorption and centrifugation process.

The same experiments were repeated with particles AS8 in order to check for the influence of the surface charge density. The obtained results exhibit the same principal characteristics than the ones described so far for sample AS2. Nevertheless, some fundamental differences which can be seen very clearly by representing the initial aggregation rate k_{11} as a function of surface coverage were found. The rate constants were obtained by fitting a straight line to the onset of the inverse square root of the monomer concentration. This procedure, however, is not entirely justified since the curves used for the fit (see upper right corner of Fig. 1) are not exactly straight lines and deviate slightly from the linear behaviour. This demonstrates that the constant kernel solution is not suitable to describe the measured flocculation processes under these experimental conditions. Nevertheless, it can be used to obtain a reasonably good estimation for the initial rate constant.

As can be seen in Fig. 2, the aggregation rate observed for sample AS8 remains almost constant for the lesser and intermediately covered particles and decreases for the highly covered particles. The value, measured at low surface coverage coincides with the value observed for the untreated particles. In this case, it becomes obvious that the protein has little influence on the aggregation. At higher surface coverage, however, the adsorbed protein molecules inhibit flocculation and the flocculation rate decreases due to steric interactions. So, no clear evidence for bridging

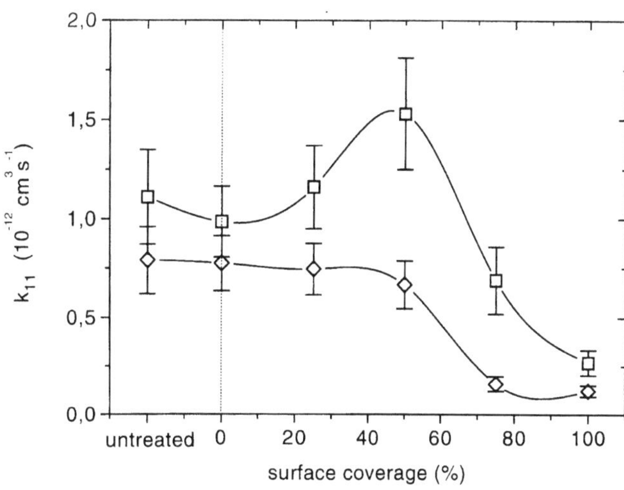

Fig. 2 Initial aggregation rate constant k_{11} as a function of surface coverage for particles AS2 (□) and AS8 (◇). The solid lines are drawn as a guide to the eye

flocculation was found for this type of particles since one expects a pronounced maximum of the aggregation rate at intermediate surface coverage (Eq. (3)).

For the particles with the higher surface charge (sample AS2) a quite different behaviour was observed. In this case, a pronounced maximum for the initial flocculation rate was found at intermediate surface coverage. This finding gives clear evidence for bridging flocculation and indicates that the conformation of the adsorbed protein should differ from the one found for system AS8. Therefore, we conclude that the particle surface charge is an important parameter for the conformation of the adsorbed protein molecules [15, 16] and the formation of protein bridges between the particles.

Acknowledgments Financial support by the CICYT project MAT96-1035-C03-02 and the *Gottlieb Daimler- und Karl Benz-Stiftung* is gratefully acknowledged.

References

1. Dickinson E, Eriksson L (1991) Adv Colloid Interface Sci 34:1
2. Gregory J (1989) Crit Rev Envir Control 19(3):185–230
3. Fernández-Barbero A, Cabrerizo M, Martínez R, Hidalgo-Álvarez R (1993) Progr Colloid Polym Sci 93:269–272
4. Ruehrwein RA, Ward A (1952) Soil Sci 73:485
5. Smoluchowski MV (1917) Z Phys Chem 92:129
6. Drake RL (1972) In: Hidy GM, Brock JR (eds) Topics in Current Aerosol Research. Vol 3. Pergamon Press, New York, p 201
7. LaMer VK (1966) Disc Faraday Soc 42:248
8. Molski A (1989) Colloid Polym Sci 267:371–375
9. Hogg R (1984) J Colloid Interface Sci 102:232
10. Ash SG, Clayfield EJ (1976) J Colloid Interface Sci 55:645
11. Moudgil BM, Shah BD, Soto HS (1987) J Colloid Interface Sci 119:446
12. Goodwin JW, Gearn JH, Ho CC, Ottewil RH (1974) Colloid Polym Sci 252:464
13. Fernández-Barbero A, Cabrerizo-Vílchez M, Martínez-García R, Hidalgo-Álvarez R (1996) Phys Rev E 53(5):4981–4989
14. Fernández-Barbero A, Schmitt A, Cabrerizo-Vílchez M, Martínez-García R (1996) Physica A 230:53–74
15. Elgersma AV, Zsom RLJ, Norde W, Lyklema J (1990) J Colloid Interface Sci 138:145–156
16. Martín-Rodríguez A, Cabrerizo-Vílchez M, Hidalgo-Álvarez R (1994) Colloids Surf A: Physiochem Eng Aspects 92:113–119

Progr Colloid Polym Sci (1997) 104:148–151
© Steinkopff Verlag 1997

H. Lichtenfeld
L. Knapschinsky
C. Dürr
H. Zastrow

Colloidal stability – investigations by single-particle scattering photometer

Dr. H. Lichtenfeld (✉)
L. Knapschinsky · C. Dürr · H. Zastrow
Max-Planck-Institut für Kolloid- und
Grenzflächenforschung
Rudower Chaussee 5
12489 Berlin, Germany

Abstract The single-particle light-scattering technique proves to be suitable for determination of particle size distribution with very high resolution. In particular, first steps of a coagulation process can be determined very sensitively. Rapid coagulation led to good correspondence with theory in the case of spherical oxide particles. For quasi-spherical particles (with edges and corners) satisfactory experimental results have been obtained by extension of the theory. In all the experiments a decay of aggregates could be proved. However, theoretical determined velocity constants at slow coagulation could not be confirmed by experiments. Coagulation takes place in a wider electrolyte region as expected.

Key words Stability – single-particle light scattering – coagulation – ferric oxide dispersions

Introduction

To investigate the stability behavior of colloid dispersions, coagulation kinetics is the right way as with the determination of velocity constants provided the interaction energy can be determined. Here we use the method of single-particle light scattering. By application of the volume scattering method, the scattered light of a large number of particles or aggregates is being measured at the same time, so it seems to be hardly possible to suppress the formation and decay of doublets and triplets by measuring techniques or by computation.

We first investigated polymer dispersions (polystyrene and polyvinyl acetate), which can be prepared to be nearly ideal spherical and monodisperse and are used for model substances in different colloid-chemical investigations. Results have been obtained [1–3] which cannot be explained by classical DLVO-theory (slow coagulation in a wide range of electrolyte concentration, reversible rapid coagulation).

It is interesting to extend these investigations to a representative of oxides. These substances possess a very different surface than the polymers mentioned.

Experimental

Our light-scattering apparatus is shown in Fig. 1. After passing a spatial filter the light of a He–Ne laser is focused by a lens system into the sample cell in such a way that the focus lines of meridional and saggital beams have little geometrical deviation. In this area of the laser focus, a suitable region in relation to light intensity, geometry and irradiance profile can be found and the particle stream is led into it. Using a dosage system the dispersion is passed through a capillary with a 0.1 mm diameter orifice at the higher end. After leaving the orifice the particle stream will be narrowed by use of hydrodynamical focusing.

The scattering volume can be determined using the Poisson distribution by experimental determination of the number of coincidences as a function of particle

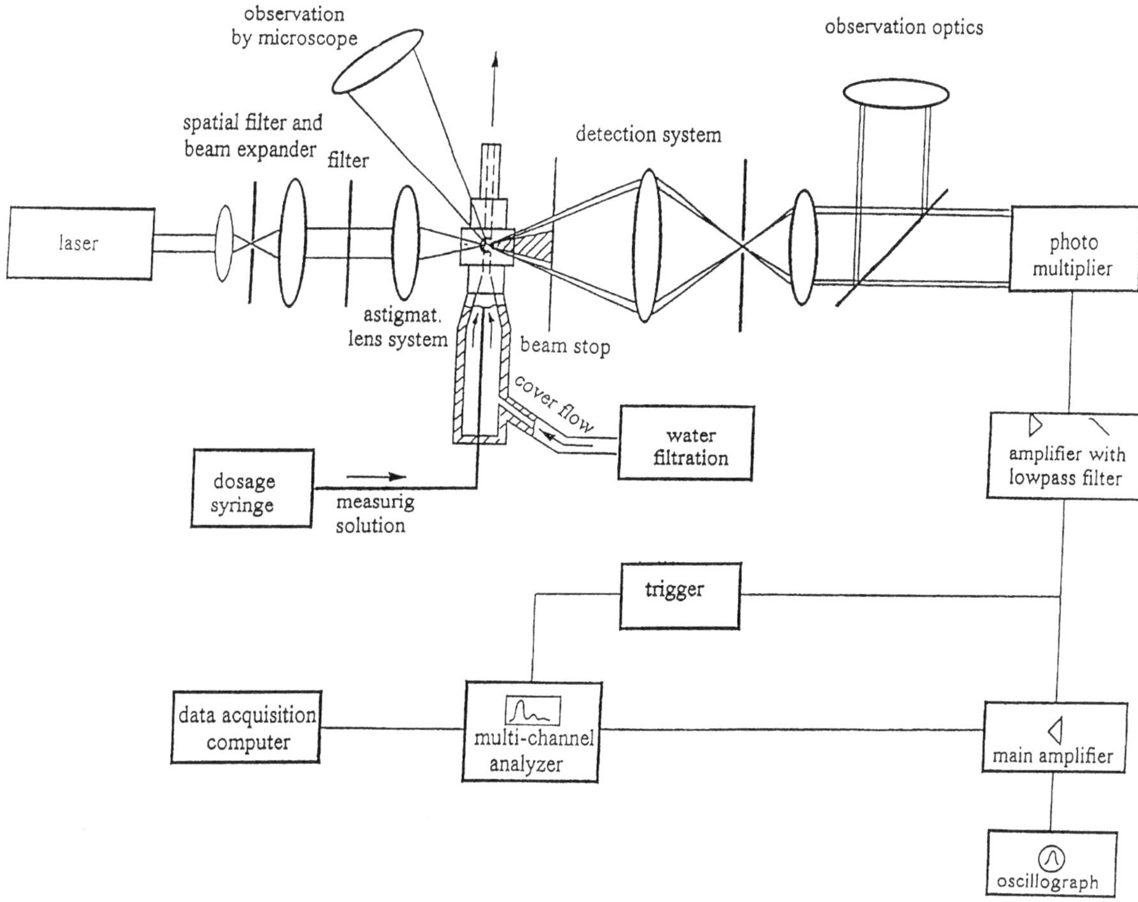

Fig. 1 Schematic diagram of our single-particle scattering photometer

concentration. It results in a scattering volume of 1.7×10^{-9} cm³. By correcting the coincidences, a particle concentration $\leq 3 \times 10^8$ cm⁻³ can be measured.

The primary beam will be blocked and pulses of scattered light generated by particles moving through the laser focus are received in an angular region of 5°–10° in the forward direction. After changing into electrical signals data acquisition takes place in a PC via a multi-channel analyser. More information is given in [3].

According to the Rayleigh–Debye–Gans-theory (RDG), the intensity of a pulse is proportional to the square of the volume of the scattering particles (for polymer particles) and hence we can get a particle-size distribution of the tested dispersion. In case of a not so wide distribution, it is possible (also for oxide particles) to distinguish between singlets, doublets and triplets – their concentration can be represented as a function of time.

An essential feature of efficiency of the apparatus is the resolving power. Here a mixture of two "monodisperse" polystyrene latices of narrow neighboring particle size

of the same concentration (0.60 and 0.61 μm) could be resolved.

In Fig. 2 the distribution behavior of a coagulating latex is shown. The detectable particle radius extends from ca. 50 nm to some micrometers. Aggregation processes are described formally by reaction equations of second order. The equations will be solved by the Runge–Kutta method for the aggregation and deaggregation constants k_{ij} and b_{ij}. Calculated curves are compared with the experimental points $z_i(t)$ to get the best fit by varying the constants.

Sample preparation

In order to obtain monodisperse Fe_2O_3 particles of around 200 nm – to avoid the influence of sedimentation – we employed a combination of homogeneous nucleation and heterogeneous seed-growth. The preparation was in accordance with the method of Matijevic and Scheiner [4]

Fig. 2 Rapid coagulation after 60 min

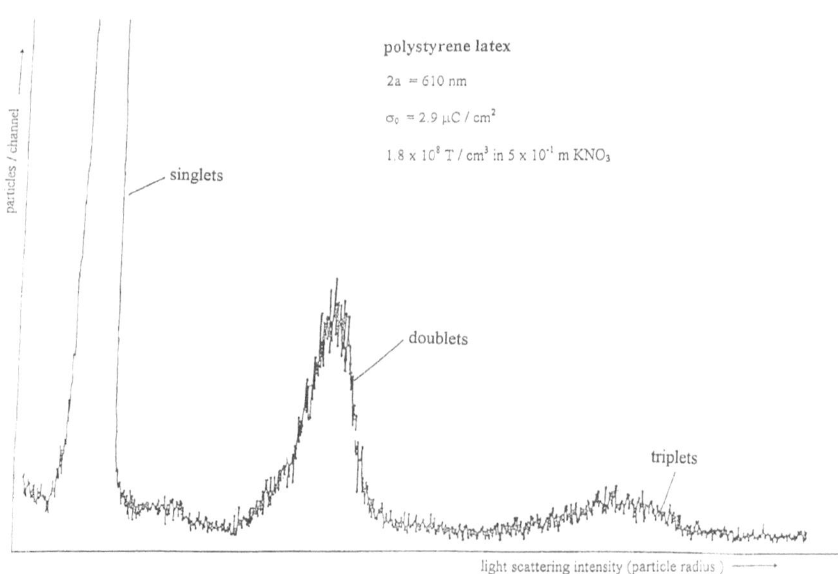

and Penners [5]. These sols had to be dispersed intensively to destroy all existing aggregates in the tested solution. Potassium nitrate was used as the coagulating electrolyte.

Results

The theoretical determination of velocity constants k_{11} can be carried out by considering hydrodynamical interaction [6] according to the following relation:

$$k_{11} = \frac{8\pi D_0 a}{\int_0^\infty \frac{\beta(u)}{(2+u)^2} \exp\{[V_R(u) + V_A(u)]/kT\}\, du}$$

$$= 8\pi D_0 aN \,, \tag{1}$$

k_{ij}:   velocity constant of formation of a doublet from two spherical single particles

D_0:   diffusion coefficient of particles at strong dilution

a: particle radius

$V_R(u)$: electrostatic repulsion energy between two particles

$V_A(u)$: van der Waals attractive energy between two particles

k: Boltzmann constant

T: temperature

N: dimensionless velocity constant

$\beta(u)$: hydrodynamic retardation

$$u = \frac{r - 2a}{a} \,,$$

r: centre distance of two particles.

$V_R(u)$ can be determined by the Derjaguin formula [7] or by the expanded formula of McCartney and Levine [8]. For $V_A(u)$ the van der Waals' attraction is to be used.

By application of the often used Hamaker constant of 5×10^{-21} J for polystyrene or PVAC we get a value of 6×10^{-12} cm^3/s at the fast coagulation which corresponds with the experimental values [2, 3]. Not so with Haematite (α–Fe$_2$O$_3$)-sols. Here the experiments have been explained only by extending the theory by Shilov et al. [9].

The above-mentioned investigations on rapid coagulation have to be relaxed for an electrolyte concentration of 0.5 mol/l. To analyse the transition to slow coagulation and the slow coagulation as a function of electrolyte concentration, the electrolyte concentration has been reduced stepwise. With the calculation of k_{11} according Eq. (1), ψ_δ-potentials have to be inserted. It is impossible to get them by measurement techniques – only ζ-potentials can be determined ($\zeta \leq \psi_\delta$).

The potentials, mentioned in Table 1, are determined by electrophoresis (Zeta-Sizer 4, Malvern Instruments) from electrophoretic mobility. According to Overbeek and Wirsema [10], the determination of potentials at the first three high electrolyte concentrations according to the Smoluchowski equation (2nd column) is – because of low potentials and high κa values (3rd column in the table) – correct, the error is not greater than 1 mV. For the next three concentrations in Table 1 it is sufficient to apply the Henry equation with the same margin of error. The measuring error is 1 mV. Theoretical and experimental velocity constants are compared as dimensionless N-numbers in Table 2 ($N = 2$ relating to $k_{11} = 12.2 \times 10^{-12}$ cm^3/s). The first three of these concentrations lead according to DLVO-theory (Eq. (1)) to rapid

Table 1 Determination of ζ-potentials

c [mol/l]	ζ [mV]	κa	ζ nach Henry [mV]
0.5	-8	246	
0.2	-21	156	
0.1	-32	110	
0.05	-42	78	-44
0.04	-45	70	-47
0.02	-55	49	-58

Table 2 Comparison of theoretical and experimental values for N

c of KNO_3 [mol/l]	N Theory (Eq. 1)	N Experiment
0.5	1.50	1.79
0.2	1.49	1.73
0.1	1.48	1.62
0.05	1.16	1.39
0.04	4×10^{-7}	1.33
0.02	10^{-41}	Stability

coagulation. Differences in the values of theoretical velocity constants are small. The (constant) difference between experiment and theory could be explained in [9], but not the other differences between the experimentally determined constants of different electrolyte concentrations. If an absolute error of $\pm 5\%$ is assumed on experimentally determined velocity constants k_{ij} and so the proof of a deviation from DLVO-theory is pending, the situation changes drastically with a further reduction of electrolyte concentration. These differences cannot be explained by other Hamaker constants and higher potentials ($\psi_\delta < \zeta$ is impossible). Changing the Hamaker constant would shift the problem only on the electrolyte scale. The assumption that the electrostatics does not have a strong influence, as expected according DLVO-theory, on those equations and parameters used here remains.

References

1. Sonntag H, Shilov V, Gedan H, Lichtenfeld H, Dürr C (1986) Colloids Surfaces 20:303
2. Lichtenfeld H, Sonntag H, Dürr C (1991) Colloids Surfaces 54:267
3. Lichtenfeld H, Knapschinsky L, Sonntag H, Shilov V (1995) Colloids Surfaces A 104:313
4. Matijevic E, Scheiner F (1978) J Colloid Interface Sci 63:509
5. Penners NGH (1985) Dissertation, Wageningen University
6. Honig EP, Roebersen GJ, Wiersema PHJ (1971) J Colloid Interface Sci 63:97
7. Derjaguin BV, Landau LD (1941) Acta Physiochim URSS 14:633
8. McCartney LN, Levine SJ (1969) Colloid Interface Sci 30:345
9. Shilov V, Lichtenfeld H, Sonntag H (1995) Colloids Surfaces A 104:321
10. Overbeek J Th G, Wiersema PH (1967) In: Bier M (ed) Electrophoresis, Vol II, Ch 1. Academic Press, New York

Progr Colloid Polym Sci (1997) 104:152–154
© Steinkopff Verlag 1997

S.U. Egelhaaf
P. Schurtenberger

The micelle-to-vesicle transition as observed by time-resolved scattering experiments

Dr. S.U. Egelhaaf (✉)
Large Scale Structures Group
Institut Laue-Langevin
B.P. 156
38042 Grenoble Cedex 9, France

P. Schurtenberger
Institut für Polymere
ETH Zürich
8092 Zürich, Schweiz

Abstract Amphiphilic molecules spontaneously self-assemble in solution to form a variety of aggregates. The understanding of the equilibrium properties of these aggregates, such as their shape and size, has made significant progress. However, only little is known about the existence of non-equilibrium or metastable states which form during structural transitions and the kinetics of their formation. Aqueous mixtures of lecithin and bile salt exhibit a transition from worm-like mixed micelles to vesicles, which spontaneously occurs upon dilution. This transition is studied using time-resolved light and small-angle neutron scattering. It is demonstrated that the temporal evolution of the aggregate structures can be followed and detailed information even on molecular length scales can be obtained.

Key words Light scattering – small-angle neutron scattering – micelles – vesicles – non-equilibrium phenomena

Equilibrium properties

Aqueous mixtures of lecithin and bile salt are important examples of multicomponent systems [1–4]. They not only serve as interesting model systems, but are also of great relevance in biology and physiology as well as in pharmaceutical applications.

Depending on the bile salt-to-lecithin ratio and the total lipid concentration, they form various structures [3]. Upon dilution of a mixed micellar stock solution the shape of the micelles changes from small spherical to large semi-flexible and locally cylindrical structures. This is caused by the decrease of the bile salt-to-lecithin ratio in the aggregates due to dilution, which appears to lower the spontaneous curvature H_0 of the mixed micelles. Endcaps are therefore avoided and the micelles are forced to grow. At even higher dilutions, the resulting value of H_0 favours the formation of locally lamellar structures and a spontaneous micelle-to-vesicle transition is induced. Beyond the micellar phase boundary, vesicles coexist with cylindrical micelles. In this region the sizes of the vesicles and micelles remain constant and only their relative amount changes. At still higher dilutions a single phase of relatively monodisperse vesicles exists, whose size decreases upon dilution.

Micelle-to-vesicle transition: time dependence and intermediate structures

While this transition has been observed in a number of surfactant systems, several aspects of it remain highly controversial. The exact sequence of the structural evolution and the existence of intermediate structures, such as connected networks or perforated bilayers, is still not conclusively determined. Not even the question whether vesicles represent true equilibrium structures is unambiguously solved. Based on the well established phase

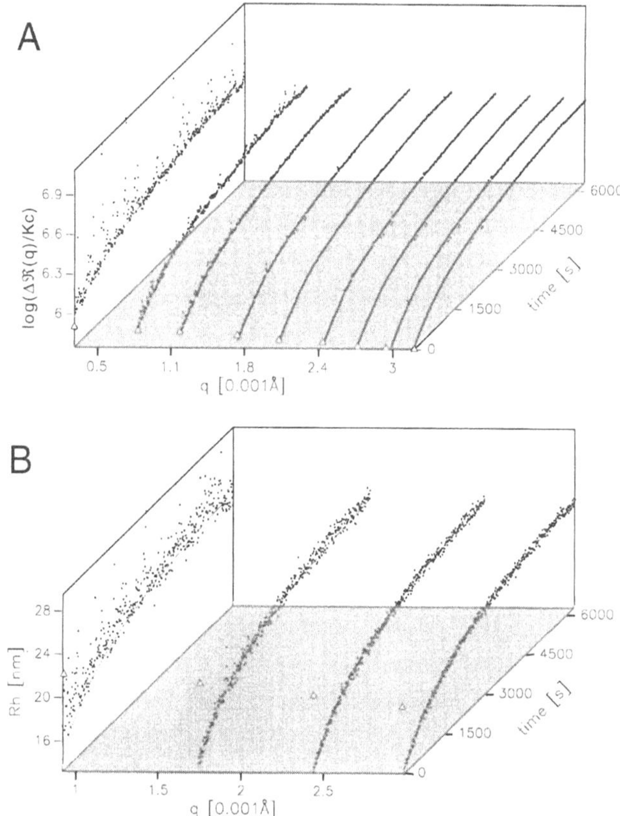

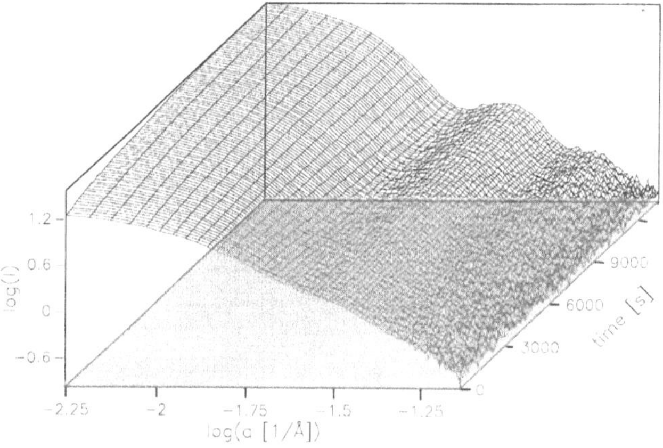

Fig. 2 Temporal evolution of the scattering intensity $I(q)$ as a function of scattering vector q for a micelle-to-vesicle transition as observed by SANS

Fig. 1 Temporal evolution of the reduced Rayleigh ratio $\Delta\Re(q)/Kc$ (A) and hydrodynamic radius $R_h(q)$ (B) as a function of scattering vector q for a micelle-to-vesicle transition as observed by LS. The data obtained from the initial, equilibrated solution is represented by open triangles

behaviour, we thus started an investigation of the kinetic aspects of the micelle-to-vesicle transition using time-resolved light scattering (LS) and small-angle neutron scattering (SANS). The transition is induced by rapidly diluting a micellar solution to a concentration corresponding to a vesicular solution.

The LS experiments were performed on a dedicated fibre-optics-based multiangle instrument [5], which simultaneously collects the scattered light at nine angles. Furthermore, the use of single and few mode fibres permits an optimization of the experimental conditions for both dynamic and static LS at the same time [6] and therefore simultaneous measurements can be performed. The thus obtained time-dependencies of the reduced Rayleigh ratio $\Delta\Re(q)/Kc$ and hydrodynamic radii $R_h(q)$ as a function of scattering vector q are shown in Fig. 1. The data of the initial, equilibrated micellar solution (Fig. 1, open tri-

angles) are significantly different from the first measurement after dilution, indicating the initial and very rapid formation of intermediary structures. Their $\Delta\Re(q)/Kc$ and $R_h(q)$ can be determined by a fit, resulting in structural information on the intermediates. From these relatively long-lived intermediary structures, vesicles are formed according to a first-order kinetics with a time-constant of about 1.2 h (the fit is shown as the solid line in Fig. 1). The vesicles might form either directly or via second, short-lived intermediates, which then very rapidly transform into vesicles and are therefore not observable.

Due to the limited range of scattering vectors q, LS allows for rather indirect conclusions on the involved aggregate structures only. This limitation can be overcome by SANS, which covers a q-range that is ideal for such studies [4]. Therefore we performed SANS experiments at D22 (ILL), which offers the unique possibility for time-resolved studies of this transition due to its very high neutron flux and the large q-range covered in a single instrumental setting. A measurement time of 1 min already results in sufficiently good statistics, although the samples typically have total surfactant concentrations of only 1 mg/ml. Figure 2 provides an example of the temporal evolution of the q-dependence of the scattered intensity $I(q)$ during the course of a dilution-induced micelle-to-vesicle transition. The SANS data confirm the conclusions drawn from LS. Moreover, they now provide us for the first time with information on molecular length scales that allow for a detailed analysis of the structural evolution during this important transition.

References

1. Hjelm RP, Thiyagarajan P, Alkan-Onyuksel H (1992) J Phys Chem 96: 8653–8661
2. Long MA, Kaler EW, Lee SP (1994) Biophys J 67:1733–1742
3. Egelhaaf SU, Schurtenberger P (1994) J Phys Chem 98:8560–8573
4. Pedersen JS, Egelhaaf SU, Schurtenberger P (1995) J Phys Chem 99: 1299–1305
5. Egelhaaf SU, Schurtenberger P (1996) Rev Sci Instr 67:540–545
6. Gisler T, Rüger H, Egelhaaf SU, Tschumi J, Schurtenberger P, Ricka J (1995) Appl Opt 34:3546–3553

Progr Colloid Polym Sci (1997) 104:155–156
© Steinkopff Verlag 1997

T. Hellweg
D. Langevin

Droplet microemulsions: Determination of the bending elastic constants by photon correlation spectroscopy

Dr. T. Hellweg (✉) · D. Langevin
Centre de Recherche Paul Pascal – CNRS
Av. du Docteur Schweitzer
33600 Pessac, France

Abstract The bending elasticity of a surfactant film at the interface oil–water in a microemulsion system can be described by the spontaneous curvature and two elastic constants κ and $\bar{\kappa}$. In this study we derive the sum $2\kappa + \bar{\kappa}$ for the system n-octane/$C_{10}E_5$/water from the polydispersity of the relaxation rate distribution found by a CONTIN analysis of the intensity correlation functions. A value of $1.2kT$ is found.

Key words Microemulsion size polydispersity – dynamic light scattering – surfactant film bending elasticity

Microemulsion are thermodynamically stable mixtures of oil and water, stabilized by surfactants and in some cases additionally by cosurfactants [1, 2]. The microstructures formed in these systems are frequently droplets, either of oil or water, covered by a surfactant mono-layer, dispersed in the continuous phase. In microemulsions containing comparable amounts of oil and water also bicontinuous structures may occur.

The stability of these different phases is mainly determined by the elastic properties of the amphiphilic film. In the case of droplet microemulsions the theory of curvature elasticity has been used for the description of the shape fluctuation of the drops [3, 4]. The knowledge of the elastic constants is necessary for an understanding of the properties of microemulsions. In recent years specially nonionic surfactants of the type alkyloligoglycolether ($C_n E_j$) have been used in studies of the behavior of microemulsions, because they can be used to form ternary microemulsions and no additional cosurfactant has to be added [5, 6]. In this study we investigate a microemulsion containing n-octane/$C_{10}E_5$/water on the water continuous side of the phase diagram (dispersed volume fraction $\Phi = 0.04$).

The well defined microstructures in microcmulsions are formed, because the surfactant layer has a preferred curvature C_0. A positive value of C_0 indicates a curving tendency towards water, a negative value, a curving tendency towards oil. The droplet radius for a given system can be predicted by

$$R = 3\Phi/[S]\Sigma , \tag{1}$$

where $[S]$ is the surfactant concentration, Φ the dispersed-phase volume fraction and Σ the area in the monolayer occupied by a surfactant molecule [2]. The area occupied per surfactant molecule is relatively constant and the monolayer can be regarded as approximately incompressible [1]. Under conditions of microemulsion formation, one important contribution to the free energy of the systems is the curvature energy of the surfactant layer. The curvature energy can be described by

$$F = \tfrac{1}{2}\kappa(C_1 + C_2 - 2C_0)^2 + \bar{\kappa}C_1C_2 . \tag{2}$$

Here, κ and $\bar{\kappa}$ are the bending and saddle-splay modulus. C_1 and C_2 are the principal curvatures of the surfactant film. In the case of droplet microemulsions $C_1 = C_2$.

The knowledge of the bending elastic constants is necessary for an understanding of the properties of microemulsions on a molecular level. If one looks at surfactant interfaces using experimental methods, which are able to resolve fluctuations of the interface it is possible to calculate the bending stiffness of the layer. For droplet microemulsions it is possible to describe the fluctuations in terms of spherical harmonics [7, 3]. This approach leads to the derivation of a quantitative connection between the polydispersity p of the microemulsion drops and the bending elastic constants [8].

$$p^2 = \frac{\langle R^2 \rangle}{\langle R \rangle^2} = \frac{kT}{8\pi(2\kappa + \bar{\kappa}) + 2kTf(\Phi)} \, . \tag{3}$$

R is the radius of the drops and $f(\Phi)$ an entropic term which is given by

$$f(\Phi) = \ln \Phi + \frac{1 - \Phi}{\Phi} \ln(1 - \Phi) \, . \tag{4}$$

The polydispersity p can be calculated from the moments of the distribution function of relaxation rates accessible by photon correlation spectroscopy. In this study we used CONTIN to compute the distribution functions [10].

The samples were investigated in the single-phase region but close to the emulsification failure boundary. In this region of the phase diagram only droplet structures are found in the emulsions. The volume fraction of oil and surfactant was 0.02, each. The phase transition (2 to 1) occurs at 304 K and we determined the polydispersity at 304, 305 and 306 K. The polydispersities in this temperature range stay constant and we calculate a value of $1.2kT$ for the sum of the elastic constants. For the same system $2\kappa + \bar{\kappa}$ has been calculated from SANS data and from interfacial tension measurements [9]. From SANS a value of $1.81kT$ is obtained. The interfacial tension measurements lead to $1.6kT$. Within the accuracy of the different methods these values are in the same order of magnitude. The hydrodynamic radius determined for the microemulsion drops in the regarded temperature range is 7.6 nm.

References

1. De Gennes PG, Taupin C (1982) J Phys Chem 86:2294–2304
2. Langevin D (1992) Annu Rev Phys Chem 43:341–369
3. Milner ST, Safran SA (1987) Phys Rev A 36:4371–4379
4. Borkovec M (1989) J Chem Phys 91:6268–6281
5. Kahlweit M, Strey R (1985) Angew Chem 97:655–669
6. Strey R (1994) Colloid Polymer Sci 272:1005–1019
7. Safran SA (1983) J Chem Phys 78:2073–2076
8. Gradzielski M, Langevin D, Farago B (1996) Phys Rev E 53:3900–3919
9. Gradzielski M, Langevin D, Sottmann T, Strey R, to be published
10. Provencher SW (1982) Comput Phys Com 27:213–217

Progr Colloid Polym Sci (1997) 104:157–159
© Steinkopff Verlag 1997

A hard sphere microemulsion

U. Olsson
P. Schurtenberger

U. Olsson (✉)
Physical Chemistry 1
Center for Chemistry
and Chemical Engineering
Lund University
P.O. Box 125
221 00 Lund, Sweden

P. Schurtenberger
Institut für Polymere
ETH Zentrum
8092 Zürich, Switzerland

Abstract We present data from an investigation of an oil-in-water nonionic three-component microemulsion system under conditions where spherical droplets with a radius of approximately 80 Å form. The structural and dynamic properties of the microemulsion have been studied using a combination of small-angle neutron scattering, static and dynamic light scattering, pulsed-gradient NMR self-diffusion and low shear viscosity measurements. We demonstrate that these liquid-like droplets have properties which to a very good approximation mimic those of classical hard sphere suspensions over a large range of volume fractions.

Key words Microemulsion – hard sphere – compressibility – diffusion – viscosity

Introduction

Microemulsions are thermodynamically stable liquid mixtures of water, oil and surfactant. While being macroscopically homogeneous, they are locally structured into polar and apolar domains separated by a surfactant-rich dividing surface. Due to the many ways of dividing space, microemulsions may show a large variation in microstructure. Under certain conditions, it is possible to stabilize spherical droplets of, say, oil in water with a low polydispersity and concentration invariant size. The conditions are a finite but not too low spontaneous curvature of the surfactant film and that the system is saturated with the dispersed oil [1].

One such system, with the nonionic surfactant pentaethylene glycol dodecyl ether ($C_{12}E_5$), water and decane, has recently been investigated in detail [2–4]. Spherical oil droplets were prepared with a surfactant-to-oil ratio $\phi_s/\phi_o = 0.815$, where ϕ_s and ϕ_o are the surfactant and oil volume fraction, respectively, over a large range of droplet volume fractions $\phi = \phi_s + \phi_o$. The surfactant consists of a dodecyl alkyl chain connected to an oligo ethylene oxide block of five ethylene oxide units. The microemulsion particles can be considered as spherical oil droplets of (hydrocarbon) radius r_{hc} covered by a dense brush of end-grafted penta ethylene oxide chains, where the grafting density is approximately 45 Å^2 per chain. This is a situation quite analogous to sterically stabilised "solid" colloid particles.

Small angle neutron scattering

Figure 1 shows the small-angle neutron scattering data (obtained at Risø National Laboratory, Denmark) from a dilute sample of $\phi = 0.02$ where the oil and water has been contrast-matched resulting in coherent scattering from the surfactant film alone [4]. A fit to the data, shown as a solid line, corresponds to the form factor of a spherical shell of radius $r_{hc} = 75$ Å and a relative polydispersity $\sigma/r_{hc} = 0.16$. This polydispersity contains contributions from both size (volume) and shape polydispersity [4].

U. Olsson and P. Schurtenberger
Hard sphere microemulsion

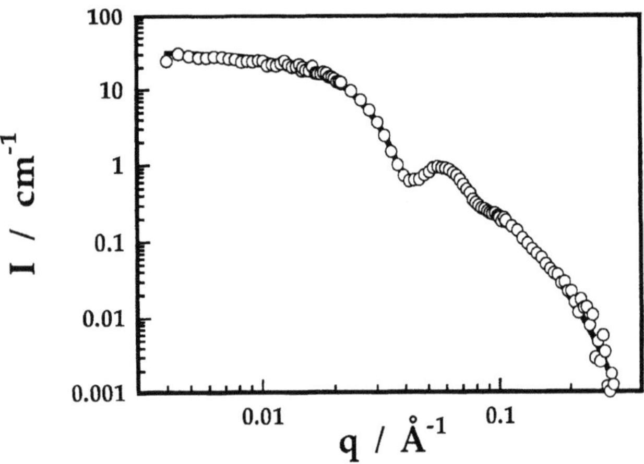

Fig. 1 SANS spectrum from a sample of $\phi = 0.02$, where the scattering length density of the oil is matched to that of water. The solid line is the best two parameter fit with a form factor of a shell yielding the radius $r_{hc} = 75\,\text{Å}$ and Gaussian relative standard deviation $\sigma/r_{hc} = 0.16$. Data taken from ref. [4]

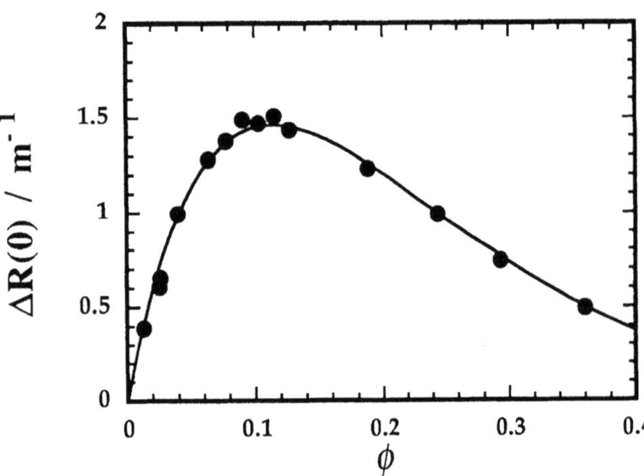

Fig. 2 Variation of the excess Rayleigh ratio, ΔR, extrapolated to zero scattering vector with the droplet volume fraction, ϕ. The solid line is the best two-parameter fit using the Carnahan–Starling equation for the osmotic pressure. The two fitted parameters were $r_{HS} = 86\,\text{Å}$ and $r_{hc} = 76\,\text{Å}$. Data taken from ref. [2]

Their relative contributions are not accurately known; however, contrast match experiments indicate that it is the shape polydispersity which dominates.

Static light scattering

In a static light-scattering experiment the effective structure factor, $S(0)$, at zero scattering vector was obtained from the extrapolated forward scattering. In Fig. 2 is shown the variation of the excess Rayleigh ratio $\Delta R(0)$ extrapolated to zero scattering vector with the volume fraction of droplets, ϕ [2]. In the monodisperse case, $S(0)$ is linked to the osmotic compressibility for which, in the case of hard spheres, an accurate expression exists due to Carnahan and Starling [5]. The experimental data were fitted with the Carnahan–Starling equation for $S(0)$ using two adjustable parameters, namely the hard sphere radius, r_{HS}, and the hydrocarbon radius, r_{hc}. The fit, shown in Fig. 2 as a solid line, results in an almost perfect match with the data. The parameters were found to be $r_{HS} = 86\,\text{Å}$ and $r_{hc} = 76\,\text{Å}$, where the latter is in excellent agreement with the SANS data. From r_{HS} we also obtain the corresponding hard sphere volume fraction $\phi_{HS} = 1.14\phi$. The fact that we are able to describe the system so well assuming monodisperse spheres indicates that polydispersity effects are minor.

Low shear viscosity

The low shear viscosity, η, was measured using capillary and, at higher concentrations, a cone-plate rheometer [3]. The two techniques gave equivalent results at intermediate concentrations. The variation of the relative viscosity η/η_0, where η_0 is the water solvent viscosity, with the hard sphere volume fraction ϕ_{HS} is shown in Fig. 3. For comparison we have also plotted data from van der Werff and de Kruif [6] for hard sphere silica dispersions of three different sizes. As can be seen, there is a perfect agreement between the microemulsion and silica data. The solid line in Fig. 3 shows the Quemada expression [7]

$$\eta/\eta_0 = (1 - \phi_{HS}/\phi_m)^{-2} \tag{1}$$

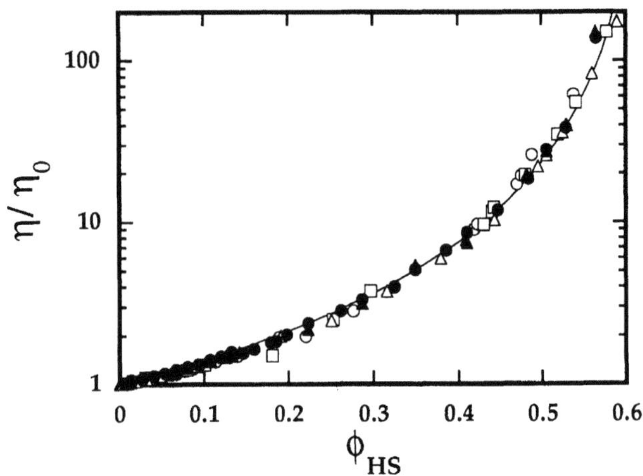

Fig. 3 Variation of the relative low shear viscosity η/η_0 with the hard sphere volume fraction ϕ_{HS}. Samples from the microemulsion (data taken from ref. [3]) was measured in a capillary (●) or in a cone and plate rheometer (▲). Shown with open symbols are data of different radii of coated silica spheres in oil, taken from ref. [6]. The solid line shows the prediction of Eq. (1)

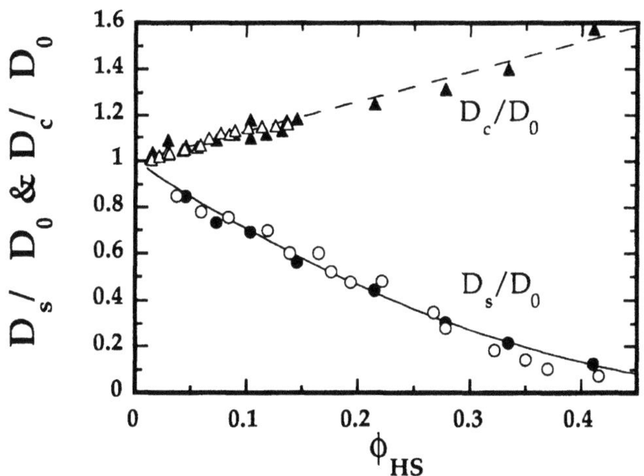

Fig. 4 Variation of the normalised collective (D_c/D_0) (▲) and long time self-diffusion (D_s/D_0) (●) coefficients with the hard sphere volume fraction ϕ_{HS}. All filled symbols refer to microemulsion data (taken from ref. [2]). D_c/D_0 data shown as open triangles correspond to silica spheres, taken from ref. [8]. D_s/D_0 data shown as open circles correspond to the self-diffusion of traces of silica spheres in a dispersion of pmma spheres (data taken from ref. [9]). The broken line is the equation $D_c/D_0 = 1 + 1.3\phi_{HS}$. The solid line shows the relation $D_s/D_0 = (1 - \phi_{HS}/0.63)^2$

$D_0 = 2.0 \times 10^{-11}$ m^2 s^{-1} (using heavy water, D$_2$O, as solvent and 23.5 °C) a hydrodynamic radius of $R_H = 95$ Å is obtained from the Stokes–Einstein relation. In Fig. 4, we also show, for comparison, D_c data on silica dispersions taken from Kops–Werkhoven and Fijnaut [8] and D_s data from van Megen and Underwood measured on traces of silica particles in dispersion of poly (methylmetacrylate) spheres [9]. As is seen, there is a good agreement, both in the collective and self-diffusion data. D_c increases approximately linearly with ϕ_{HS}, $D_c/D_0 \approx 1 + 1.3\phi_{HS}$, up to high volume fractions.

It is known from simple liquids that there is a correlation between D_s and η, where the product $D_s\eta$ is approximately constant upon variations in pressure or temperature [10]. A similar correlation was also found in a colloidal hard sphere system [11], with $D_s/D_0 = (\eta/\eta_0)$, which implies that the variation of D_s/D_0 can be described by the inverse of the relation describing η/η_0:

$$D_s/D_0 = (1 - \phi_{HS}/\phi_m)^2 . \tag{2}$$

This equation is shown as a solid line in Fig. 4 and as is seen, it provides a good description of the self-diffusion data.

with $\phi_m = 0.63$. As is seen, the Quemada relation provides a very accurate description of the data.

Diffusion

The concentration dependence of the collective (D_c) and self-diffusion (D_s) coefficients of the droplets are presented in Fig. 4, where we have plotted D_c/D_0 and D_s/D_0 as a function of ϕ_{HS} [2]. Here, D_0 denotes the diffusion coefficient extrapolated to infinite dilution. Due to the liquid nature of the droplets their NMR spectrum is in the motional narrowing regime with long transverse relaxation times. For this reason, the long time self-diffusion coefficients are conveniently and accurately measured using the pulsed gradient spin-echo (PGSE) technique. D_c was obtained from dynamic light scattering. From

Conclusion

As been shown above, it is possible to prepare a hard sphere microemulsion up to a rather high volume fraction ($\phi \approx 0.5$) of droplets. The main advantages of the microemulsion as a model hard sphere system are that it is easily prepared from commercially available components and that, due to the liquid nature, self-diffusion coefficients can be conveniently and accurately measured with the NMR PGSE technique. The droplets do crystallise at higher concentrations. However, while both volume and interfacial area are incompressible, they change their shape in the vicinity of the disorder–order transition.

Acknowledgments Collaboration with Håkan Bagger-Jörgensen, Kell Mortensen and Marc Leaver on parts of the work presented here is kindly acknowledged.

References

1. Safran SA (1994) Statistical Thermodynamics of Surfaces, Interfaces, and Membranes. Addison-Wesley, Reading, MA
2. Olsson U, Schurtenberger P (1993) Langmuir 9:3389–3394
3. Leaver MS, Olsson U (1994) Langmuir 10:3449–3454
4. Bagger-Jörgensen H, Olsson U, Mortensen K, Langmuir (accepted)
5. Carnahan NF, Starling KE (1969) J Chem Phys 51:635
6. van der Werff JC, de Kruif CG (1989) J Rheol 33:421–454
7. Quemada D (1977) Rheol Acta 16:82–94
8. Kops-Werkhoven MM, Fijnaut HM (1981) J Chem Phys 74:1618
9. van Megen W, Underwood SM (1989) J Chem Phys 91:552
10. Packhurst Jr HJ, Jonas J (1975) J Chem Phys 63:2698, 2705
11. van Bladeren A, Peetermans J, Maret G, Dhont JKG (1992) J Chem Phys 96

Progr Colloid Polym Sci (1997) 104:160–162
© Steinkopff Verlag 1997

A.R. Denton
H. Löwen

Unusual phase behaviour from peculiar pair potentials: A density-functional-perturbation study

Submitted to Proceedings of the International Workshop on Optical Methods and the Physics of Colloidal Dispersions (30.9.1996–1.10.1996, Mainz, Germany)

Dr. A.R. Denton (☒)
Institut für Festkörperforschung
Forschungszentrum Jülich GmbH
52425 Jülich, Germany

H. Löwen
Institut für Theoretische Physik II
Universität Düsseldorf
40225 Düsseldorf, Germany

Abstract Applying a density-functional-based perturbation theory, we examine the phase behaviour of model colloidal systems characterized by hard core and additional piecewise-constant repulsive or attractive interactions. For sufficiently narrow square-shoulder repulsions, the theory predicts coexistence between expanded and condensed isostructural (f.c.c.) crystals, in quantitative agreement with available simulation data. A square-well attraction of equal width results in a solid–solid critical point at lower density and higher temperature and a significantly wider coexistence region. Adding a constant long-ranged attraction leads to a novel phase diagram exhibiting a *quadruple* point, where two fluid phases and two solid phases coexist simultaneously.

Key words Colloids – phase behaviour – density – functional theory

Recent computer simulations [1, 2] and theoretical studies [3–8] of systems interacting via extremely short-range pair potentials have produced convincing evidence for a first-order isostructural phase transition between expanded and condensed solids, the corresponding phase diagram exhibiting three-phase coexistence between a single fluid phase and the two solids. Possible physical manifestations of such systems are uncharged colloidal particles mixed with non-adsorbing polymer, or charge-stabilized colloidal suspensions, whose macroions interact via electrostatic and van der Waals forces [9].

Here we employ a combination of density-functional (DF) theory and thermodynamic perturbation theory to study the phase behaviour of model systems characterized by short- and long-range piecewise-constant interactions. Our goal is to isolate the influence of specific features of the pair potential on the stability of solid–solid transitions and to explore the intriguing possibility of *four*-phase (vapour–liquid–solid–solid) coexistence. We focus in particular on pair potentials composed of a hard core of diameter σ plus a repulsive square-shoulder or attractive square-well of width δ and magnitude ε. Since ε scales with temperature, these systems are completely characterized by the parameter δ/σ. We further consider the influence of long-range attraction, in the simplest conceivable way, by adding to the square-well a "van der Waals" tail, i.e. a constant attraction of infinite range, but infinitesimal strength $-a\varepsilon\sigma^3/V$, where V is the volume. This introduces an additional dimensionless parameter a, which gauges the strength of the attraction.

The relevant theoretical quantity is the Helmholtz free energy functional $F[\rho]$, a functional of the spatially varying one-particle density $\rho(\mathbf{r})$. For pair potentials with a steeply repulsive short-range interaction, thermodynamic perturbation theory [10, 11] accurately approximates $F[\rho]$ by decomposing the full pair potential $\phi(r)$ into a repulsive short-range reference potential $\phi_o(r)$ and relatively weak long-range perturbation $\phi_p(r)$. In our case, $\phi_o(r)$ is the hard-sphere (HS) pair potential and $\phi_p(r)$ is a step function of amplitude ε and range $\sigma + \delta$.

To first order in the perturbation,

$$F[\rho] \simeq F_{HS}[\rho] +$$

$$\frac{1}{2}\int d\mathbf{r}\int d\mathbf{r}'\, \rho(\mathbf{r})\,\rho(\mathbf{r}')g_{HS}(\mathbf{r},\mathbf{r}')\,\phi_p(|\mathbf{r}-\mathbf{r}'|)\,, \tag{1}$$

where $g_{HS}(\mathbf{r},\mathbf{r}')$ is the pair distribution function of the HS solid. The HS solid free energy $F_{HS}[\rho]$ separates naturally into an (exactly known) ideal-gas term and an excess term $F_{ex}[\rho]$. For the latter, we use the modified weighted-density approximation (MWDA) [12]:

$$F_{ex}^{MWDA}[\rho]/N = f_{HS}(\hat\rho)\,, \tag{2}$$

where $f_{HS}(\hat\rho)$ is the excess free energy per particle of the *uniform* HS fluid evaluated at a weighted density

$$\hat\rho \equiv \frac{1}{N}\int d\mathbf{r}\int d\mathbf{r}'\, \rho(\mathbf{r})\,\rho(\mathbf{r}')\,w(|\mathbf{r}-\mathbf{r}'|;\hat\rho)\,, \tag{2'}$$

defined as a *weighted* average of the physical density with respect to a weight function $w(r)$, which in turn is specified by the requirement

$$\left(\frac{\delta^2\,F_{ex}^{MWDA}[\rho]}{\delta\rho(\mathbf{r})\,\delta\rho(\mathbf{r}')}\right)_{\rho(\mathbf{r})\to\rho} = -\,k_B T c_{HS}^{(2)}(|\mathbf{r}-\mathbf{r}'|;\rho)\,, \tag{3}$$

$c_{HS}^{(2)}(|\mathbf{r}-\mathbf{r}'|;\rho)$ being the two-particle Ornstein–Zernike direct correlation function. The fluid-state input functions f_{HS} and $c_{HS}^{(2)}$ we take from the analytic solution of the Percus–Yevick equation for hard spheres [10].

Following Likos et al. [4], we approximate $g_{HS}(\mathbf{r},\mathbf{r}')$ by a unit step function, thus properly excluding self-correlation but otherwise neglecting pair correlations. This mean-field approximation is expected to be quite reasonable in the high density solid, where two-particle correlations are determined largely by the highly non-uniform one-particle density. It is also supported by Monte Carlo simulations of the HS crystal [13] and by DF theory calculations for the Lennard–Jones system [14]. At the solid densities of interest here, the density distribution is parametrized to reasonable accuracy by the isotropic Gaussian ansatz,

$$\rho(\mathbf{r}) = \left(\frac{\alpha}{\pi}\right)^{3/2}\sum_{\mathbf{R}} e^{-\alpha|\mathbf{r}-\mathbf{R}|^2}\,, \tag{4}$$

the sum running over lattice sites $\mathbf{R}$ of an f.c.c. crystal. Minimization of the functional $F[\rho]$ with respect to the single variational parameter α determines the free energy of the solid.

For the fluid phase, we use the uniform (constant ρ) limit of Eq. (1) [11]:

$$f(\rho) \simeq f_{HS}(\rho) + 2\pi\rho\int_0^\infty dr\, r^2\, g_{HS}(r)\phi_p(r)\,, \tag{5}$$

taking for the HS fluid functions f_{HS} and $g_{HS}(r)$ the essentially exact Carnahan–Starling and Verlet–Weis expressions [10], respectively. The sole effect of the van der Waals tail is to add a structure-independent term proportional to $(-a\rho^2/2)$ to the free energy per volume of both the fluid and the solid phases.

For a given pair potential, the fluid and solid free energies per volume are computed as a function of average reduced density $\rho\sigma^3$ at fixed reduced temperature $k_B T/\varepsilon$. Sufficiently short-range interactions ($\delta\ll\sigma$) induce inflection in the curve of F/V vs. ρ that can result in solid–solid phase coexistence. The densities of coexisting phases are established by means of a Maxwell common tangent (or equal area) construction, ensuring equality of the chemical potentials and pressures. Repetition of the procedure for a series of temperatures systematically maps out the phase diagram in the $T-\rho$ plane.

Predictions of the theory for isostructural solid–solid coexistence in the square-shoulder system are presented in Fig. 1, together with recently available Monte Carlo simulation data [2]. With increasing shoulder width, the solid–solid critical point shifts to lower densities and the coexistence region widens, while the critical temperature remains relatively constant. Evidently, the theory slightly overestimates the coexistence densities, but otherwise is in excellent agreement with simulation.

Figure 2 displays our results for the full phase diagram of the square-well system with van der Waals tail. The long-range attraction proves to have only a minor effect

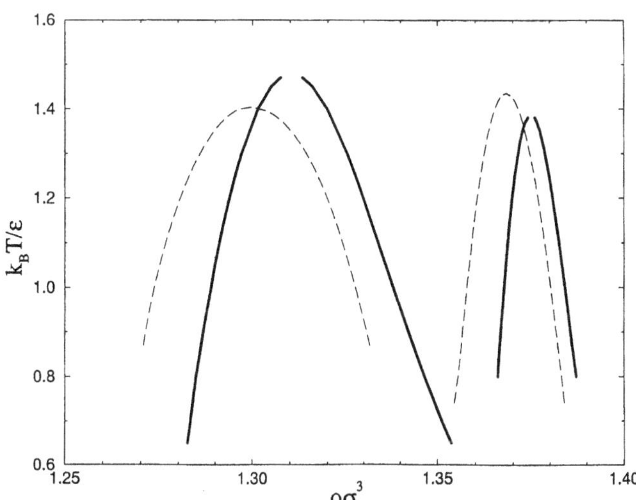

Fig. 1 Phase diagram of temperature vs. density (reduced units) for the square-shoulder system, exhibiting coexistence between two isostructural (f.c.c.) solids. Solid curves are theoretical predictions, dashed curves the corresponding simulation data [2] for shoulder widths $\delta/\sigma = 0.03$ (right-most curves) and $\delta/\sigma = 0.08$ (left-most curves)

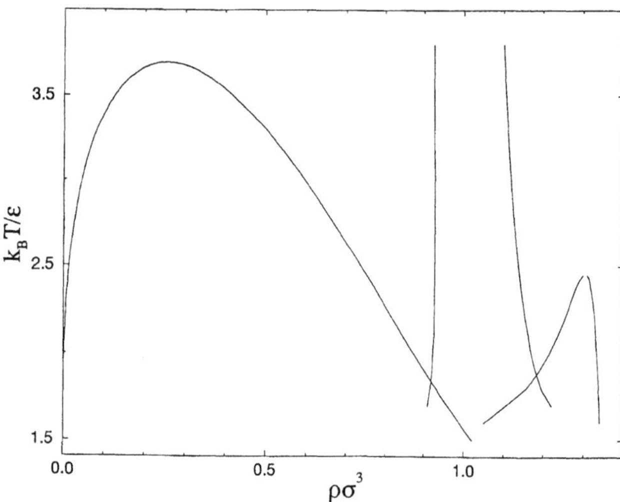

Fig. 2 Phase diagram for the square-well system with constant long-range attractive tail, exhibiting a quadruple point. From left to right, curves represent vapour–liquid, liquid–solid, and solid–solid phase coexistence for the case of well width $\delta/\sigma = 0.03$ and van der Waals parameter $a = 40$

on solid–solid coexistence, which is seen to differ considerably from the square-shoulder case. For the same step width, $\delta/\sigma = 0.03$, the square-well critical point occurs at lower density and higher temperature and the coexistence region is significantly wider. Furthermore, the transition remains stable over a narrower range of δ than for the square-shoulder system [2]. In the absence of long-range attraction, the theory predicts preemption by the liquid–solid transition for δ/σ as small as 0.05–0.06. For such short-range potentials, the vapour–liquid critical temperature falls well below the liquid–solid–solid triple point temperature, in which case only the liquid can coexist with the two isostructural solids. Inclusion of the van der Waals tail, however, raises the vapour–liquid critical point and the vapour–liquid–solid triple point. For a sufficiently strong attraction, as illustrated in Fig. 2, the two triple points can coincide, resulting in a *quadruple* point at which the two fluid phases coexist with the two solid phases.

Summarizing, using a relatively simple density-functional-perturbation theory, we have demonstrated that sufficiently short-range piecewise-constant interactions–whether attractive *or* repulsive–can result in unusual phase behaviour distinguished (at sufficiently low temperature) by a first-order isostructural solid–solid transition. Adding a weak constant long-range attraction to a square-well pair potential has little effect on the solid–solid transition, but strongly enhances the vapour–liquid transition. The resulting phase diagram exhibits *four* phases (vapour, liquid, expanded- and condensed-solids), which in this model can simultaneously coexist at a quadruple point. Whether or not the remarkable phase behaviour predicted for these model systems can be observed experimentally is not yet clear. The present study, however, may help to resolve the issue by guiding the parametrization of more realistic pair potentials (e.g., DLVO) that better model the interactions in real colloidal systems. Work along these lines is in progress.

Acknowledgment We gratefully acknowledge Drs. A.M. Bohle and C.N. Likos for helpful discussions, and Dr. P. Bolhuis for kindly providing simulation data for the square-shoulder system prior to publication.

References

1. Bolhuis P, Frenkel D (1994) Phys Rev Lett 72:2211; (1995) Phys Rev E 50:4880
2. Bolhuis P, Frenkel D (1996) J Phys, to appear
3. Tejero CF, Daanoun A, Lekkerkerker HNW, Baus M (1994) Phys Rev Lett 73:752
4. Likos CN, Németh ZS, Löwen H (1994) J Phys Condens Matter 6:10965
5. Németh ZS, Likos CN (1995) J Phys Condens Matter 7:L537
6. Likos CN, Senatore G (1995) J Phys Condens Matter 7:6797
7. Rascón C, Mederos L, Navascués G (1995) J Chem Phys 103:9795
8. Kincaid JM, Stell G, Goldmark E (1976) J Chem Phys 65:2172
9. Pusey PN (1991) In: Hansen J-P, Levesque D, Zinn-Justin J (eds) Liquids, Freezing and Glass Transition. North-Holland, Amsterdam, pp 763–942
10. Hansen JP, McDonald IR (1986) Theory of Simple Liquids, 2nd ed. Academic, London
11. Weeks JD, Chandler D, Andersen HC (1971) J Chem Phys 54:5237; Andersen HC, Chandler D, Weeks JD (1976) Adv Chem Phys 34:105
12. Denton AR, Ashcroft NW (1989) Phys Rev A 39:4701
13. Weis J-J (1974) Mol Phys 28:187; Kincaid JM, Weis J-J (1977) Mol Phys 34:931
14. Mederos L, Navascués G, Tarazona P, Chacón E (1993) Phys Rev E 47:4284

Progr Colloid Polym Sci (1997) 104:163–165
© Steinkopff Verlag 1997

D.S. Horne
I.R. McKinnon

DWS behaviour in gelling systems: Preliminary comparison with rheological measurements over sol–gel transition in latex-doped gelatin

Dr. D.S. Horne (✉)
Hannah Research Institute
Ayr KA6 5HL, Scotland, United Kingdom

I.R. McKinnon
Department of Chemistry
Monash University Clayton
Clayton, Victoria, Australia

Abstract DWS measurements have been made of latex particle mobility in a gelatin solution as the solution is cooled through the sol–gel transition and comparisons made with visco-elastic properties of the same solution obtained using a sensitive rheometer. The DWS correlation functions show the square root dependence on channel time predicted for measurements made with a back-scattering configuration. Over the temperature range in which the gelatin solution viscosity remains independent of shear rate, the DWS relaxation time is proportional to the ratio of dynamic viscosity to absolute temperature (η/T). Viscosity, however, becomes shear rate dependent as the solution gels and the correlation breaks down. Through the sol–gel transition, over the measured phase angle range 85–5°, a correlation is observed between DWS relaxation time and the reciprocal of the tangent of that phase angle, a measure of the viscoelastic relaxation behaviour. Beyond this region, DWS relaxation time passes through a maximum, associated with which is a decline in correlation function amplitude. It is suggested that this behaviour is associated with the freezing out of longer-lived relaxation modes in the gel.

Key words Diffusing wave spectroscopy – light scattering – gelation – rheology

Introduction

By treating the transport of light as a diffusion process, diffusing wave spectroscopy (DWS) extends the traditional techniques of dynamic light scattering to strongly multiple scattering, opaque media [1, 2]. A particularly simple stretched exponential analytical form for the correlation function is obtained for backscattering geometry and is applicable to our system incorporating a bifurcated bundle of optical fibres [3].

The relaxation time measured by DWS for particle mobility in a fluid system is a function of both particle size and particle–particle interactions. This precludes the direct translation of mobility into particle size, the most prevalent use of traditional dynamic light scattering. How-ever, it has been demonstrated that dispersion viscosity provides an adequate correction factor for interaction effects, allowing the use of DWS techniques for size determination in opaque fluid emulsions [4]. The ultimate manifestation of hindrance in particle–particle mobility is gelation. Here we report preliminary observations on a comparison of viscoelastic and DWS behaviour encountered through the sol–gel transition of a cooling gelatin solution doped to a 1% volume fraction with 0.1 μm latex particles.

Experimental

Our light scattering set-up for DWS measurements, incorporating a bifurcated optical fibre bundle and multi-time base autocorrelator (Malvern 7032, 256 channels) has been described previously [3]. Data analysis procedures and verification of the DWS approach for this particular configuration are also included therein [4]. Linear viscoelastic shear moduli were measured during gel formation in a Bohlin CVO rheometer by applying an oscillatory stress and measuring the resultant strain. A double-gap Couette geometry was employed. Rheological data were obtained at oscillation frequencies of 0.04, 0.08, 0.15 and 0.3 Hz in successive measurements at approx. 2.5 min intervals as the gelatin solution was cooled, data collection being made over the third oscillation period in every case.

The gelatin solution was prepared by dispersing 2.5 g gelatin (Sigma Type II from porcine skin, Bloom No. 175) in 47.5 g of a 1% volume fraction suspension of polystyrene latex particles (Sigma LB1, diam. 114 nm) at 55 °C. 4 ml of the solution was transferred to a sample container set in a thermostatted block pre-heated to 50 °C. After insertion of the heated DWS probe and an equilibration period, the power to the heating element was switched off and the system allowed to cool to ambient temperature. During this time, which lasted some 8 h due to the large thermal inertia of the system, correlation functions accumulated over 30 s were stored continuously for subsequent analysis. 30 ml of the remaining hot solution was poured into the double-gap geometry fitted to the rheometer and cooled using a temperature gradient profile which attempted to parallel that measured during the cooling of the latex-doped gelatin solution undergoing DWS measurements.

Results

The DWS measurements show relaxation time, τ_{DWS}, increasing during the cooling period, passing through a maximum of approx. 1.8 s in the neighbourhood of 23.6 °C and declining thereafter to a value about 100X the relaxation time measured at 50 °C (Fig. 1). The correlation function amplitude remains largely constant until it shows the plotted decrease over the same temperature range as τ_{DWS} exhibits its maximum (Fig. 1). Similar behaviour has been observed previously in firm acid milk gels [5].

During cooling down to 24.5 °C, the only detectable viscoelastic change in the gelatin solution is in the dynamic viscosity, G^*/ω, where $G^* = \sqrt{|G'^2 + G''^2|}$ and ω is the frequency of oscillation. This rises monotonically, independent of frequency, down to approx. 24.3 °C (data not shown). Its value is dominated completely by the viscous shear modulus, G''. Below 24.3 °C, a dependence of viscosity on oscillation frequency clearly emerges as the doped gelatin solution gels. The gel point is clearly seen in the

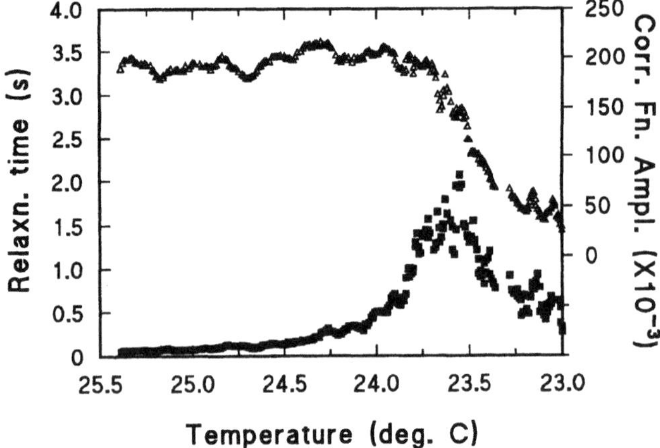

Fig. 1 DWS relaxation time (■) and correlation function amplitude (△) measured in back-scattering mode on cooling 5% (wt/wt) gelatin solution doped with 1% (v/v) 0.1 μ polystyrene latex

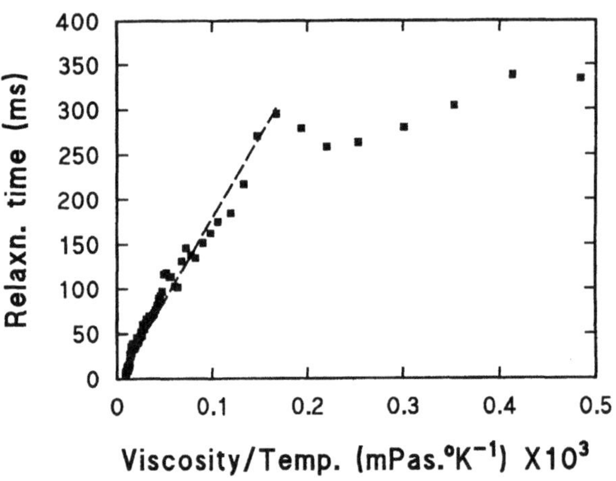

Fig. 2 Relationship between measured DWS relaxation time and the ratio, obtained at the same temperature, of rheological dynamic viscosity and temperature (°K) over the temperature range 50.0–24.0 °C. Dashed line is calculated linear fit for points from 50 °C to 24.2 °C

rapid rise in the elastic modulus, G', which causes it to overtake the viscous component (both measured with $\omega = 0.08$ Hz) at approximately 24.1 °C.

Down to 24.3 °C in this system where τ_{DWS} has risen to around 300 ms. a linear correlation is observed between τ_{DWS} and the shear-independent solution viscosity divided by absolute temperature (Fig. 2). This is as expected for a dispersion where the particles are undergoing a hindered Brownian motion and has been documented in previous studies on emulsions [4] and cooling milk [5]. Below 24.3 °C, where shear-rate-dependent behaviour is observed in the viscosity value, the linearity in the correlation is lost (Fig. 2).

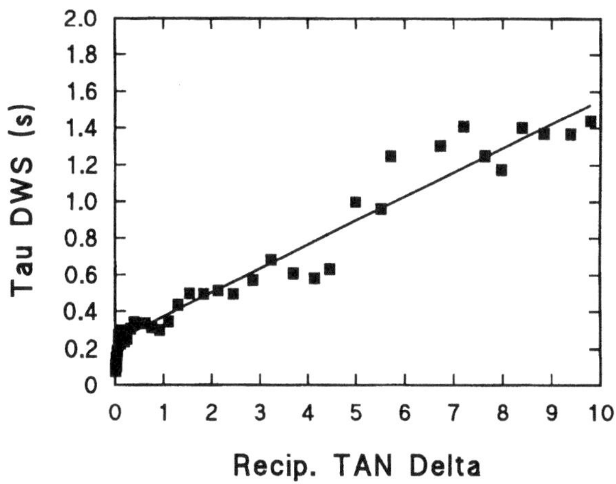

Fig. 3 Correlation between τ_{DWS} and reciprocal TAN δ, both measured at the same temperature, as 5% gelatin solution cools through sol–gel transition. The line is the calculated linear fit to points within the temperature range from 24.3 °C to 23.8 °C and measured phase angles from 85 °C to 5°

Through the rheologically observed gel point, over the temperature range 24.3–23.8 °C, the minimum set by the termination of the rheological measurements at this point, a linear correlation is found between τ_{DWS} and the reciprocal of the tangent of the rheological phase angle (Fig. 3). Thus, whereas in a fluid system hydrodynamic effects, manifested in the viscosity, were the principal source of hindrance to particle motion, rendering τ_{DWS} proportional to η^*, in the viscoelastic regime, τ_{DWS} is now governed by the ratio of elastic-to-viscous components, becoming dominated by the growth of elastic modulus as gelation proceeds. In rheological terms, the reciprocal of the tangent of the phase angle is a measure of the stress relaxation time, of the time dependence of the response of the gelling protein to the applied oscillating force field. In a simple Maxwellian dashpot and spring model, it would be the time constant for a monoexponential decay. Here it is more likely that a distribution of relaxation pathways exists and that their constrained relaxation would lead to a stretched exponential behaviour. A plausible explanation for the maximum observed in τ_{DWS} is that eventually the longer-lived contributions to the distribution become frozen-out, leaving only the faster relaxations, their loss coinciding with the drop in correlation function amplitude indicating a decline in the number of mobile light scattering entities.

References

1. Pine DJ, Weitz DA, Chaikin PN, Herbolzheimer E (1988) Phys Rev Lett 60:1134–1137
2. Weitz DA, Pine DJ (1993) In: Brown W (ed) Dynamic Light Scattering: The Method and Some Applications. Oxford University Press, Oxford, pp 652–720
3. Horne DS (1989) J Phys D 22:1257–1265
4. Horne DS, Davidson CM (1993) Colloid Surf A: Physicochem Eng Aspects 77:1–8
5. Horne DS (1991) In: Schmitz KS (ed) Photon Correlation Spectroscopy: Multicomponent Systems. SPIE Bellingham, Washington, pp 166–180

Progr Colloid Polym Sci (1997) 104:166–167
© Steinkopff Verlag 1997

M.-O. Ihm
F. Schneider
P. Nielaba

Phase transitions in nonadditive hard disc systems: a Gibbs ensemble Monte Carlo study

M.-O. Ihm · F. Schneider
Dr. P. Nielaba (✉)
Institut für Physik
Universität Mainz
55099 Mainz, Germany

Abstract We study the properties of a model fluid in two dimensions with Gibbs ensemble Monte Carlo (GEMC) techniques, in particular we analyze the entropy-driven phase separation in case of a nonadditive symmetric hard disc fluid. By a combination of GEMC with finite size scaling techniques we locate the critical line of nonadditivities as a function of the system density, which separates the mixing/demixing regions and compare with a simple analytical approximation.

Key words Phase transitions – Gibbs ensemble – finite size scaling – nonadditive hard discs

Phase transitions in systems with purely repulsive interaction have got much attention in the last years [1–8]. Besides studies of three-dimensional systems with nonadditive hard spheres, the properties of two-dimensional hard discs [5], soft spheres [7], hard hexagons [8] and of the multicomponent Widom Rowlinson model [2] were investigated recently.

The Gibbs ensemble Monte Carlo (GEMC) method [9, 10] has been successful in predicting coexistence densities in a variety of different classical and quantum [11] systems undergoing a phase separation; for a review see Ref. [12]. In these studies the simulation is done in two boxes in parallel, which have their own periodic boundary conditions, acceptance rules for particle exchanges between the boxes and volume changes (conserving the total volume) are designed such that the pressure and the chemical potential in the two boxes are equal. For parameters where the system is in the coexistence region and far away from the critical point the correlation length ξ is eventually smaller than the linear system size. Due to the inherent GEMC procedure the interfacial free energy of the system is minimized by phase separation into the two simulation boxes. However, as noted by Mon and Binder [13], near the critical point the correlation length may exceed the linear system size and thus density fluctuations on all length scales can be expected. As a result the GEMC boxes often change their identity. In particular, in the case of purely repulsive interaction the free energy only consists of entropy. In this case the interfacial free energy between the different phases is quite small and thus the conditions on the GEMC procedure are extreme. On the other hand, the block analysis finite size scaling method [14] has been proven to be useful to obtain critical point parameters and coexistence densities even for fluid systems [15].

In this paper we report on a combination of GEMC and block analysis techniques for a system of nonadditive hard discs [16, 17]. We obtain the phase diagram of the system and compare with a simple analytical approximation; a 20% deviation is found.

We consider a system of hard discs (of diameter d) with N_A particles of type A and N_B particles of type B and interaction potential:

$$U(r_{12}) = \begin{cases} 0: & r_{12} > d_{s_1 s_2}, \\ \infty: & r_{12} < d_{s_1 s_2}. \end{cases} \tag{1}$$

r_{12} is the distance of two particles, $s_1, s_2 \in \{A, B\}$ are their species and $d_{AA} = d_{BB} = d$, $d_{AB} = d + \Delta/2$. The total number of particles N and the total volume V is fixed and thus the average density $\rho^* = \rho * d^2 = N d^2 / V$. Due to the additional

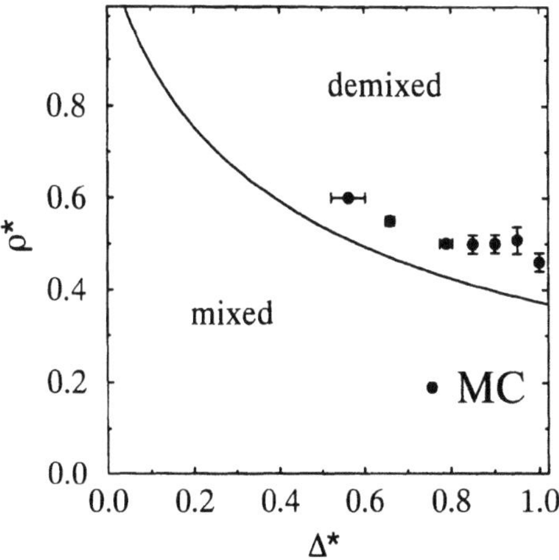

Fig. 1 Phase diagram; plot of critical density versus critical nonadditivity. Symbols: Critical points from MC simulations; full line: critical line from Eq. (2)

repulsion between A and B-type particles we expect a phase separation into a A-rich and a B-rich fluid phase for large values of $\Delta > \Delta_c$ and fixed total density. For $\Delta = 0$ we have a pure hard disc fluid with no phase separation in the fluid phase.

In order to get a rough idea about the critical line l_c, defined by $\rho_c^*(\Delta_c)$ we computed [16, 17] l_c in analogy to a study in three dimensions [18] by convex envelope arguments for the free energy and arrive at the compact expression (see Fig. 1):

$$\rho_c^*(\Delta_c) = \frac{4}{\pi}\left(1 - \sqrt{\frac{\Delta_c}{d + \Delta_c}}\right). \qquad (2)$$

In order to locate the critical values Δ_c as a function of ρ^* we perform GEMC simulations [16, 17]. The GEMC

results were obtained with $N = 512$ particles and about 10^6 Monte Carlo steps, where each step consisted of 400 attempted moves, 20 particle exchange and 2 volume change attempts, cell occupancy lists were successfully used to speed up the procedure.

Since the phase separation is driven by entropy we expect only a small interfacial free energy in case of phase coexistence and thus the conditions for the GEMC procedure are extreme. The critical points were obtained by inspection of number-difference histograms $P_L(N_A - N_B)$ on different length scales L obtained by subdivision of the simulation boxes of sizes V_1 and V_2 ($V_1 + V_2 = V$) into smaller subsystems of size $L \times L$. For $\Delta < \Delta_c$ the distributions are all singly peaked, for larger Δ we obtain a single peak structure of $P_L(N_A - N_B)$ for large L and a double peak structure for small L. An analysis of these histograms with the fourth-order cumulants [14]

$$U_L = 1 - \langle(N_A - N_B)^4\rangle_L / 3\langle(N_A - N_B)^2\rangle_L^2 \qquad (3)$$

allows a determination of critical points, due to the cumulants L-invariance at the critical point, since $U_L(L/\xi) \approx U_L(0)$. The GEMC results for the critical line are presented in Fig. 1. Similar to the prediction of Eq. (2), ρ_c is a decreasing function of Δ_c, however at a given density the GEMC results for Δ_c are about 20% larger than the predictions of Eq. (2), calling for better approximations.

In summary by a combination of GEMC and finite-size scaling techniques we obtain results for the critical line for nonadditive symmetric hard discs even in cases where the interfacial free energy is purely entropic.

Acknowledgments F.S. thanks the DFG for the support Ni-259/6-1,2, P.N. thanks the DFG for support (Heisenberg foundation). The computations were partly carried out on the *CRAY YMP* of the *HLRZ* at Jülich.

References

1. Biben T, Hansen JP (1991) Phys Rev Lett 66:2215; (1991) J Phys, Condens Matter 3:F65
2. Lebowitz JL, Mazel A, Nielaba P, Samaj L (1995) Phys Rev E 52:5985
3. Jung J, Jhon MS, Ree FH (1995) J Chem Phys 102:1349
4. Amar JG (1989) Mol Phys 67:739
5. Gazzillo D, Pastore G (1989) Chem Phys Lett 159:388
6. Gazzillo D (1991) J Chem Phys 95:4565
7. Mountain RD, Harvey AH (1991) J Chem Phys 94:2238
8. Frenkel D, Louis AA (1992) Phys Rev Lett 68:3363

9. Panagiotopoulos AZ (1987) Mol Phys 61:813
10. Smit B, De Smedt Ph, Frenkel D (1989) Mol Phys 86:931; Smit B, Frenkel D (1989) Mol Phys 68:951
11. Schneider F, Nielaba P (1995) Phys Rev E 51:5162
12. Panagiotopoulos AZ (1992) Molec Simul 9:1; Smit B (1993) In: Allen MP, Tildesley DJ (eds) Computer Simulation in Chemical Physics. Kluwer, Dordrecht, p 173
13. Mon KK, Binder K (1992) J Chem Phys 96:6999

14. Binder K (1981) Z Phys B 43:119
15. Rovere M, Heermann DW, Binder K (1988) Europhys Lett 6:585; Rovere M, Nielaba P, Binder K (1993) Z Phys B 90:215
16. Ihm MO (1993) Diplomarbeit, Mainz
17. Schneider F, Ihm MO, Nielaba P (1994) In: Landau DP, Mon KK, Schüttler HB (eds) Computer Simulation Studies in Condensed Matter Physics VII. Springer, Berlin, p 188
18. Melnyk TW, Sawford BL (1975) Mol Phys 29:891

Progr Colloid Polym Sci (1997) 104:168–172
© Steinkopff Verlag 1997

Diffusion of colloids at short times

M. Watzlawek
G. Nägele

M. Watzlawek[1] (✉) · G. Nägele
Fakultät für Physik
University of Konstanz
78434 Konstanz, Germany

Present address:
Institut für Theoretische Physik II
Heinrich-Heine-Universität
Universitätsstr. 1
40225 Düsseldorf, Germany

Abstract We study the combined effects of electrostatic and hydrodynamic interactions (HI) on the short-time dynamics of charge-stabilized colloidal spheres. For this purpose, we calculate the translational and the rotational self-diffusion coefficients, D_s^t and D_s^r, as function of volume fraction ϕ for various values of the effective particle charge Z and various concentrations n_s of added 1–1 electrolyte.

Our results show that the self-diffusion coefficients in deionized suspensions are less affected by HI than in suspensions with added electrolyte. For very large n_s, we recover the well-known results for hard spheres, i.e. a linear ϕ-dependence of D_s^t and D_s^r at small ϕ. In contrast, for deionized charged suspensions at small ϕ, we observe the interesting non-linear scaling properties $D_s^t \propto 1 - a_t \phi^{4/3}$ and $D_s^r \propto 1 - a_r \phi^2$. The coefficients a_t and a_r are found to be nearly independent of Z. The qualitative differences between the dynamics of charged and uncharged particles can be well explained in terms of an effective hard sphere (EHS) model.

Key words Translational diffusion – rotational diffusion – hydrodynamic interaction – charge-stabilized colloidal suspensions

Introduction

Since several years, the effect of HI on the short-time self-diffusion coefficients of hard sphere suspensions has been investigated in detail by various authors [1–5]. For the calculation of the first and second virial coefficients of D_s^t and D_s^r in an expansion in terms of the volume fraction ϕ, both the influence of two-body and three-body HI was taken into account. At small ϕ, the currently established results for the normalized diffusion coefficients H_s^t and H_s^r are given by [1, 4]

$$H_s^t = \frac{D_s^t}{D_0^t} = 1 - 1.831\phi + 0.88\phi^2 + \mathcal{O}(\phi^3) \tag{1}$$

and [2, 5]

$$H_s^r = \frac{D_s^r}{D_0^r} = 1 - 0.630\phi - 0.67\phi^2 + \mathcal{O}(\phi^3), \tag{2}$$

respectively. Here, D_0^t and D_0^r are the Stokesian diffusion coefficients for a colloidal sphere of radius a dispersed in a solvent of viscosity η.

The possibility to express H_s^t and H_s^r in terms of a power series in ϕ arises from the fact that hard sphere suspensions at small ϕ can be considered as dilute both with respect to the particle hydrodynamics and to the microstructure. For charge-stabilized suspensions, however, this is not possible in general [6, 7]. Especially deionized, i.e. salt-free suspensions exhibit pronounced spatial correlations even at very small ϕ, so that these systems are diluted only as far as the HI is concerned. The corresponding radial distribution function $g(r)$ has a pronounced ϕ-dependence, a well-developed first maximum, and it shows a correlation hole, i.e. a spherical region with zero probability for finding another particle, which usually extends over several particle diameters [6, 7]. In contrast, the $g(r)$ of hard spheres is nearly a unit step function $g(r) \simeq \Theta(r - 2a)$ for $\phi \leq 0.05$. Therefore, the calculation of H_s^t and H_s^r at small ϕ is more demanding for charged suspensions than for hard spheres, because for the charged particles it is necessary to use distributions functions generated from computer simulations or integral equation methods [6, 7].

We will show subsequently that it is essentially the presence of the correlation hole for charged suspensions, which causes large and interesting differences in the short-time diffusion of charged and uncharged suspensions.

Calculation of H_s^t and H_s^r

In the following, we shortly summarize the main expressions needed to calculate H_s^t and H_s^r for charge-stabilized suspensions. A more detailed description of the method used by us for the calculation of short-time diffusion coefficients is given in refs. [7, 8].

As shown in refs. [5, 9], both H_s^t and H_s^r can be measured using depolarized dynamic light scattering (DDLS) from suspensions of optically anisotropic colloidal spheres. On the time scales, which are accessible by DDLS, the theoretical expression for H_s^t is given by [3]

$$H_s^t = \frac{1}{3D_0^t} \langle \mathrm{Tr} \, \mathbf{D}_{11}^{tt} (\mathbf{r}^N) \rangle \, . \tag{3}$$

The corresponding expression for H_s^r is obtained from Eq. (3) by simply replacing the superscript "t" by "r". The hydrodynamic diffusivity tensors $\mathbf{D}_{11}^{tt}$ and $\mathbf{D}_{11}^{rr}$ relate the force/torque exerted by the solvent on an arbitrary particle 1 with its translational/angular velocity [3, 10]. Due to the many-body character of HI, both tensors depend on the particle configuration $\mathbf{r}^N = (\mathbf{r}_1, \cdots, \mathbf{r}_N)$ of all N interacting particles, and in principle the full N-particle distribution function is needed to perform the ensemble average $\langle \cdots \rangle$. $\mathrm{Tr} \, \mathbf{D}_{11}^{tt}$ denotes the sum over the diagonal elements of $\mathbf{D}_{11}^{tt}$.

For an appropriate evaluation of Eq. (3), we use a rooted cluster expansion [2, 5], which leads to a "hydrodynamic virial expansion" of H_s^t:

$$H_s^t = 1 + H_{s1}^t \phi + H_{s2}^t \phi^2 + \mathcal{O}(\phi^3) \, . \tag{4}$$

Here, the coefficient H_{s1}^t is given by an integral over the product of $g(r)$ with a translational hydrodynamic mobil-

ity function, which depends only on the distance r of two spheres [1, 2, 7, 8]. The second coefficient H_{s2}^t accounts for three-body HI. For evaluating H_{s2}^t, one needs therefore an expression for the static triplet correlation function $g^{(3)}(\mathbf{r}, \mathbf{r}')$ which appears as part of the integrand of a three-fold integral.

A similar analysis is used to calculate H_s^r, leading to results which involve now rotational hydrodynamic two-body and three-body mobility functions [4, 5, 7, 8].

The results for H_s^t and H_s^r depicted in Eqs. (1) and (2) were derived from Eq. (3) by using in H_{s1}^t and H_{s1}^r the $g(r)$ of hard spheres evaluated up to linear order in ϕ, whereas the vanishing density form of $g^{(3)}(\mathbf{r}, \mathbf{r}')$ was used in calculating the coefficients H_{s2}^t and H_{s2}^r. In these results, exact two-body HI is accounted for H_{s1}^t and H_{s1}^r, whereas only the leading long-distance contribution to the three-body mobility functions was used for the calculation of H_{s2}^t and H_{s2}^r [1, 2, 4, 5].

For charged suspensions, however, it is not possible to use in Eq. (3) low-order virial expressions of the two-body and three-body static distribution functions. In this study, we use instead results for $g(r)$, which are obtained from the rescaled mean spherical approximation (RMSA), as applied to the one-component macrofluid model of charge-stabilized colloidal suspensions [6]. The effective pair potential $u(r)$ acting between two particles is modelled by the repulsive part of the famous DLVO-potential, i.e. $\beta u(r) = K \exp[-\kappa(r - 2a)]a/r$, for $r > 2a$. Here, $\beta = (k_B T)^{-1}, K = Z^2(L_B/a)(1 + \kappa a)^{-2}, L_B = \beta e^2/\varepsilon$, and ε denotes the dielectric constant of the solvent. The screening parameter κ is given by $\kappa^2 = L_B [3|Z|\phi/a^3 + 8\pi n_s]$, where n_s is the concentration of added 1–1 electrolyte, and the counterions are assumed to be monovalent [6, 7]. Moreover, we use Kirkwood's superposition approximation for $g^{(3)}(\mathbf{r}, \mathbf{r}')$, inserting again the RMSA-$g(r)$. Further details concerning the numerical calculation of $H_{s1}^t, H_{s1}^r, H_{s2}^t$, and H_{s2}^r are given in refs. [7, 8].

Results and discussion

We focus first on the short-time diffusion coefficients of deionized charged suspensions, i.e. where $n_s = 0$. Our results for H_s^t and H_s^r are shown in Figs. 1 and 2. The used system parameters are typical for systems which have been under experimental study [11]. Obviously, the effect of HI on the self-diffusion coefficients is less pronounced for charged suspensions than for hard spheres at the same ϕ. Furthermore, we find a quite different volume fraction dependence of H_s^t and H_s^r for charged and uncharged particles. Whereas for hard spheres the ϕ-dependence of H_s^t and H_s^r is linear at small ϕ (cf. Eqs. (1) and (2)), we

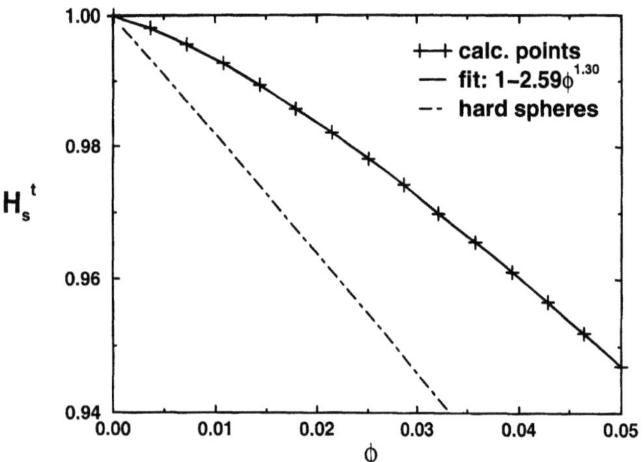

Fig. 1 Normalized short-time translational diffusion coefficient H_s^t for a deionized charged suspension with $Z = 200$, $a = 45$ nm, $T = 294$ K, and $\varepsilon = 87.0$. Also shown is the result for hard spheres according to Eq. (1)

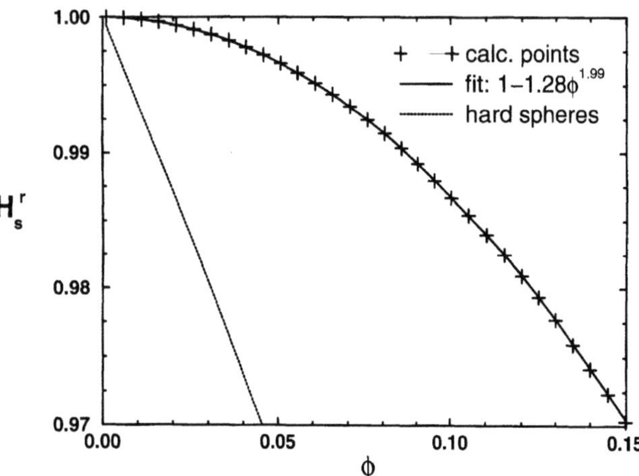

Fig. 2 Normalized short-time rotational diffusion coefficient H_s^r for a deionized suspension with system parameters as in Fig. 1. Also displayed is the result for hard spheres according to Eq. (2)

obtain from a least-squares fit of our numerical results (shown as crosses in Figs. 1 and 2) the following results for deionized charged suspensions for $0 \leq \phi \leq 0.05$ [8]:

$$H_s^t = 1 - a_t \phi^{1.30}, \quad a_t = 2.59 , \tag{5}$$

$$H_s^r = 1 - a_r \phi^{1.99}, \quad a_r = 1.28 . \tag{6}$$

The coefficients a_t and a_r are found to be nearly independent of the effective particle charge when $Z \geq 200$ [7, 8]. Note from Fig. 2 that Eq. (6) constitutes the best fit function for $H_s^r(\phi)$ even in the extended interval $0 \leq \phi \leq 0.15$. In case of H_s^t however, the parametric form

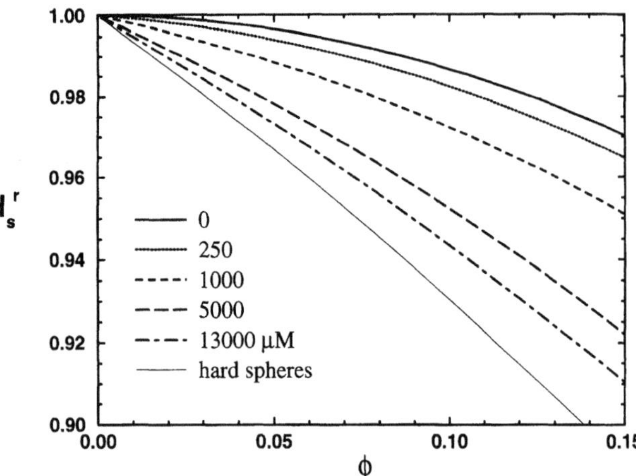

Fig. 3 Volume fraction dependence of H_s^r for various amounts of added 1–1 electrolyte, as indicated in the figure. All other system parameters as in Fig. 1

$H_s^t = a_t \phi^p$ provides no good fit for values of ϕ extending beyond 0.05 [8].

There is a simple physical explanation for the weaker influence of HI on the self-diffusion coefficients of charged suspensions as compared to uncharged ones. As already mentioned, the $g(r)$ of deionized suspensions displays a pronounced correlation hole, resulting from the strong electrostatic interparticle repulsion. Consequently, the hydrodynamic coupling between the translational or rotational motions of two spheres becomes rather small, thus giving rise to the observed weak influence of HI. Unlike charged particles, the influence particularly of the short-range part of HI is rather strong for hard sphere suspensions at small ϕ. This is due to the large probability of finding hard sphere particles at contact or close to the contact distance $r = 2a$.

Along this type of arguments, it is also possible to explain the differences of $H_s^t(\phi)$ and $H_s^r(\phi)$ in deionized suspensions and suspensions with nonvanishing n_s. We only show here the results of our calculations of $H_s^r(\phi)$ for example. From these results in Fig. 3, we notice that H_s^r becomes more and more affected by HI when n_s is increased. For very large n_s, H_s^r of charged particles approaches the result for hard sphere suspensions, obtained semianalytically in ref. [5] (cf. Eq. (2)).

This finding is easily explained by noticing that the extension of the correlation hole decreases with increasing n_s, leading to a stronger hydrodynamic coupling of the particles. Upon addition of electrolyte, the electrostatic repulsion of the particles becomes more and more screened and short-ranged, resulting in a pure hard-core repulsion for $n_s \rightarrow \infty$ [6–8]. Therefore, the microstructure of the suspension gradually transforms to that of hard spheres, with H_s^r approaching the parametric form given in Eq. (2). We further note that the radial distribution function $g(r)$ corresponding to Fig. 3 exhibits a small correlation hole even for $n_s = 13$ mM. This leads to the small differences of our results for $n_s = 13$ mM and for hard spheres in Fig. 3.

We mention that our results for H_s^t show similar trends, i.e., a gradual transformation of the ϕ-dependence of H_s^t from Eqs. (5) to (1) with increasing n_s [8].

In the remainder of this article, we focus on the qualitatively different ϕ-dependencies of H_s^t and H_s^r found in case of deionized charged and uncharged suspensions. For an intuitive physical explanation, we use an effective hard sphere model (EHS model) [6, 7], describing the actual $g(r)$ as a unit step function $g_{\text{EHS}}(r) = \Theta(r - 2a_{\text{EHS}})$. The EHS radius $a_{\text{EHS}} > a$ accounts in a crude fashion for the correlation hole, observed in the actual $g(r)$. We identify $2a_{\text{EHS}} = r_m$, where r_m is the position of the first maximum of $g(r)$. It is now crucial to notice that r_m shows an interesting scaling property when $n_s = 0$. Due to the strong electrostatic repulsion, r_m has the same ϕ-dependence as the average geometrical distance $\bar{r}$ between two spheres. Hence,

$$a_{\text{EHS}} \propto r_m \propto \bar{r} = a\sqrt[3]{4\pi/3}\,\phi^{-1/3} \,. \tag{7}$$

Using the approximation $g_{\text{EHS}}(r)$ of $g(r)$, it is easy to calculate the coefficients H_{s1}^t and H_{s1}^r in an approximative way. By using far-field expansions of the hydrodynamic two-body mobility functions [7, 8, 10], one obtains the following results from the leading terms of these expansions:

$$H_{s1}^t = -\frac{15}{8}\left(\frac{a}{a_{\text{EHS}}}\right) + \mathcal{O}(a_{\text{EHS}}^{-3}) \,, \tag{8}$$

$$H_{s1}^r = -\frac{5}{16}\left(\frac{a}{a_{\text{EHS}}}\right)^3 + \mathcal{O}(a_{\text{EHS}}^{-5}) \,. \tag{9}$$

This leads together with Eqs. (4) and (7) to the expressions

$$H_s^t = 1 - A^t\phi^{4/3} + \mathcal{O}(\phi^2), \quad A^t > 0 \,, \tag{10}$$

$$H_s^r = 1 - A^r\phi^2 + \mathcal{O}(\phi^{8/3}), \quad A^r > 0 \,, \tag{11}$$

with exponents which are in good agreement with our numerical findings given in Figs. 1 and 2 (cf. Eqs. (5) and (6)).

Therefore, we have shown by a simple analytic calculation based on the EHS model that the observed differences in the functional forms of $H_s^t(\phi)$ and $H_s^r(\phi)$ between charged and uncharged suspensions are mainly caused by the leading terms of the hydrodynamic two-body mobility functions in combination with the scaling property

$r_m \propto \phi^{-1/3}$, valid for deionized suspensions. The higher-order terms in the hydrodynamic far-field expansions only give rise to minor corrections to the observed scaling properties depicted in Eqs. (5) and (6). These terms become increasingly important for larger volume fractions $\phi \geq 0.05$ (cf. Eqs. (10) and (11) in the EHS model).

When electrolyte is added to the suspension, Eq. (7) becomes invalid because of the enhanced screening of the direct particle interactions. This causes a change in the functional behaviour of $H_s^t(\phi)$ and $H_s^r(\phi)$, as can be seen both from the EHS model and from our numerical results (cf. Fig. 3 in case of H_s^r) [7, 8].

Using the EHS model, it is also possible the motivate the nearly Z-independence of a_t and a_r in Eqs. (5) and (6). Since r_m is nearly independent of Z for $Z \geq 200$, the EHS model predicts charge independent results for the short-time diffusion coefficients of deionized suspensions, in agreement with our numerical results.

We mention that it is also possible to deal with H_{s2}^t and H_{s2}^r within the EHS model, giving further insight in the volume fraction dependence of the diffusion coefficients of deionized suspensions [7, 8]. It is then possible to explain qualitatively the surprising fact that $H_s^r(\phi)$ is well parametrized up to $\phi = 0.15$ by the functional form $H_s^r = 1 - a_r\phi^2$, obtained in the EHS model by using only the leading term in the far-field expansion of the rotational two-body mobility functions [7, 8].

Conclusion

We have presented calculations of the translational and rotational short-time self-diffusion coefficients for charge-stabilized suspensions. The self-diffusion coefficients of charged suspensions are less affected by hydrodynamic interactions than the corresponding coefficients of hard spheres. As a major result we have found substantially different volume fraction dependencies of H_s^t and H_s^r for (deionized) charged and uncharged suspensions. The observed differences are well explained in terms of an effective hard sphere model by observing the big differences in the microstructure of suspensions of charged and uncharged particles.

We note finally that recent DDLS measurements of H_s^r in deionized suspensions of charged fluorinated polymer particles compare favourably with our results in Eq. (6) [11]. On the other hand, to our knowledge, no experimental data of H_s^t for deionized charge-stabilized suspensions are accessible so far. We further point out that the interesting qualitative differences between charge-stabilized suspensions and hard spheres exist also with respect to sedimentation [12] and long-time self-diffusion [13].

References

1. Cichocki B, Felderhof BU (1988) J Chem Phys 89:1049
2. Jones RB (1988) Physica A 150:339
3. Jones RB, Pusey PN (1991) Annu Rev Phys Chem 42:137
4. Beenakker CWJ, Mazur P (1983) Physica A 120:388
5. Degiorgio V, Piazza R, Jones RB (1995) Phys Rev E 52:2707
6. Nägele G (1996) Phys Rep 272:215
7. Watzlawek M, Nägele G, Physica A 235, to appear
8. Watzlawek M, Nägele G, University of Konstanz, submitted
9. Degiorgio V, Piazza R, Bellini T (1994) Adv Coll Int Sci 48:61
10. Jones RB, Schmitz R (1988) Physica A 149:373
11. Bitzer F, Palberg T, Leiderer P, University of Konstanz, private communication
12. Thies-Weessie DME, Philipse AP, Nägele G, Mandl B, Klein R (1995) J Coll Int Sci 176:43
13. Nägele G, Baur P, University of Konstanz, submitted

Progr Colloid Polym Sci (1997) 104:173–176
© Steinkopff Verlag 1997

E. Dubois
V. Cabuil
F. Boué
J.C. Bacri
R. Perzynski

Phase transitions in magnetic fluids

E. Dubois · Dr. V. Cabuil (✉)
Laboratoire de Physicochimie Inorganique
Case 63
4 place Jussieu
75252 Paris Cedex 05, France

F. Boué
Laboratoire Léon Brillouin
CNRS-CEA
CE Saclay
91191 Gif-Sur-Yvette Cedex, France

J.C. Bacri · R. Perzynski
Laboratoire Acoustique et Optique
de la Matière Condensée
URA CNRS 800
Case 78
4 place Jussieu
75252 Paris Cedex 05, France

Abstract Experimental results concerning phase transitions in magnetic fluids are presented. Magnetic fluids are colloidal dispersions of nanometric magnetic particles which are stabilized either through electrostatic repulsions between charged particles in water, or through steric repulsions between surfactant coated particles in oil. In the both cases, fluids are synthesized in order to have repulsions which strongly dominate the interparticles interactions balance. SANS experiments performed on aqueous magnetic fluids show that the system is strongly repulsive and not sensitive to a decrease of temperature. In this case, phase transitions are not observed in a reasonable range of temperature or of applied magnetic field. In aqueous systems, SANS measurements show that repulsions are gradually screened by addition of salt, and "gas–liquid"-like transitions are observed decreasing the temperature. The temperature threshold is related to a screening parameter taking into account the ionic strength and the particles diameter. The same kind of behavior is observed for oily systems: steric repulsions have to be decreased through addition of a bad solvent of the surfactant chains in order to observe "gas–liquid" transitions if temperature is decreased.

Key words Magnetic fluids – phase transitions – colloidal dispersions – neutron scattering

Introduction

Magnetic fluids are colloidal dispersions of nanometric magnetic particles in a liquid [1]. Each particle is a monodomain, and thus a permanent magnetic dipole [2]. Stability of these dispersions depends on the balance of the different interactions between particles and is very important as soon as magnetic fluids are used for technical applications [3]. In such dispersions, "gas–liquid"-like transitions are observed if parameters as temperature or intensity of an applied magnetic field are modified. The composition of the liquid carrier monitors the range of stability of the colloid [4–6]. Experimental determination of phase transition thresholds for chemically synthesized ferrofluids [7] is presented and connected to neutron scattering experiments in the monophasic state. Electrostatically stabilized colloids will be mainly considered, but preliminary results on sterically stabilized suspensions will be introduced.

Samples

The colloids used here are maghemite nanoparticles (γ-Fe_2O_3) obtained through a chemical synthesis [7]. The process leads to polydisperse particles. Their size

Table 1 A Size characteristics of the samples as determined by magnetization measurements; A, B, C, D, E are dispersed in water, F in cyclohexane. B For sample B, comparison of SANS size measurements to their calculated values from distribution of 1A

A							B		
Name	A	B	C	D	E	F	B	SANS	Calculated
d_0 [nm]	6.2	7.1	9.1	9.2	12	7	d_w [nm]	8.3	7.8
σ	0.15	0.15	0.2	0.15	0.15	0.2	R_g [nm]	3.6	4.1

distribution is well described by a lognormal law, characterized by two parameters: d_0 and σ[8], typically $d_0 = 7$ nm and $\sigma = 0.4$. The size distribution is narrowed by a size sorting process described elsewhere [6]. The characteristics of the particles used in this work are given in Table 1A. Two kinds of dispersions are synthesized: aqueous or oily dispersions. In aqueous dispersions at a pH value of 7, particles are coated by citrate ions which ensure negative surface charges and electrostatic repulsions [9]. Surface charge density is constant with particles size and equal to 1.4 charge/nm² which corresponds to 0.22 C/m². In oily dispersions, particles are coated with oleic acid ensuring steric interparticle repulsions.

Experiments

Small-angle neutron scattering experiments (SANS) are performed for aqueous magnetic fluids on PAXE spectrometer in the LLB (CEN Saclay, France). The wavelength is 10 Å and the scattering vector q ranges from 8×10^{-3} to 7×10^{-2} Å^{-1}. The intensity I can be written as $I(q) = (b_1 - b_2)^2 F(q) g(q)$, where $b_1 = 6.96 \times 10^{10}$ and $b_2 = -0.53 \times 10^{10}$ cm^{-2} are the densities of the diffusion lengths for maghemite and water, Φ is the volume fraction of particles, $F(q)$ is the form factor and $g(q)$ the interparticle correlation function. For low particles concentrations, $g(q) \to 1$, the spectra give the form factor and allow size determination. Two kinds of particle sizes may be deduced: a weight-averaged diameter $d_w = d_0 \exp(4.5\sigma^2)$ and a radius of gyration $R_g = 0.5d_0 \exp(7\sigma^2)$ [10]. At higher concentrations, spectra reflect the interparticle interactions.

Phase transition thresholds are determined for aqueous and oily systems by an optical microscope observation. In water, an increase of ionic strength (addition of NaCl) or a decrease of temperature T induces transitions [6, 11]. In cyclohexane, an addition of ethylalcohol or a decrease of T induces transitions. The magnetic fluid is observed in a cell 100 μm thick. Its temperature can be monitored between $-15°$ and 40 °C. In both systems, the threshold corresponds to the nucleation of small black droplets of a dense liquid phase, dispersed in a dilute phase.

Results

Aqueous dispersions

SANS experiments are performed on sample B in the monophasic state. Figure 1A shows the intensity versus q for three different initial volume fractions of particles. At

Fig. 1 Neutron scattering experiments on sample B. (a) $I = f(q)$ at [NaCl] = 0 for several volume fractions. (+) $\Phi = 0.01$, ($\times$) $\Phi = 0.04$, and $\Phi = 0.11$ at [NaCl] = 0 and different temperatures, ($\diamond$) $T = 8$ °C, ($\square$) $T = 25$ °C, ($\bullet$) $T = 30$ °C. (b) $\Phi = 0.11$: – at [NaCl] = 0, at different temperatures: ($\diamond$) $T = 8$ °C, ($\square$) $T = 25$ °C, ($\bullet$) $T = 30$ °C, – for several NaCl concentrations at $T = 25$ °C: ($\times$) [NaCl] = 0.11 M, (+) [NaCl] = 0.22 M ($\triangle$) [NaCl] = 0.37 M. At 25 °C, the macroscopic transition threshold is around 0.4 mol/L ($\kappa d_0 \sim 19.5$) and the sample with [NaCl] = 0.37 mol/L separates between 17 and 15 °C

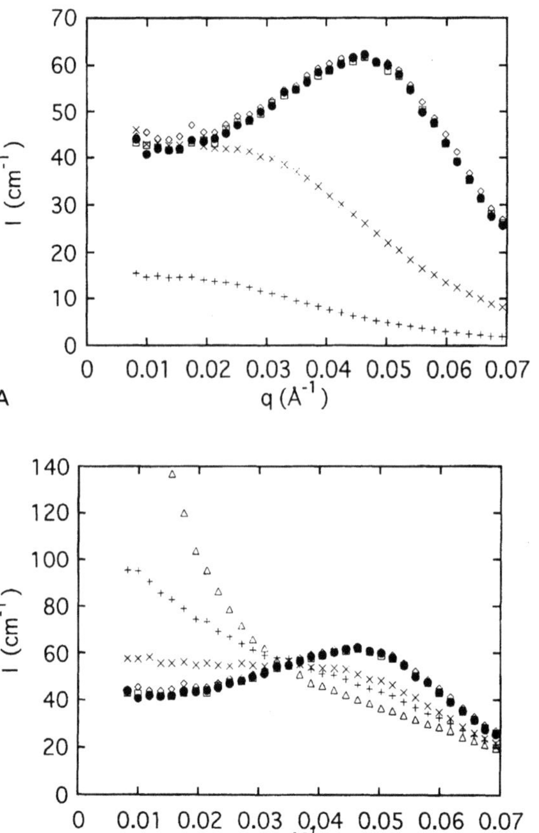

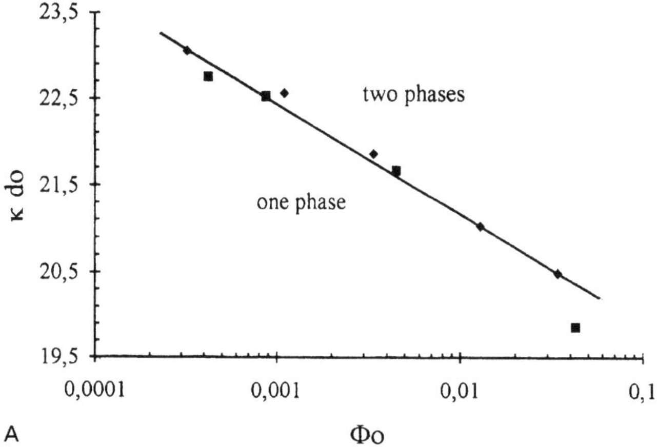

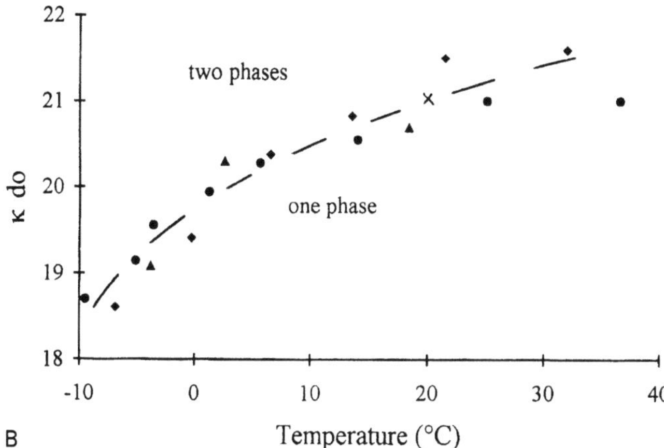

Fig. 2 A κd_0 versus Φ_0 for samples B (■) and C (♦) at room temperature; B) κd_0 versus temperature for samples A (●), D (▲), E (♦) and C (×) at $\Phi_0 = 0.01$

$\Phi = 1\%$, interactions are negligible: particle sizes determinations are given in Table 1B. At higher concentrations, $I(q = 0)$ expresses a strongly repulsive regime characterized by

$$\frac{\partial}{\partial \Phi}\left(\Phi\frac{\partial \mu}{\partial \Phi}\right) \approx 20 k_B T \ ,$$

where μ is the chemical potential of the solution. At $\Phi = 11\%$, the maximum of $I(q)$ is related to the strength of interparticles electrostatic repulsions. q_{max} corresponds to a mean distance $d = 14$ nm between particles, in good agreement with $d = d_0/\sqrt[3]{\Phi_0} = 15$ nm. At low ionic strength, this strong repulsion is not modified by a decrease of temperature as shown in Fig. 1A. As NaCl is added, the maximum of $I(q)$ disappears and the intensity at small q increases (Fig. 1B): electrostatic repulsions are progressively screened.

Phase transition thresholds are determined for particles of Table 1 dispersed in water at several volume fractions. Figure 2A concerns experiments at room

Table 2 Temperature thresholds of samples F ($\Phi = 0.014$) for several percentages of ethylalcohol. The percentage is the ratio of the ethylalcohol volume over the total volume

% Ethylalcohol	24.3	26.6	28
Temperature	$< -3\,°C$	$10\,°C$	$22\,°C$

temperature, the repulsions being screened through addition of NaCl. Figure 2B gives the ionic strength thresholds as a function of temperature. The relevant reduced parameter is κd_0, where κ^{-1} is the Debye length characteristic of the screening of electrostatic repulsions.

Cyclohexane magnetic fluids

As it is observed for aqueous systems, if repulsions are efficient, these ferrofluids are stable even at low temperature or if a magnetic field is applied; it is thus again necessary to decrease the repulsions to observe phase transitions in reasonable ranges of temperature and magnetic field. This decrease may be performed by addition of non-adsorbing polymers which modifies the interaction balance, introducing attractive depletion forces between particles [5, 12]. Another possibility is to add to the solution a bad solvent of the surfactant chains ensuring the steric repulsions. Here ethylalcohol is choosen as a bad solvent for oleic acid chain. Thresholds are given for sample F in Table 2.

Discussion and conclusion

These results point out the similar behavior of aqueous and oily magnetic fluids allowing a general description of the phase behavior of these systems. The SANS measurements and the determination of thresholds bring to the fore the fact that it is necessary to decrease the strong interparticle repulsions of the colloidal system to be able to induce a phase separation by applying a magnetic field or lowering the temperature. The way of decreasing repulsions depends on the nature of particles stabilization. For ionic particles, an increase of ionic strength screens electrostatic repulsions; for particles coated with surfactant chains, a depletion effect or a modification of the solvent quality is necessary. In electrostatic systems, the relevant parameter of the problem, firstly introduced by Victor and Hansen [13] in their theoretical work, is κd_0. It normalizes the electrostatic screening to the particle size d_0, which appears to be a parameter of paramount experimental importance. For the particles sizes investigated here and in a zero magnetic field, dipolar anisotropic interactions are negligible compared to the isotropic ones (Van der Waals and electrostatic repulsions). Oily magnetic fluids have to be studied in more detail. In future, the behavior of large particles at high-volume fraction under field will be investigated to explore the dipolar aspects of the system.

Acknowledgments This work has been supported by DRET (DGA).

References

1. Rosensweig R (1982) Pour la Science 104
2. Rosensweig R (1985) Ferrohydrodynamics. Cambridge University Press, Cambridge
3. Massart R, Bacri JC, Perzynski R (1995) Techniques de l'ingénieur D2180:1
4. Bacri JC, Perzynski R, Salin D, Cabuil V, Massart R (1989) J Colloid Int Sci 132:43
5. Cabuil V, Perzynski R, Bastide J (1994) Prog Colloid Polymer Sci 97:75
6. Massart R, Dubois E, Cabuil V, Hasmonay E (1995) J Magn Magn Mat 149:1
7. Massart R, IEEE Trans Magn (1981) MAG-17:1247
8. Cabuil V, Perzynski R (1996) In: Berkovski (ed) Magnetic Fluids and Applications Handbook Begell house inc, New York, p 14
9. Bacri JC, Perzynski R, Salin D, Cabuil V, Massart R (1990) J Magn Magn Mat 85:27–32
10. Bacri JC, Boué F, Cabuil V, Perzynski R (1993) Colloids Surf A 80: 11–18
11. Dubois E, Cabuil V, Perzynski R, Bacri JC, to be published
12. Lekkerkerker HNW, Dhont JKG, Verduin H, Smits C, van Duijneveldt JS (1995) Physica A 213:18
13. Victor JM, Hansen JP (1985) J Chem Soc Faraday Trans 2 81:43–61

Progr Colloid Polym Sci (1997) 104:177-179
© Steinkopff Verlag 1997

H. Graf
H. Löwen
M. Schmidt

Cell theory for the phase diagram of hard spherocylinders

H. Graf · H. Löwen (✉) · M. Schmidt
Institut für Theoretische Physik II
Heinrich-Heine-Universität Düsseldorf
Universitätsstraße 1
40225 Düsseldorf, Germany

H. Löwen
Institut für Festköperforschung
Forschungszentrum Jülich
52425 Jülich, Germany

Abstract A cell theory is proposed to obtain the full phase diagram of hard spherocylinders involving isotropic, nematic, smectic as well as plastic and aligned crystalline phases. Despite its conceptual and numerical simplicity this free-volume theory yields the correct topology of the phase diagram in semi-quantitative agreement with recent computer simulations.

Key words Hard spherocylinders – phase transitions – cell theory

Hard spherocylinders with an orientational degree of freedom represent a standard model for liquid crystals [1] which has the advantage to be simple since the only parameters characterizing the system are the number density ρ and the aspect ratio $p = L/D$. Here, L is the length of the cylindrical part and D is the sphere and cylinder diameter. The density can suitably be scaled by the closed packed density ρ_{CP} leading to the dimensionless quantity $\rho^* = \rho/\rho_{CP}$, $0 \leq \rho^* \leq 1$. A further advantage is that, for $p = 0$, the well-known system of (isotropic) spheres is recovered and that the opposite limit $p \to \infty$ leads to the analytical Onsager solution for the isotropic-nematic transition. Recently, the full phase diagram in the whole parameter space spanned by ρ^* and p was obtained by computer simulation [2]; parts of it were known from former work [3, 4]. The resulting phase diagram has a pretty rich topology involving isotropic, nematic, smectic and plastic crystalline phases as well as aligned crystals with different stacking sequences.

The aim of this paper is to propose a cell theory for the full phase diagram of hard spherocylinders. This theory is similar in spirit to the free-volume approach of ref. [5] where a system of completely aligned spherocylinders was studied. However, as an essential part of our theory, we also incorporate the orientational degrees of freedom by assuming an effective shape of the particles gained from an orientational average. Our theory gives, for the first time,

a stable plastic crystal (rotator solid) and an AAA-stacked solid whose stabilities were not investigated in previous density functional calculations [6]. It has the further advantage of being conceptionally and numerically simpler than sophisticated and basically uncontrolled density functional approximations. Surprisingly, our cell approach gives the correct topology of the phase diagram in semi-quantitative agreement with the simulation data [2].

Let us now briefly describe our theory. In order to locate phase coexistence we have to know the Helmholtz free energy f per particle in any phase which is under consideration. For the *isotropic* (fluid) phase we use the well-known analytical expression of scaled particle theory [7]. In the remaining phases we approximately split the system into cells containing one particle. The cell size is chosen in such a way that the particles do not feel the interaction with their neighbours. In this case, the total free energy f splits naturally into a part f_{rot} stemming from the orientational degrees of freedom and another part f_{cm} resulting from the motion of the centre-of-mass coordinate: $f = f_{rot} + f_{cm}$.

Calculating f_{rot} we consider the angular distribution function $g(\omega)$ which is rotational symmetric around a fixed director ω_0. Here, ω and ω_0 are unit vectors. Particularly, we use a Maier–Saupe form for $g(\omega)$, i.e., $g(\omega) = (1/\mathcal{N})\exp[\alpha P_2(\omega \cdot \omega_0)]$, where $P_2(x) = (3x^2 - 1)/2$ is the second Legendre polynomial, $\mathcal{N}$ is to guarantee correct normalization $\int f(\omega)\,d^2\omega = 1$ and α is a variational parameter which is yet to be determined. Then the rotational free energy per particle reads $f_{rot} = \int d^2\omega\, g(\omega)\ln[g(\omega)]$.

The key idea for calculating f_{cm} is to map the system onto *completely oriented* particles with an *effective shape* which in general differs from the original spherocylindric shape. This effective shape should depend on the angular distribution function $g(\omega)$ in order to take orientational correlations between nearest neighbours roughly into account. We then estimate the centre-of-mass part f_{cm} by assuming a cell model for the solid part and a scaled-particle approach for the fluid part of the different phases [5] in this substitute system. Finally, the choice of effective shape is optimized by choosing the variational parameter α such that the total free energy f becomes minimal with respect to α.

Let us now describe the mapping in more detail. We define a mean distance $\bar{R}$ by averaging over the orientational distribution function g as

$$\bar{R}(\theta, [f]) = D/2 + L/2 \int d^2\omega'\, g(\omega')|\omega\cdot\omega'|. \tag{1}$$

Clearly, $\bar{R}$ is a functional of g and only depends on the angle θ defined via $\cos\theta = \omega_0\cdot\omega$. The actual effective shape is now obtained as a Legendre transform of $\bar{R}$ with respect to θ. Equivalently, the effective shape is the envelope of all planes which have a distance $\bar{R}(\theta, [f])$ from the centre-of-mass of the spherocylinder and whose normal forms an angle θ with the director ω_0. It is instructive to consider two special cases: For fully aligned spherocylinders, we have $g(\omega) = \delta(\omega - \omega_0)$ and the effective shape coincides with the original spherocylindric shape. Second, for an isotropic distribution $g(\omega) = 1/4\pi$, we get $\bar{R} = D/2 + L/4 = $ const. The Legendre transform is again a sphere of the same radius $R_{mean} = D/2 + L/4$ and the resulting effective shape is a sphere of radius R_{mean}. Note that – by construction – this is the *mean radius* R_{mean} of the spherocylinder which is obviously smaller than its *maximal* radius, $R_{max} = D/2 + L/2$. Therefore, the volume corresponding to the effective shape is in general smaller than $4\pi R_{max}^3/3$ as resulting from a fully rotated spherocylinder, hence taking roughly orientation correlation effects between nearest neighbour into account.

We finally propose a variational principle for the free energy. In the original version of the cell model [8], the free energy per particle is bounded from above by $-k_B T\ln(V_{cell})$ where V_{cell} is the free cell volume in suitable units. Hence, the best upper bound is achieved by maximizing V_{cell} within the given constraints of fixed average density and geometry [9]. Of course, since we invoke scaled particle approximations and work with an ef-

fective shape, our results are no longer true upper bounds to the exact free energy. As a further approximation, we carry over the variational principle in order to optimize the actual free energy values in the different phases.

Let us now describe the theoretical treatment of the different phases in more detail. The *plastic crystal* is a rotator solid with positional but without orientational order. In this case, the distribution function $g(\omega)$ is constant. As we demonstrated above, the effective system consists of spheres of radius R_{mean} and the cell theory reduces to that of spheres. The *aligned solid with* ABC *stacking* has both positional and orientational order, exhibiting a distorted fcc-structure. The free-volume cell is estimated to be a rhombic dodecahedron corresponding to an undistorted fcc hard sphere crystal. In the *aligned solid with* AAA *stacking*, on the other hand, the free-volume cell is a hexagonal prism. In the *smectic*-A *phase* the free energy splits into a contribution of a one-dimensional cell model for a rod of length $2\bar{R}(\theta = 0, [g])$ and the scaled particle contribution of a two-dimensional liquid of hard discs with diameter $2\bar{R}(\theta = \pi/2, [g])$. Note that the total free energy is then minimized with respect to the layer spacing and the orientation distribution. Finally, the scaled particle free energy of parallel hard spherocylinders [10] of diameter $D_n^* = 2\bar{R}(\theta = \pi/2, [g])$ and length $L_n^* = 2\bar{R}(\theta = 0, [f]) - D_n^*$ is used to calculate f_{cm} of the *nematic phase*. In order to compensate lack of configuration space in the cell approach we add a constant of $-1.8k_B T$ to

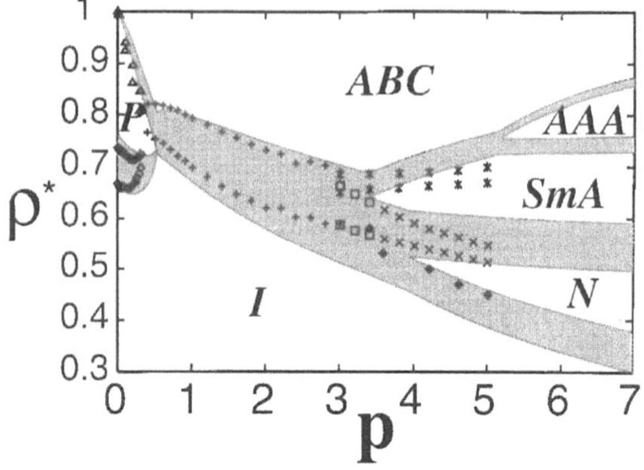

Fig. 1 Phase diagram of hard spherocylinders in the $(p-\rho^*)$-plane. The shaded area is the coexistence region calculated within our theory and the dots are the simulation data [2]. There is an aligned ABC-solid, an aligned AAA-solid, a plastic crystal (P), an isotropic fluid (I) and a nematic (N) and smectic-A (SmA) phase. The meaning of the symbols for the simulation data are: (+) I-ABC transition, (◇) I-P transition, (□) I-SmA transition, (♦) I-N transition, (×) N-SmA transition, (·) SmA-ABC transition, (▲) P-ABC transition

f_{cm} in the plastic crystal and in the aligned solids and consistently a third of this constant in the smectic-A phase. A second constant of $-2.25k_B T$ is added in the nematic phase. These two constants can be regarded as fitting parameters.

The results for the phase diagram are shown in Fig. 1 in the range of $0 \leq p \leq 7$ and compared against the simulation data [2]. The theory predicts correctly all the stabilities of the different phases, exhibits all qualitative trends and is in semiquantitative agreement with the exact data although the density jumps are overestimated.

We are at present calculating the phase diagram of spherocylinders in an external field coupling to the orientational degrees of freedom. Here, we obtained critical points between phases of same geometry.

References

1. Frenkel D (1991) In: Hansen JP, Levesque D, Zinn-Justin J (eds) liquids Freezing and Glass Transition. North-Holland, Amsterdam; Vroege GJ, Lekkerkerker HNW (1992) Rep Prog Phys 55:1241; Löwen H (1994) Phys Rev E 50:1232
2. Bolhuis PG (1996) PhD thesis, University of Utrecht. Bolhuis P, Frenkel D, to be published
3. Veerman JAC, Frenkel D (1990) Phys Rev A 41:3237
4. McGrother SC, Williamson DC, Jackson G (1996) J Chem Phys 104:6755
5. Taylor MP, Hentschke R, Herzfeld J (1989) Phys Rev Lett 62:800; Sharlow MF, Selinger RLB, Ben-Shaul A, Gelbart WM (1995) J Phys Chem 99:2907
6. Poniewierski A, Holyst R (1990) Phys Rev A 41:6871
7. Barker JA, Henderson D (1976) Rev Mod Phys 48:587
8. Kirkwood JG (1950) J Chem Phys 18:380; Wood WW (1952) J Chem Phys 20:1334
9. Löwen H, Schmidt M, this issue; Schmidt M, Löwen H (1996) Phys Rev Lett 76:4552
10. Cotter MA, Martire DE (1970) J Chem Phys 52:1909

Progr Colloid Polym Sci (1997) 104:180–182
© Steinkopff Verlag 1997

S.M. Clarke
A.R. Rennie

Structures of spherical particles dispersed in density matched media under oscillatory shear

S.M. Clarke · Dr. A.R. Rennie (✉)
Polymers and Colloids Group
Cavendish Laboratory
Madingley Road
Cambridge, CB3 0HE, United Kingdom

Abstract Small-angle light scattering has been used to investigate the structure of sterically stabilised colloidal particles under oscillatory shear. Strain amplitude is significant and evidence for 'string' phases under certain conditions is presented.

Key words Colloids – oscillatory shear – light scattering

Introduction

X-ray, neutron and light scattering have been employed to investigate the structure of particulate dispersions under shear [1, 2]. This information can be used to interpret rheological measurements or test predictions of theoretical models of flow. Scattering work has mostly used dispersions of spherical particles, both charge and sterically stabilised, under steady shear. Various structures for sterically stabilised dispersions under shear have been discussed. These include distorted liquids, sliding layers of particles and 'strings' in which the particles move in lines along the flow direction [2]. Recently, anisotropic particles have also been described [3, 4]. The effects of oscillatory shear have however received less attention [5, 6].

This paper presents small-angle light scattering data from a dispersion (volume fraction 0.4) of 1.5 μm diameter polymethylmethacrylate (PMMA) spherical particles sterically stabilised [7] in a mixture of cycloheptyl bromide and Decalin under oscillatory shear. The particles are suspended in a media that is almost matched in refractive index to the particles. This method allows the bulk of concentrated samples to be investigated without multiple scattering. In addition a mixture of solvents was chosen such that the suspending medium matches the density of the particles. Density matching avoids effects which arise due to sedimentation or creaming of the particles [8] and implies that the volume fraction is equal to the weight fraction.

The experimental apparatus is described elsewhere [9, 10]. The essential features include, a collimated laser beam incident on the sample in a plate–plate shear cell, the scattered light is diverted with a number of optical components onto a charged coupled device detector. This type of detector has the advantages of rapid, two-dimensional data collection and provides numerical values for subsequent quantitative analysis of the intensity. The oscillatory shear was applied by turning the bottom plate. The acceleration of the moving plate is fast such that over most of the cycle, the velocity, and therefore the strain rate, $\dot{\gamma}$, is constant. The strain rate profile is a square wave. To allow comparison of results with other materials and models, the shear rate is expressed in terms of the Peclet number, Pe, which is defined as $Pe = 6\pi\dot{\gamma}\eta a^3/8kT$, where η is the viscosity and a the particle diameter. Strain amplitude and rate can both be varied in this apparatus.

Results

Figure 1a presents the scattering patterns from the dispersion of PMMA particles in the plate–plate cell under static conditions. This pattern has a beam stop in the centre to prevent saturation of the detector by the primary beam. The ring of intensity in this pattern is characteristic of a liquid-like structure. Figure 1b presents the scattering pattern from the same dispersion as Fig. 1a under an oscillatory shear with a strain amplitude of 1.6 and a

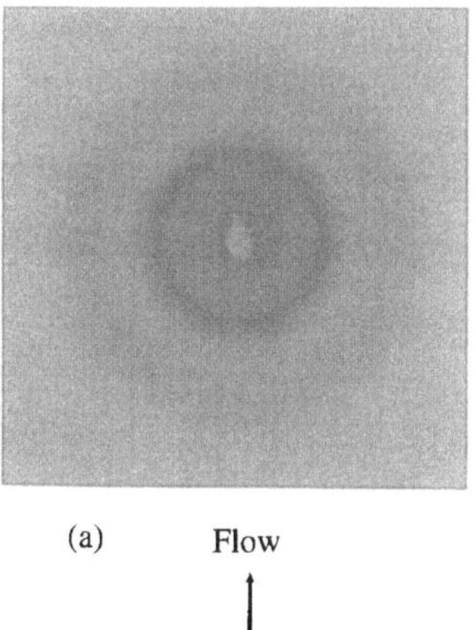

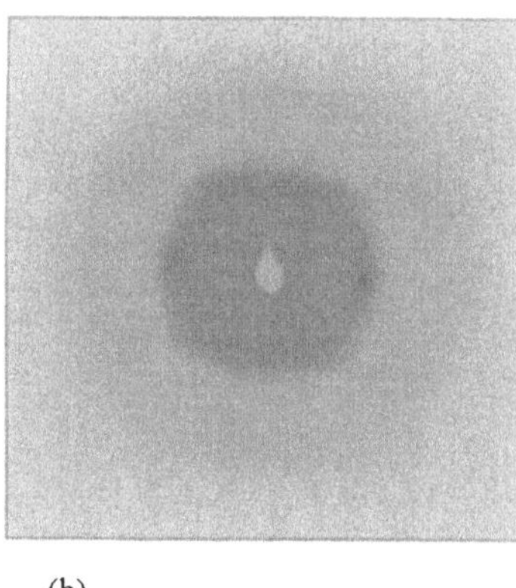

Fig. 1 Small-angle light scattering patterns from a dispersion of PMMA particles in a plate–plate shear cell under **A** static conditions and **B** oscillatory shear with a strain amplitude of 1.6 and a strain rate corresponding to a Peclet number of 170. The flow and vorticity directions are indicated in the figure

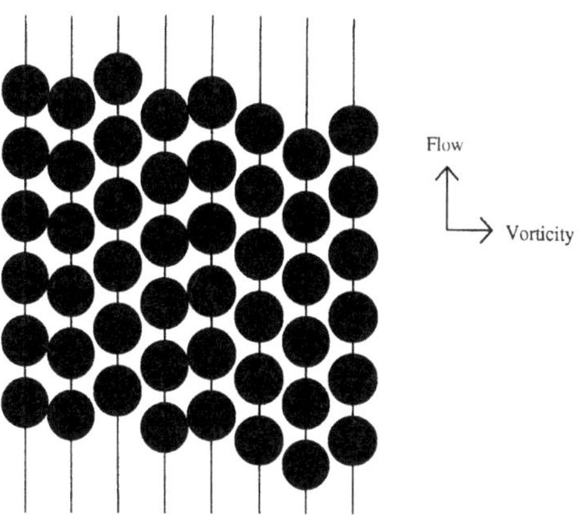

Fig. 2 Schematic illustration of the "string" phase showing lines or strings of particles. The flow and vorticity directions are indicated in this figure. A description of this structure is given in the text

Peclet number of 170. This pattern consists of lines across the detector in the vorticity direction and a series of spots in the vorticity direction on the vorticity axis and is characteristic of a "string" phase.

A quantitative discussion of the liquid-like structure of Fig. 1a in terms of structure factors for hard spheres [11] has been given previously [10]. The scattering from lines of particles which constitute the "string" phase has been described [12]. The lines across the pattern (Fig. 1b) correspond to a characteristic spacing of particles within each of the strings. The spots, on the vorticity axis in the pattern, arise from a characteristic spacing between the strings normal to the axis of the strings.

Quantitative calculations of the structure giving rise to this pattern will be presented elsewhere [13]. A schematic illustration of the structure is given in Fig. 2. The separation of particles along each string is 1.7 μm and the separation of the strings is 1.4 μm. The density of the particles in this structure constrains the packing, however, there can exist only weak correlations between positions of particles in neighbouring strings. Extensive order of strings relative to each other would give rise to modulations of intensity along the lines in the scattering pattern that are not observed.

Discussion and conclusions

Small angle light scattering from dispersions of spherical particles in a density matched media has been shown to provide evidence for structures induced by oscillatory shear. In this contribution the interpretation of an experimental scattering pattern is demonstrated for one flow geometry. "String" phases have been described in simulations, mainly under conditions of continuous shear and without interparticle hydrodynamic interactions [14]. The present and other recent work [10] indicate that the string phase is not found for these materials under continous shear but only under particular conditions of oscillatory motion.

References

1. Clarke SM, Rennie AR (1996) Current Opinion in Colloid and Interface Science 1:34

2. Ackerson BJ (1996) Current Opinion in Colloid and Interface Science 1:450

3. Clarke SM, Convert P, Rennie AR (1996) Europhys Letts 35:233

4. Ramsay JDF, Lindner P (1994) J Chem Soc Farad Trans 90:2001
5. Ackerson BJ (1990) J Rheol 34:553
6. Yan YD, Dhont JKG, Smits C, Lekkerkerker HNW (1994) Physica A 202:68
7. Antl L, Goodwin JW, Hill RD, Ottewill RH, Owens SM, Papworth S (1986) Colloid Surf 17:67
8. Clarke SM, Rennie AR (1996) Farad Discuss 104 in press
9. Tromp RH, Rennie AR, Jones RAL (1995) Macromolecules 28:4129
10. Clarke SM, Ottewill RH, Rennie AR (1995) Adv Colloid Interface Sci 60:95
11. Ashcroft NW, Lekner J (1966) Phys Rev 145:83
12. Hosemann R (1954) Naturwiss 19:440
13. Clarke SM, Rennie AR, in preparation
14. Brady JF (1996) Current Opinion in Colloid and Interface Science 1:472

Progr Colloid Polym Sci (1997) 104:183–186
© Steinkopff Verlag 1997

J. Vogel

Is there scaling of the intermediate scattering function without hydrodynamic interactions? Brownian rods on a 1d-ring

This work has been carried out as part of the IICM-Programme funded by the Commission of the European Community

Dr. J. Vogel (✉)
Department of Physics and Astronomy
James Clerk Maxwell Building
The King's Buildings
The University of Edinburgh
Edinburgh EH9 3JZ, United Kingdom

Abstract With the help of a Monte Carlo simulation of a one-dimensional system of hard rods on a ring we calculated the wave vector and time-dependent intermediate scattering function $f(k, t)$. We were looking for a scaling behavior for this function like the one found in recent light scattering experiments by P.N. Segrè and P.N. Pusey in dense hard sphere systems. For high densities we do not find perfect scaling but a collapse of the various large scattering vector functions $f(k, t)$ around the peak of the structure factor to a narrow range in time.

Key words Scaling – intermediate scattering function – Brownian hard rod dispersion

Introduction

Recently, an interesting scaling for the intermediate scattering function $f(k, t)$ [1] of a dense colloidal dispersion of hard spheres of radius a was found [2]. Plotting $\ln f(k, t)$ divided by its initial slope $D_S(k)k^2$ as a function of time t for various scattering vectors k shows that all curves beyond $ka = 2.5$ fall upon each other not only for short times but at all times. Here $D_S(k) = D(k, t \to 0)$ is the short-time limit of the collective diffusion coefficient $D(k, t)$,

$$D(k, t) = -\frac{\partial \ln f(k, t)}{k^2 \partial t} \ . \tag{1.1}$$

This is surprising since in this regime of wave vectors the static structure factor $S(k)$ shows its strongest variations. It also implies that the short-time and long-time diffusion constants $D(k, t \to \infty)$ only differ by a constant factor, independent of the scattering vector. The underlying physical picture to this surprising result is still unclear but it is argued that the local cage of neighboring particles around a given particle reduces its diffusive behavior essentially to self diffusion.

We like to understand the physical picture which leads to this scaling behavior. Is hydrodynamic interaction between the particles essential to see it? Is it just an effect of the surrounding cage of particles?

To clarify these questions we perform a Monte Carlo simulation of a system of hard rods of length $2a$ on a one-dimensional ring of length L. Hydrodynamic interaction between the Brownian particles was omitted. If the scaling exists then some hints should remain even in one dimension. For the simulation we use a method proposed by Cichocki and Hinsen [3]. At each Monte Carlo step one particle i, $i = 1, \ldots, N$, was picked randomly and moved by a distance δx drawn from a Gaussian distribution of variance $\langle \delta x\, \delta x \rangle_{\Delta t} = 2D_0 N \Delta t$. Here D_0 denotes the single particle diffusion constant. Overlap configurations were discarded. Configurations of all N-particles were stored at equidistant time intervals. Typically, 2^{19} different configurations were used in calculating the time averages needed for the density auto-correlation functions. The time step used was $10^{-4} t_0$. Both, the transformation into wave vector space and the calculation of the time-averaged density auto-correlation

$$f(k, t) = \frac{\langle \rho\rho \rangle_{k, t}}{\langle \rho\rho \rangle_{k, t=0}} \tag{1.2}$$

function were done with the help of fast Fourier transform algorithms.

Results of simulation

In the simulation the intermediate scattering function $f(k, t)$ of the N-particle system was calculated for several discrete scattering vectors $k_n a = 2\pi a n/L$, with $n = 8, \ldots, 18$. The size of the ring was chosen to be $L = 32a$. Simulations were performed with 3, 10, 12 and 14 particles, corresponding to volume fractions $\phi = 0.1825, 0.625, 0.75$ and 0.875, respectively. The $k_n a$ values chosen above cover the region around the first peak of the static structure factor, see Fig. 1a–d. The form of the structure factor agrees well with Wertheim's solution of the Percus–Yevick equation [4] transformed into the k-space.

Now we switch to dynamics, the temporal decay of the intermediate scattering function $f(k, t)$. Time was measured in units of $t_0 = a^2/D_0$, about half the time a particle takes to diffuse over a distance comparable to its diameter $2a$. With increasing volume fraction Fig. 2 shows that the intermediate scattering function spreads out in time. Wave vectors around the peak of the structure factor decay much slower. The low-density sample, Fig. 2a, a pecularity of the one-dimensional structure factor being only proportional to $\sin(2ka)/ka$, changes this common $3d$ result.

Without hydrodynamic interactions the initial decay rate, the short-time diffusion coefficient, is entirely determined by the static structure factor:

$$D_{\mathrm{eff}}(k) = \frac{D_0}{S(k)} \ . \tag{2.1}$$

To see this behavior verified one has to go back to times of the order of 10^{-4}–$10^{-3} t_0$. The Figs. 2a–d now compare the decay of the intermediate scattering function at different densities both, unscaled and scaled, with the initial slopes $D_{\mathrm{eff}}(k)k^2$. For the lowest density in Fig. 2a, as can be expected, the scaling is perfect. We only see short-time diffusion within the time t_0. For the higher volume fractions, Figs. 2b–d, the particles have collided several times with each other. The range over which the intermediate scattering function decays for various scattering vectors around the peak of the structure factor initially spreads out over two decades in time. After the scaling procedure was done, this goes down to a mere factor of two for the highest volume fraction $\phi = 0.875$. Very much like in the three dimensional measurements of P.N. Segrè and P.N. Pusey the small scattering vectors below the peak of the structure factor do not participate in the scaling but are set off. Unlike in this experiment in our one-dimensional case

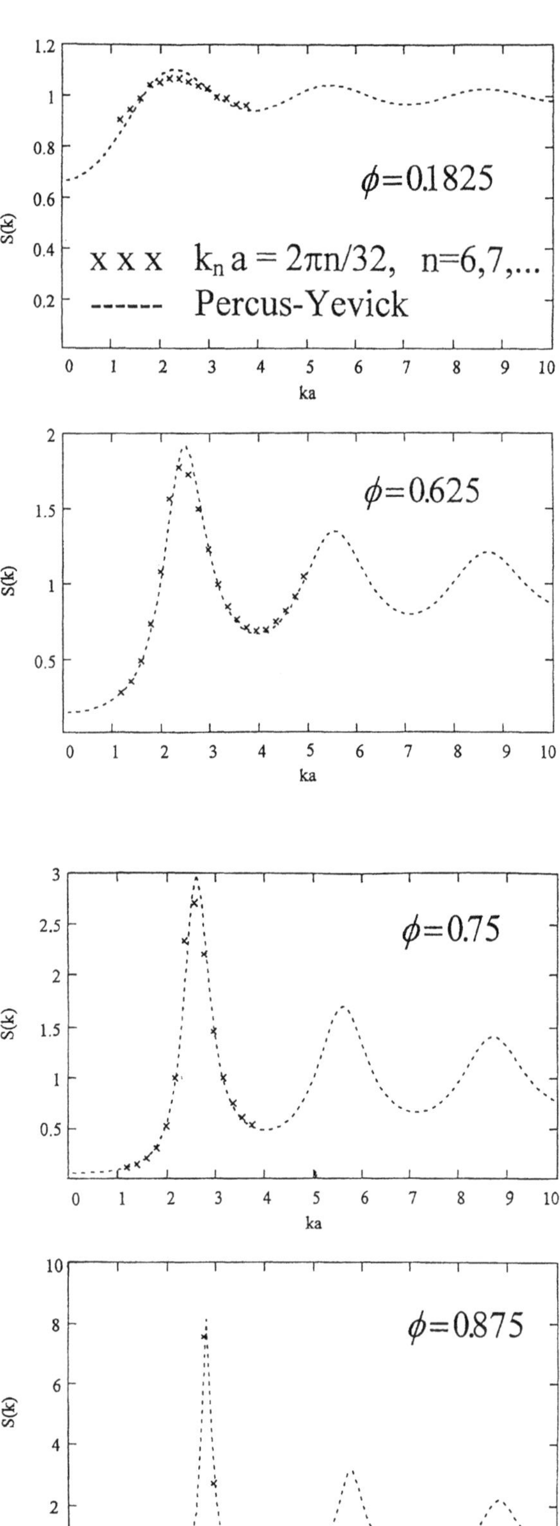

Fig. 1 The four plots (a)–(d) show the static structure factor for the four volume fractions $\phi = 0.1875, 0.625, 0.75$ and 0.875 (symbols) together with Wertheim's solution of the PY equation (dashed line)

they are shifted to larger times. For the highest densities the ratio between short-time and long-time diffusion is already about a factor of two.

Conclusion

We performed a Monte Carlo simulation of an one-dimensional system of Brownian particles on a ring. The logarithm of the intermediate scattering function $f(k, t)$, when scaled with its initial slope, collapses from a range of two magnitudes in time down to about a factor of two. Due to the small particle number within the model the time-dependent diffusion coefficients oscillate in time and cannot be compared with the experiment.

For a final judgement of the physical mechanism leading to scaling, a mechanism which seems to be independent of hydrodynamic interaction effects, one has to go to higher particle numbers and perform a more elaborate simulation in higher dimensions. This work is in progress.

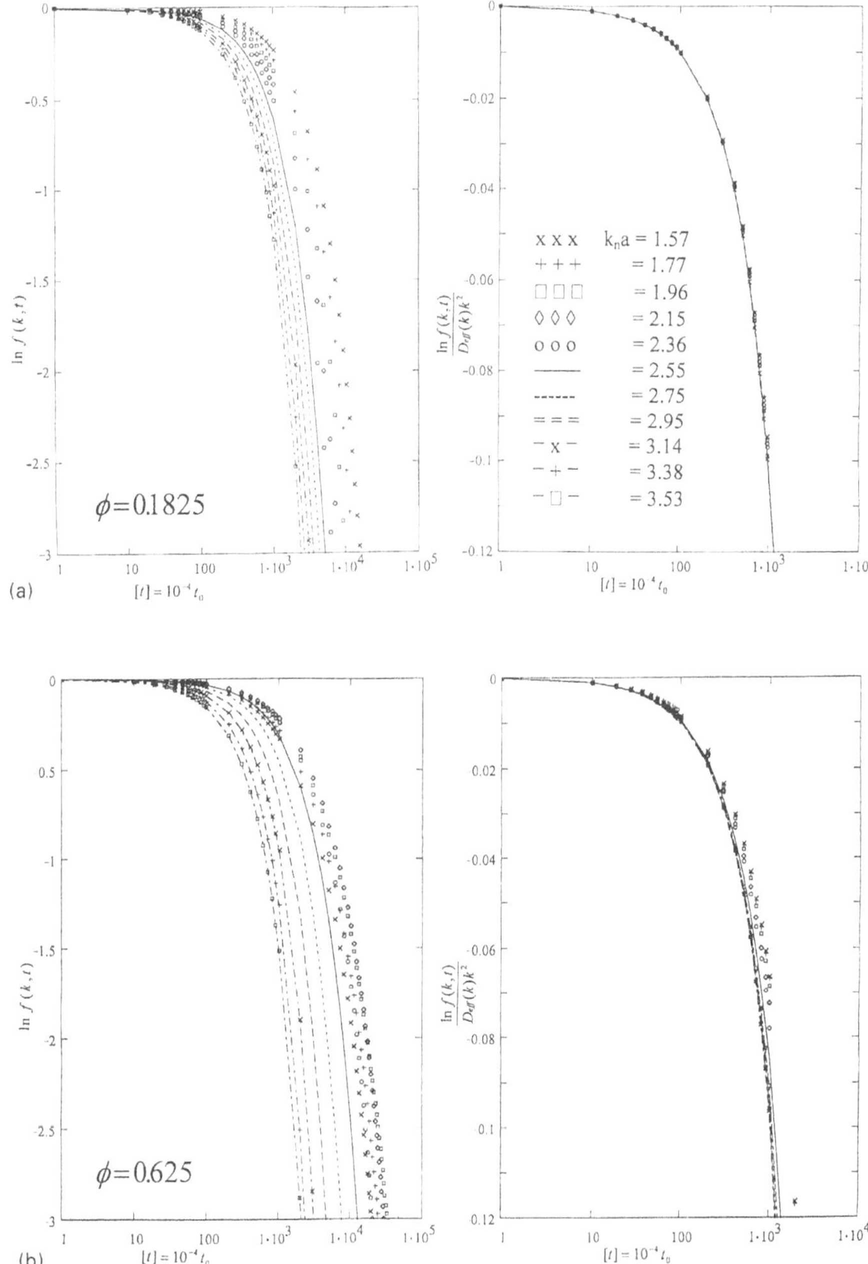

Fig. 2 The four double-plots (a)–(d) show on the left-hand side the intermediate scatterings function $f(k, t)$ for the four volume fractions $\phi = 0.1875, 0.625, 0.75$ and 0.875 as a function of time measured in units of t_0. The parameter is the scattering vector $k_n a$, with $n = 8$–18, see text. The right-hand sides of the four plots (a)–(d) show the scaled intermediate scattering function $\ln f(k, t)/(k^2 D_S(k))$ again as a function of time measured in units of t_0

Fig. 2 Continued

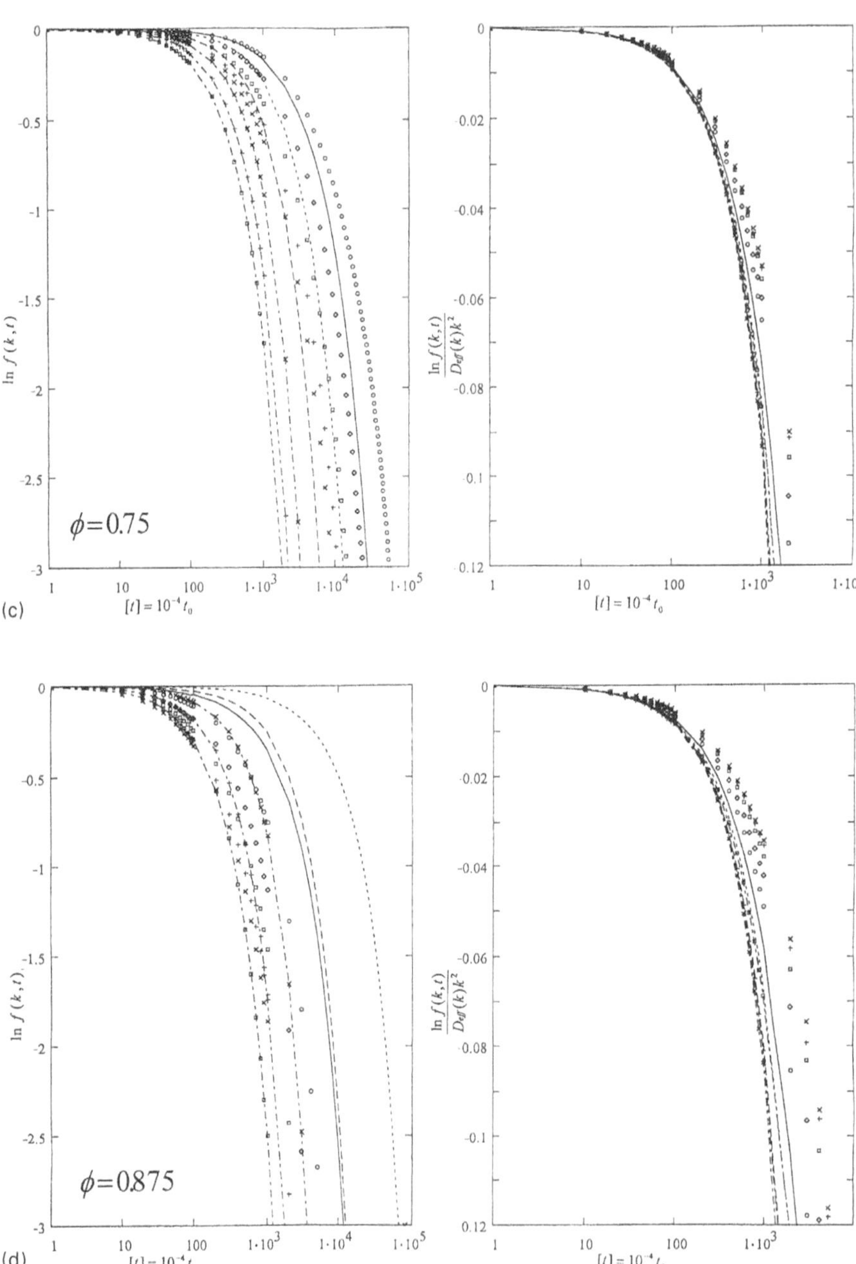

Acknowledgments I wish to thank Wilson Poon, Peter Pusey and Phil Segrè for many discussions. Especially Joachim Wittmer tried hard to convince me how not to perform Monte Carlo simulations.

References

1. For a review see e.g.: Pusey PN (1991) In: Hansen JP, Levesque D, Zinn-Justin J (eds) Liquids, freezing and glass transition. Les Houches 1989, Session LI, pp 763–992. Elsevier, Amsterdam

2. Pusey PN, Segrè PN, Behrend OP, Meeker SP, Poon WCK (1995) Dynamics of concentrated colloidal suspensions, preprint, November; Physica A, in press; Segrè PN, Pusey PN (1996) Phys Rev Lett 77:771; Segrè PN, Pusey PN (1996) The dynamics of hard sphere colloidal suspensions, preprint, September

3. Cichocki B, Hinsen K (1990) Physica A 166:473

4. Wertheim MS (1964) J Math Phys 5:5

Progr Colloid Polym Sci (1997) 104:187–190
© Steinkopff Verlag 1997

R. Sigel
G. Strobl

Static and dynamic light scattering from the nematic wetting layer in an isotropic liquid crystal

R. Sigel (✉) · G. Strobl
Fakultät für Physik
Universität Freiburg
Hermann-Herder-Straße 3
79104 Freiburg, Germany

Abstract By the use of the evanescent wave technique the fluctuations of the nematic wetting layer in an isotropic liquid crystal have been investigated by static and dynamic light scattering. As the homeotropically oriented layer is birefringent, its critical angles of total internal reflection for the ordinary and extraordinary rays differ from that of the bulk. For a suitable selected angle of incidence there is a propagating extraordinary ray in the layer which becomes evanescent in the bulk. In this case the scattering of the layer is enhanced and therefore well accessible to the measurement. It shows up to be strongly dependent on the scattering vector component parallel to the interface, in contrast to the bulk scattering with no marked dependence on the scattering vector. At angles of incidence with evanescent waves in bulk and layer, a broad long time decay shows up, which seems to be typical for evanescent measurements and can be described with an algebraic functional dependence.

Key words Light scattering – evanescent wave – liquid crystals – wetting – interface – fluctuations

Introduction

By the use of the evanescent wave scattering geometry shown in Fig. 1, it is possible to examine interfaces by the light scattering technique [1]. For a plane wave impinging on an interface at an angle φ to the surface normal, the normal wave vector component $k_{i,z}$ of the light refracted under the angle φ' can be calculated from Snells law:

$$k_{i,z} = \frac{2\pi n_2}{\lambda}\cos(\varphi') = \frac{2\pi n_2}{\lambda}\sqrt{1 - \frac{n_1}{n_2}\sin^2(\varphi)} \ . \tag{1}$$

If the index of refraction n_1 of the ambient is larger than the index n_2 of the medium beyond the interface, $k_{i,z}$ becomes imaginary when

$$\varphi > \varphi_0 := \arcsin\left(\frac{n_2}{n_1}\right) . \tag{2}$$

Under this condition of total internal reflection (TIR), the refracted ray is no longer propagating perpendicular to the interface but becomes the exponentially decaying evanescent wave, as can be seen by the phase factor:

$$\exp[ik_{i,z}z] = \exp[-|k_{i,z}|z] \ . \tag{3}$$

The evanescent wave, the extent of which is comparable to the wavelength of light is used as the illuminating light of a scattering experiment, in order to detect only fluctuations close to the interface. To achieve TIR in the experiment the sample is kept in direct contact with a highly refractive hemispherical lens.

Concerning liquid crystals, an interesting interface phenomenon is the nematic wetting in the isotropic phase, i.e., close to the nematic isotropic phase transition a nematic surface layer forms the boundary of an isotropic bulk. Because of the lack of a suitable experimental technique for observing the dynamics of this layer, only the equilibrium properties have been investigated so far (e.g. [2]). The evanescent scattering technique fills this gap, as is discussed in this paper.

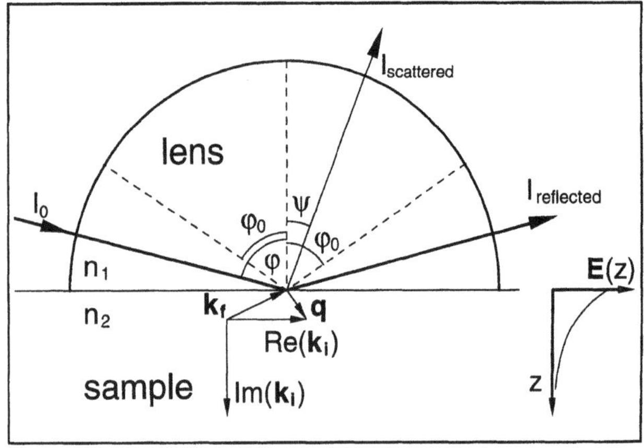

Fig. 1 Evanescent wave scattering geometry. The sample is in direct contact with a highly refractive hemispherical lens. The normal scattering vector component of the illuminating evanescent wave is imaginary. The scattered ray is refracted when leaving the sample

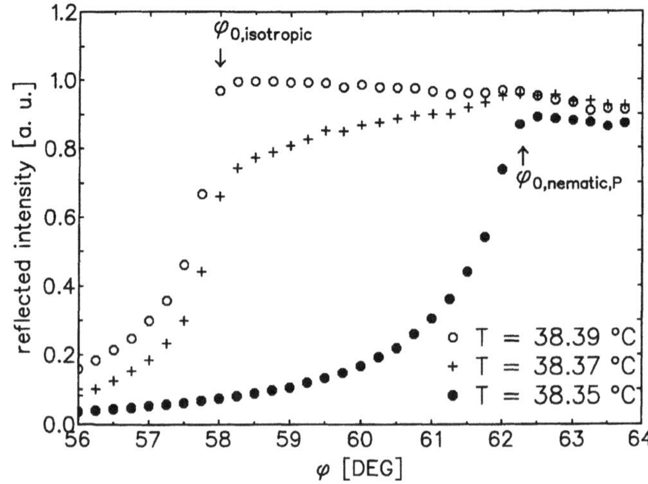

Fig. 2 By reflectivity measurements, the critical angle of TIR as well as the phase transition temperature can be determined

Experiment

The experiments were performed with the liquid crystal CB8 on a lens coated with lecithin, leading to a orientation in the nematic phase with the optical axis perpendicular to the interface (homeotropic orientation). The light source was an argon ion laser used at a wavelength of 488 nm. The lens consists of the glass Schott LaSF N9.

Results

The critical angle of TIR was determined by measurement of the dependence of the reflected intensity on the angles of incidence, as shown in Fig. 2. As the illuminating light is polarized with the electric field vector horizontally (H-polarization, parallel to the plane of reflection), the refracted wave is the extraordinary ray in the low temperature nematic phase. The very small biphasic temperature range of the phase transition indicates a pure sample.

The static scattering for varying φ in Fig. 3 is measured for a constant value $q_x = 0.0203 \text{ nm}^{-1}$ of the scattering vector component parallel to the interface by a suitable choice of the detection angle. There are two steps located at $\varphi_b \approx 58°$ and at $\varphi_l^e \approx 62°$ corresponding to the critical angles of Fig. 2. At φ_b, the drop in intensity is caused by a shrinking of the scattering volume connected with the transition of the illuminating light from a plane wave to an evanescent wave. Up to φ_l^e there remains an enhancement of the scattered intensity which can be traced back to an amplification mechanism within the refractive index profile, which ceases at the layer critical angle for TIR. This mechanism will be discussed elsewhere. The effect can only

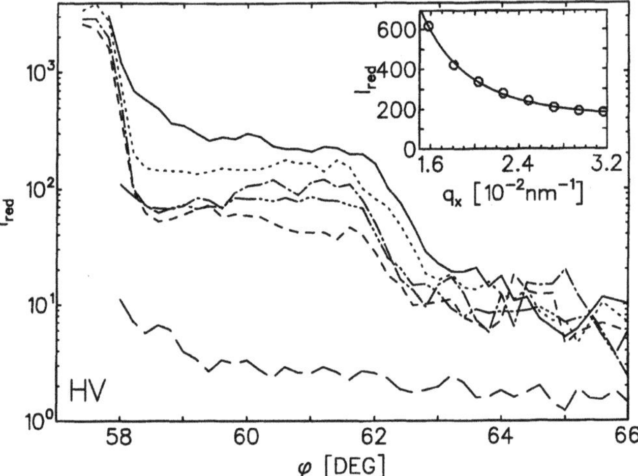

Fig. 3 Dependence of the measured scattered intensity on the angle of incidence for fixed $q_x = 0.0203 \text{ nm}^{-1}$. The different curves correspond to different temperatures: from top to bottom 38.39, 38.43, 38.48, 38.58, 38.70, and 38.80 °C. The data of the small figure in the upper right corner are measured at $T = 38.41°C$ and $\varphi = 59.2°$. The measured intensities are divided by the toluene cuvette value, so only geometric factors are needed for full reduction to Rayleigh ratios

be observed in a temperature range of about 0.3 K above the bulk phase transition. There is no similar effect for incident vertically polarized light. In this case, the light is an ordinary ray in the layer and the corresponding critical angle is below the bulk value; so, there is no amplification.

The correlation functions for $\varphi_b < \varphi < \varphi_l^e$ can be described by KWW functions, as shown in Fig. 4. The observed strong dependence of the static and dynamic light scattering on q_x is in contrast to the almost q independent behavior of the isotropic bulk; however, it resembles the

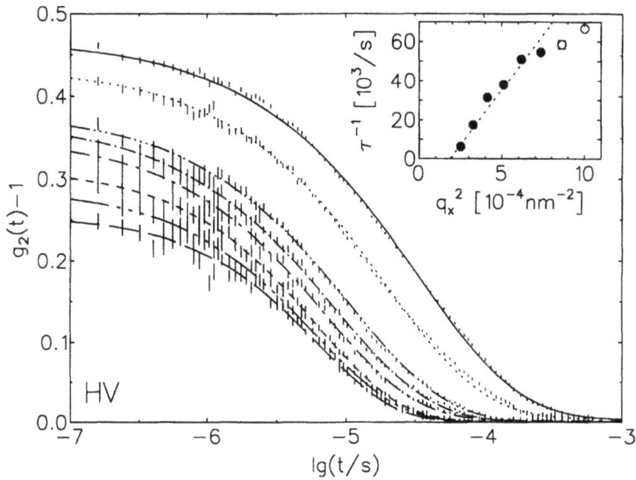

Fig. 4 The DLS data, here shown for $T = 38.41\,°\text{C}$ and $\varphi = 59.6°$, are strongly dependent on q_x. In units of $10^{-2}\,\text{nm}^{-1}$, from top to bottom the intensity correlation functions correspond to q_x values of 1.58, 1.81, 2.03, 2.26, 2.49, 2.71, 2.93 and 3.16. They can be fitted by KWW functions. The inset displays the q_x dependence of the corrected relaxation time. The error bars used in the fit are calculated using the error model of Schätzel [3]

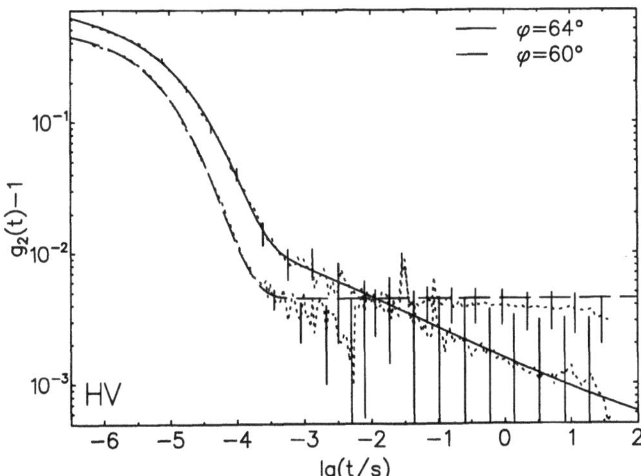

Fig. 5 Comparison of the intensity time correlation function below and above φ_l^e, measured at $q_x = 0.025\,\text{nm}^{-1}$ and $T = 38.54\,°\text{C}$. Discussion in the text. The calculated error bars are plotted only for one out of ten data points

scattering of the nematic phase. The error bars to the correlation data are calculated using the error model of Schätzel [3].

A comparison of DLS measurements below and above φ_l^e is shown in Fig. 5. While below φ_l^e a reasonable experimental base line is reached, there is a broad long time

decay above φ_l^e. This seems to be typical for evanescent wave scattering measurements: as for diffusing latices, experimentally and theoretically similar results were obtained, which were described as algebraic decay [1]. This decay is caused by the scattering of fluctuations in the inhomogeneous evanescent field. Here, they can be fitted by the empirical formula

$$g_1(t) = a\frac{\exp(t/\tau)^\beta(1 - \delta) + \delta}{1 + (t/\kappa\tau)^\nu} + c \qquad (4)$$

with $\kappa = 1$ fixed. In (4) only the part $(1 - \delta)$ decays as a KWW function, the other part decays algebraically with exponent ν. The trends of the relaxation time τ derived by fits of (4) show a similar behavior as the KWW description below φ_l^e, while ν has values around 0.2. The obtained intercept a, i.e., the extrapolation to $\tau = 0$, is rather high, it takes values between 0.8 and 1.4. This is a weak point of (4).

The dynamic VH measurements can be described in the same way and show a similar behavior, starting already from φ_b. Experimental results derived in HH polarization behave like the presented HV data.

Discussion

As the layer has a nematic-like order, the fluctuations compare with those of the nematic phase. The sudden disappearance of the enhanced scattering for $\varphi_b < \varphi < \varphi_l^e$ on heating of the sample might be an indication of a surface phase transition (the prewetting transition), predicted by Sheng [4]. As by this phase transition the birefringence of the layer gets much weaker, the step of the enhancement at φ_l^e gets close to and eventually merges to the step at φ_b. Furthermore, the surface fluctuations are much weaker, in the sense like anisotropy fluctuations of the isotropic bulk phase are much weaker than those of the nematic phase.

The comparison of the experimental data to model calculations will be presented in a forthcoming paper.

Conclusions

The nematic wetting layer in an isotropic liquid crystal shows up in evanescent wave scattering measurements with angles of incidence in between the critical angles for TIR from the bulk and of the extraordinary ray in the layer: in a narrow temperature range above the bulk phase transition, an enhanced scattered intensity is observed. The strong dependence of the static and dynamic light scattering of the layer on the scattering vector is in contrast to the isotropic bulk behavior.

References

1. Lan KH, Ostrowsky N, Sornette D (1986) Phys Rev Lett 57:17–20
2. Immerschitt S, Koch T, Stille W, Strobl G (1992) J Chem Phys 96:6249–6256
3. Schätzel K (1993) In: Brown W (ed) Dynamik Light Scattering. Clarendon Press, Oxford, pp 76–148
4. Sheng P (1982) Phys Rev A 26:1610–1617

Progr Colloid Polym Sci (1997) 104:191–193
© Steinkopff Verlag 1997

M. Schmidt
S. Krieger
D. Johannsmann

Film formation of latex dispersions observed with evanescent dynamic light scattering

M. Schmidt · Dr. D. Johannsmann (✉)
Max-Planck-Institute for Polymer Research
P.O. Box 3148
55021 Mainz, Germany

S. Krieger
Hoechst AG
G832
65926 Frankfurt, Germany

Abstract We have applied evanescent dynamic light scattering (DLS) to study the film formation of polymer latex dispersions. A dynamical glass transition is observed. When the particles fuse, the scattering rate decreases. After fusion, a new fast process ($\tau \sim 1$ ms) is observed, which is not present in bulk DLS data. We attribute the process to discontinuous relaxation of surface induced stresses (transient micro-cracks).

Key words Evanescent dynamic light scattering – colloidal dispersions – film formation

Motivation

During film formation, polymer latices experience an irreversible structural change [1]. The particles come into contact, fuse and form a uniform film (coalescence). The correlation of material and process parameters on the one hand and behavior of the film during coalescence and in its final state on the other hand are of high technical relevance. Typical material parameters are size distribution, charge, surfactant, and glass temperature of the latices. The important process parameters are temperature and humidity during film formation. The drying speed, the (nonlinear) rheological behavior of the latex during application, and the toughness, stiffness, and homogeneity of the final product are parameters of interest to the engineer.

In this work, we address the dynamic behavior during film formation. Evanescent light scattering [2] has two advantages for these investigations:

● Multiple scattering is largely eliminated because the thickness of the scattering volume (0.2–5 μm) is much less than the photon mean free path even for dense suspensions [3].
● The evanescent light probes the dynamics at the interface.

The drawbacks are:

● Because the scattering vector q is complex, the scattering amplitude is no longer strictly proportional to the Fourier transform of the contrast [4].
● The local dynamic behavior close to the interface may depend on the distance from the interface so that the scattering volume may cover different dynamical regimes. This problem can be addressed by using a particle size in the range of the smallest accessible penetration depth (~ 200 nm), so that at sufficiently high angles of incidence, only the first monolayer of latices is probed.

When water evaporates, the densification of latex particles in the film will lead to a slow down of its dynamic behavior. A glass transition is observed [5, 6]. When the particles coalesce, the time scale of the α-process goes to infinity and the scattering rate decreases because the film becomes clear.

Of special importance are the boundary conditions at the substrate surface. Because the rigid substrate prohibits lateral displacement, the densification is unidirectional. The surface induces stress during film formation. These stresses will be strongest at the substrate surface. If the material cannot adjust to the boundary conditions, (micro-) cracks will form.

Experimental procedure and results

The films were prepared by casting the dispersions (poly-methyl-methacrylate/butyl-acrylate, $d = 93$ nm) onto a high index glass slide optically connected to a semi-cylinder prism. At ambient conditions, drying is complete after 5–6 min. For the purpose of DLS measurements, the film formation was slowed down to several hours by limiting the exchange of air above the film surface. Quasi-static conditions were established during a sampling time of 3000 s. Measurements were performed at room temperature (23.5 °C), above the minimum film formation temperature MFT of 20 °C.

Figure 1A shows a set of autocorrelation functions (ACFs) during the initial densification over a time period of several hours. Two separate processes are observed. The time constant of the slower process diverges as the film

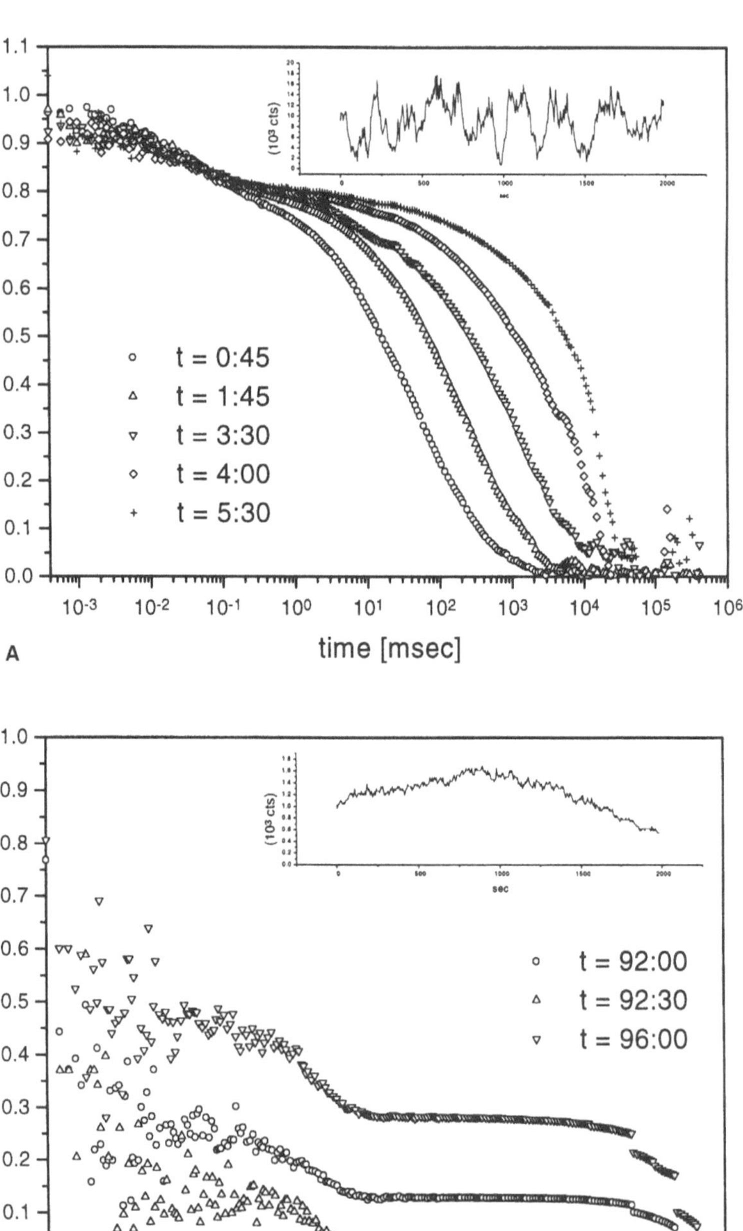

Fig. 1A Autocorrelation functions ($\theta = 90°$) as measured by evanescent DLS during initial densification over a time period of 6 h. The dynamics reflects a dynamical glass transition. The inset shows a typical time photon count signal, with strong fluctuations during sampling time.
B Evanescent autocorrelation functions taken 4 d later. A new process with a decay time in the ms range is found. The photon count rate is much less than in the initial phase

formation proceeds. We identify the processes with the α- and the β-process of an underlying glass transition [5, 6]. After fusion, both the fast and slow processes become invisible due to a lack of optical contrast, while a new process with a decay time in the ms range is observed for some samples (Fig. 1B). We never observed such a fast process in standard bulk light scattering under identical conditions. The amplitude of the fast process strongly fluctuates. For some films, no fast process was observed at all.

Tentatively, we connect the surface process to shear stress imposed by the unidirectional compression. Because the rigid surface prevents lateral shrinkage, a test volume does not only shrink, but also becomes oblate. The shear stress may relax discontinuously, forming "microcracks". Since the scattering rate does not increase with time, the cracks seem to heal on the time scale of the experiment. The time scale of crack formation should be in the ms range.

We could not establish a clear correlation between the occurrence of the fast process and the minimum film formation temperature (MFT). The process was observed for both (poly-methyl-methacrylate/butyl-acrylate) and poly-vinyl acetate. The conditions for the occurrence of the fast process have to be established in more detail.

Conclusions and outlook

We have applied evanescent dynamic light scattering to investigate the film formation of aqueous latex dispersions. The technique avoids multiple scattering and probes the interfacial dynamics mainly. The latex dispersion undergoes a glass transition. In the glassy state the scattering rate decreases due to particle fusion. After fusion, a new fast process ($\tau \sim 1$ ms) is observed. Presumably, it is connected to shear stress induced by unidirectional compression. Transient micro-cracks may form.

Surface stresses may lead to near-surface stress birefringence which should be observable with evanescent microscopy. Evanescent microscopy may also prove the occurrence of cracks and display their shape.

Acknowledgments This work was partly supported by the German Ministry for Education and Research (BMBF) under contract 03N3018S.

References

1. Richard J (1996) In: Recent Advances in Polymeric Dispersions, NATO ASI proceedings. Kluwer Academic, Dordrecht, to appear
2. Lan KH, Ostrowsky N, Sornette D (1986) Phys Rev Lett 57:17–20
3. Wiese H, Horn D (1991) J Chem Phys 94:6429–6443
4. Gao J, Rice SA (1989) J Chem Phys 90:3469
5. van Megen W, Underwood SM (1994) Phys Rev E 49:4206–4220
6. Bartsch E, Antonietti M, Schupp W, Sillescu H (1992) J Chem Phys 97:3950

Progr Colloid Polym Sci (1997) 104:194–197
© Steinkopff Verlag 1997

S. Neser
T. Palberg
C. Bechinger
P. Leiderer

Direct observation of a buckling transition during the formation of thin colloidal crystals

S. Neser (✉) · C. Bechinger · P. Leiderer
Fakultät für Physik
Universität Konstanz
78434 Konstanz, Germany

T. Palberg
Institut für Physik
Universität Mainz
55099 Mainz, Germany

Abstract We have investigated a colloidal suspension in a thin wedge formed by two glass plates in the presence of a lateral pressure. Starting with a single hexagonal layer, with increasing separation between the glass plates additional layers are added. This process is accompanied by a number of structural transitions necessary to maintain a high packing fraction under the given boundary conditions. Besides the well-known sequence of hexagonal and quadratic phases, we observe two new phases which are identified with the buckling and the rhombic phase recently predicted by other authors.

Key words Colloidal crystals – finite size effects – buckling – layering – phase transitions

Suspensions of small colloidal spheres are widely used model systems for crystallization phenomena in two and three dimensions [1, 2]. There are several methods to prepare two-dimensional (2D) colloidal systems, for example, by confinement of the suspension in a narrow gap between two glass plates [2] or by trapping the particles at the water–air interface [3]. In contrast to ideal 2D systems, for example, electrons in surface states above the liquid helium surface [4], in colloidal systems the dimensionality has to be reduced artificially. It is clear that such a system can only be an approximation of the ideal 2D case. On the other hand, in a colloidal system confined between two glass plates the restriction of motion perpendicular to the 2D plane can be controlled via particle–wall-interactions and the separation of the walls [5]. Therefore, such a system is ideally suited for investigations of the transition regime between 2D and 3D. Pieranski et al. [6] and van Winkle and Murray [7] investigated colloidal particles in a wedge geometry. Starting with one hexagonal monolayer at small plate separations, with increasing cell height they found a sequence of morphological transitions of the form

$$n\triangle \rightarrow (n+1)\square \rightarrow (n+1)\triangle ,$$

where n denotes the number of layers, with $\square$ and $\triangle$ corresponding to the quadratic and hexagonal phase, respectively. Koshikiya and Hachisu [8, 9] have reported some peculiar maze like patterns in the region between one hexagonal and two quadratic layers. Pansu et al. [10] pointed out that this sequence optimizes the packing fraction of the spheres in the slab and suggested a continuous transition between $n\square$ and $n\triangle$ via an intermediate phase possessing rhombic (r) symmetry. A buckling instability of the 2D hexagonal lattice has been found by Chou and Nelson [11], associated with a second-order phase transition to a buckling phase (b) as a function of increasing cell height. As a result of their Monte Carlo simulations Schmidt and Löwen [12] recently published a phase diagram for hard spheres being confined between two plates with distances ranging between one- and two-particle diameters. They obtained a phase diagram which shows a rich variety of phase transitions including the two new phases, the buckling and the rhombic phase. In this paper we present experimental evidence for both buckling and rhombic phases.

To fabricate cells of wedged geometry, first we thoroughly cleaned microscope cover slides with thicknesses of 2 and 0.17 mm, respectively. Then we put a small droplet

(10 μl) of suspension between the two slides, removed any excess liquid and sealed the sample cell along its perimeter with epoxy glue. As particles we used surfactant stabilized polystyrene spheres (Bangs Labs Lot. No. 20-PS196, 840 nm diameter by DLS) suspended in water with a volume fraction of 10% and a polydispersity of about 5%. The solidification of the epoxy sealing and capillary forces causes a bending of the thin cover slide due to mechanical stress and lead to the formation of a wedge between the two glass plates. The resulting geometry of the wedges was determined by analyzing the Newton fringe pattern appearing in the slab.

While the epoxy sealing has a small permeability for water which slowly evaporates out of the cell it is a perfect barrier for the particles. The evaporation of the water generates an – albeit very small – flow due to which most of the particles move towards the edges of the cell where the particle concentration increases and the system starts to crystallize. This corresponds to applying a lateral pressure to the system. During the evaporation process the phase boundary liquid/solid moves through the cell and a colorful striped pattern develops parallel to the interference fringes of the wedged cell. Typically, the evaporation process takes approximately 2–3 weeks.

In the following, we will show that those patterns can be interpreted in terms of a sequence of regions with hexagonal and quadratic symmetry. Additionally, we have found different intermediate phases in between, two of them being consistent with the buckling and the rhombic phase predicted by other authors [10–12]. In this paper, we will concentrate on systems which do not exceed two layers.

Starting at a cell height supporting only one hexagonal layer and moving in a direction of increasing cell height we always found the same sequence of phases:

$$1\triangle \rightarrow \text{buckling } (b) \rightarrow 2\square \rightarrow \text{rhombic } (r) \rightarrow 2\triangle \ ,$$

where the extent of the regions depends on the actual wedge profile. Figure 1A shows a scanning electron microscope picture of the buckling phase taken after complete evaporation of the solvent and removal of the top plate.

A

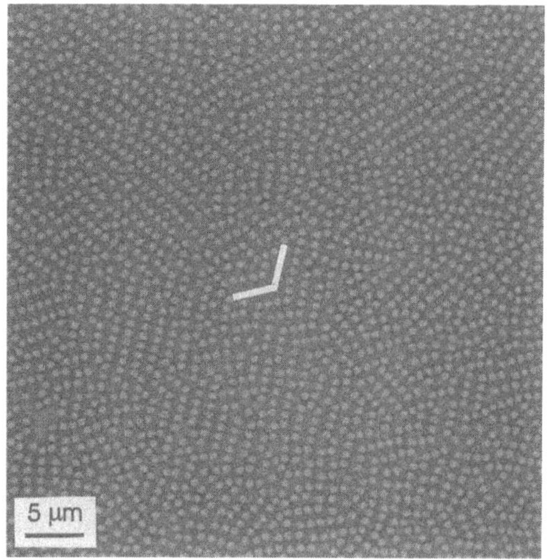

B

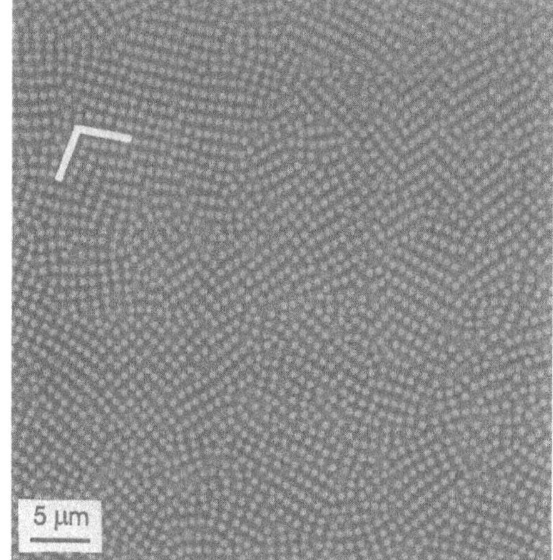

C

Fig. 1A Scanning electron microscope picture of the buckling phase. **B** Buckling phase near the phase boundary to the $1\triangle$ structure. The cell height increases from right to left. The brighter and darker particles belong to the lower and upper sublayer, respectively and the white lines are a guide to the eye to mark the kinks in the particle chains. Near the $1\triangle$ structure the kink angle is approximately 120°. **C** Buckling phase near the phase boundary to the $2\square$ structure. The cell height increases from right to left. Due to the larger separation of the two sublayers compared to Fig. 1B, only the particles of the lower sublayer are visible. The white lines are a guide to the eye marking the kinks in the particle chains. In the vicinity of the $2\square$ structure the kink angles approach 90°

The buckling phase consists of a zig-zag structure of particle chains alternating between an upper and a lower sublayer, vertically separated by a fraction of the particle diameter. Within a chain the particles are closely packed. The separation of neighboring chains belonging to the same sublayer is a function of the cell height. Figure 1B shows a region of the buckling phase near the phase boundary to the $1\triangle$ structure. The different vertical positions of the upper and the lower sublayer with respect to the focal plane of the microscope gives rise to an optical contrast which allows to distinguish between them. In Fig. 1B the vertical displacement of the sublayers is still small and the distance of two neighboring chains in the same sublayer is close to that of the corresponding distance in the hexagonal monolayer. With increasing cell height the separation of the neighboring chains decreases until it finally approaches the hard sphere radius which is the onset of the quadratic phase. Figure 1C depicts a region of the buckling phase near the $b \rightarrow 2\square$ boundary showing the considerably smaller chain separation.

The formation of the buckling phase can be understood in terms of a simple hard sphere packing model. We consider a system with a plate separation of one effective hard sphere diameter subject to lateral pressure. In that case an equilibrium structure of $1\triangle$ will result, i.e. a dense-packed single hexagonal layer. A slight increase in the plate separation will allow the system to release some of the lateral pressure by elevating or lowering particle chains. This corresponds to the buckling instability of the 2D hexagonal lattice shown by Chou and Nelson [12]. A two sublayer system is created in which the lateral chain distance in the upper and lower sublayer, respectively, decreases continuously as the plate separation increases. At the same time the packing fraction is kept at a high level. When d equals the hard sphere diameter the chains in each sublayer touch to form the $2\square$ structure. We therefore believe to have repeated the experiments of Pieranski et al. [6] and van Winkle and Murray [7] at higher volume fractions where Schmidt and Löwen predict a larger stability region of the buckling phase.

Another dominant feature of the buckling phase are the kinks in the chains. The typical kink angle is $120°$ at the onset of buckling decreasing to $90°$, when the transition to the two layer quadratic phase is approached. The rather random arrangement of the chains is in good agreement with the miniscule differences in free energy between the possible buckling patterns found by Chou and Nelson [11] and Schmidt and Löwen [12]. As already stated above, with increasing cell height the transition from the buckling to the two layer quadratic phase is reached and a further increase leads to a phase shown in Fig. 2 looking like a distorted hexagonal lattice. These distortions can best be seen following a single lattice line

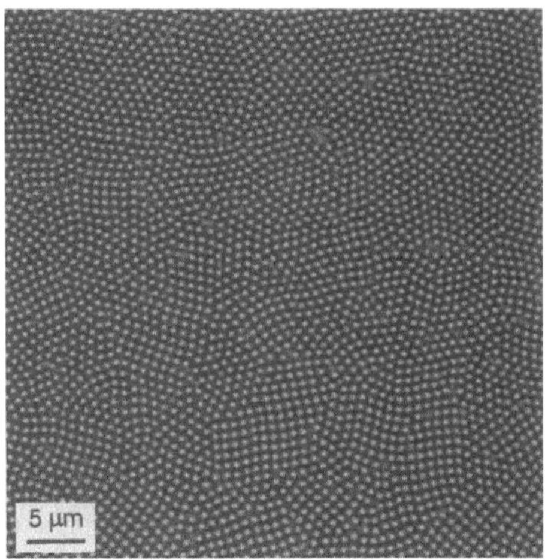

Fig. 2 Distorted hexagonal lattice suggested to be the rhombic phase. The cell height increases from right to left

with the eye. The appearance of that region together with its position between $2\square$ and $2\triangle$ structures strongly suggests it to be the rhombic phase. Finally, for sufficient wall separations the two layer hexagonal phase is reached.

It is an interesting question why the buckling and rhombic phases appear so dominantly in our experiment while being only a minor feature in the experiments of other authors [6, 7]. We suppose that the vertical displacement of chains finally leading to the buckling phase is favored in our geometry, because a lateral flow, i.e. a lateral pressure is imposed during crystallization. This is in good agreement with the results of Schmidt and Löwen, who performed simulation runs with fixed plate separation and packing fraction and observed the presence of a lateral pressure.

In conclusion, we have studied a colloidal system in wedge geometry with a small additional flow. For cell heights between one and two effective hard sphere diameters we find a rather smooth transition from a single hexagonal layer via a buckling phase to two quadratic layers and from those via the rhombic phase to two hexagonal layers. These data are consistent with the recently published theoretical phase diagram of Schmidt and Löwen [12]. Both the buckling and the rhombic phase are much more pronounced in our experiments than in those of Pieranski et al. [6] and van Winkle and Murray [7]. We attribute this, in particular, to the presence of an additional flow which forces the particles close to the hard sphere packing limit.

Acknowledgments We gratefully acknowledge financial support of the Deutsche Forschungsgemeinschaft (SFB 306 and SFB 513) and helpful discussions with Hartmut Löwen and Matthias Schmidt.

References

1. Pusey PN (1991) In: Hansen JP, Levesque D, Zinn-Justin J, Liquids, Freezing and Glass Transition: II, North-Holland, Amsterdam
2. Murray CA, van Winkle DH (eds) (1987) Phys Rev Lett 58:1200–1203
3. Pieranski P (1980) Phys Rev Lett 45: 569–572
4. Peeters FM (1987) Electrons on a liquid helium film. In: Devreese JT, Peeters FM (eds) Summerschool Proc Advanced Study Institute on the Physics of the Two Dimensional Electron Gas. Plenum, New York
5. Kepler GM, Fraden S (1994) Langmuir 10:2501–2506
6. Pieranski P et al (1983) Phys Rev Lett 50:900
7. van Winkle DH, Murray CA (1986) Phys Rev A 34:562–573
8. Ogawa T (1983) Phys Soc Jpn 52: 167–170
9. Koshikiya Y, Hachisu S (1982) Lecture in Colloid Symp Japan (Sept. 1982) (in japanese language) cited in [8]
10. Pansu B, Pieranski Pi, Pieranski Pa (1984) J Physique 45:331–339
11. Chou T, Nelson DR (1993) Phys Rev E 48:4611
12. Schmidt M, Löwen H (1996) Phys Rev Lett 76:4552

Progr Colloid Polym Sci (1997) 104:198–200
© Steinkopff Verlag 1997

D. Vollmer
J. Vollmer
R. Strey

Microemulsions:
Phase transitions and their dynamics

Dr. D. Vollmer (✉)
Institut für Physikalische Chemie
Universität Mainz
Welder-Weg 11
55099 Mainz, Germany

J. Vollmer
Fachbereich Physik
Universität-GH-Essen
45117 Essen, Germany

R. Strey
Institut für Physikalische Chemie
Universität zu Köln
50923 Köln, Germany

Abstract By differential scanning microcalorimetry we investigate temperature-induced phase transitions and their dynamics in mixtures of water, oil and a non-ionic surfactant. Special emphasis is on an investigation of the transition from a lamellar to a microemulsion phase and on the emulsification failure. The first-order phase transition from a lamellar to a microemulsion phase leads to heat changes up to $1k_B T$ per surfactant molecule. These large values for the latent heat are quantitatively described by an interfacial model which takes into account the temperature dependence of the spontaneous curvature.

The dynamics of phase separation due to an emulsification failure leads to oscillations in the specific heat when heated with constant scan speed. The period and amplitude of the oscillations depend on composition and scan speed. A square root dependence on scan speed is found for the period. The values for the heat absorbed during each oscillation are compared with those calculated from the interfacial model, and a mechanism for the origin of the oscillations is proposed.

Key words Microemulsion – calorimetry – phase transitions and their dynamics – oscillations

Mixtures of water, oil and surfactant show a large variety of phases when composition or temperature is varied [1, 2]. In particular, they form microemulsions where thermodynamically stable nanometer-sized oil and water domains are separated by a surfactant monolayer: single-phase droplet, cylindrical, bicontinuous and lamellar structures are found. Two or three of these phases may coexist with each other, or with a water or oil rich phase [3]. Although structural investigations have been performed intensively, thermal properties have barely been studied. In particular, until recently there were no measurements on the latent heat and the dynamics of phase transitions had not received much attention. In the present paper, we address these questions by differential scanning microcalorimetry. The variation of the specific heat $C_p^{\mathrm{rel}}(T)$ relative to a (irrelevant) background is measured as a function of temperature, using a MC-2

microcalorimeter (Microcal Inc.). Details on the experimental setup are given in Ref. [4].

In Fig. 1 the variation of $C_p^{\mathrm{rel}}(T)$ is given while passing the lamellar-to-droplet and the emulsification boundary (microemulsion ME to $\overline{2\phi}$) for a solution of H_2O–octane–$C_{12}E_5$ (n-dodecyl pentanethyleneglycol ether). The arrow in the inset of Fig. 1 indicates the corresponding line in a section through the Gibbs phase prisma, where the volume fraction of oil to oil-plus-water is kept fixed at 0.85. The composition of the sample, given by the volume fraction of water $\phi_w = 0.13$, of octane $\phi_o = 0.76$ and of surfactant $\phi_s = 0.11$, was chosen in such a way that water droplets are found in large regions of the high-temperature microemulsion channel. The lamellar to microemulsion transition (dashed line) is a first-order phase transition between two different single-phase structures. The emulsification boundary (solid line) marks the

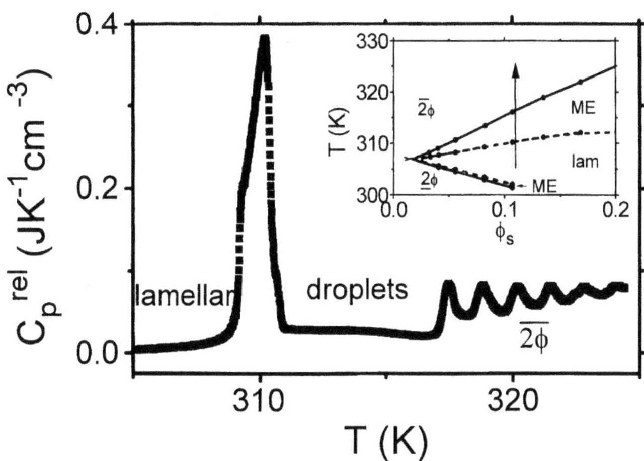

Fig. 1 Temperature-dependent variation of the specific heat $C_p^{\mathrm{rel}}(T)$ for a mixture of H_2O–Octane–$C_{12}E_5$. The inset shows a section through the phase prism at fixed volume fraction of oil to oil-plus-water, and varying volume fraction of surfactant and temperature

onset of a transition where a single phase of water droplets becomes unstable against shrinking of droplets and formation of a water-rich coexisting phase $\overline{2\phi}$. The thermogram shows a pronounced peak at 310 K and oscillations for $T \gtrsim 317$ K. Comparison with the phase diagram shows that the temperature of the maximum of the peak is close to the lamellar-microemulsion phase boundary, whereas the onset of the oscillations coincides with the emulsification boundary.

After this overview, we first discuss the peak related to the lamellar-to-microemulsion transition. The width of the peak is independent of scan speed v_s; it gives a measure of the width of the two-phase region. The area under the peak is a measure for the latent heat $\Delta Q = T \Delta S$, yielding $\Delta Q = (0.37 \pm 0.08)\,\mathrm{J\,cm}^{-3} \approx 1.1 k_\mathrm{B} T/(\text{surfactant molecule})$ and increases with surfactant concentration [2]. To model this behavior theoretically, we observe that the difference in entropy between the two phases ΔS corresponds to the difference of the temperature derivatives of the mean-field approximation to the free energies for the underlying microstructures just above (droplets: F_{drop}) and below (lamellae: F_{lam}) the transition temperature, so that

$$\Delta Q = T \Delta S = -T\left(\frac{\partial F_{\mathrm{drop}}}{\partial T} - \frac{\partial F_{\mathrm{lam}}}{\partial T}\right).$$

The free energy per unit volume can be calculated from an interfacial model due to Helfrich [5], when taking into account the temperature dependence of the radius of spontaneous curvature $c_0(T)$ [2, 6], leading to

$$F_{\mathrm{drop}} = \frac{2\kappa\phi_s}{l_s}\frac{1}{R^2}\left[(1 - Rc_0(T))^2 + \frac{\bar{\kappa}}{2\kappa}\right], \tag{1}$$

and

$$F_{\mathrm{lam}} = \frac{2\kappa\,\phi_s}{l_s}\,c_0(T)^2\,, \tag{2}$$

where $R = [3l_s(\phi_w + \phi_s/2)]/\phi_s$ corresponds to the radius of the droplets in the single phase ME; $l_s = 1.3$ nm to the length of a surfactant molecule; $\kappa \approx 0.8 k_\mathrm{B}\bar{T}$ and $\bar{\kappa} \approx -0.4 k_\mathrm{B}\bar{T}$ with $\bar{T} = 305.6$ K denote the bending modulus and the Gaussian modulus, respectively [2]; and $c_0(T) = a[1 + \bar{\kappa}/(2\kappa)](\bar{T} - T)$ with $a = 0.012\,\mathrm{K}^{-1}\,\mathrm{nm}^{-1}$ is the spontaneous curvature. The latter is the only temperature-dependent parameter in the free energies. Equations (1) and (2) permit us to calculate the latent heat of the transition

$$\Delta Q = \frac{4\kappa a T}{3 l_s^2}\frac{\phi_s^2}{\phi_w + 0.5\phi_s}\left(1 + \frac{\bar{\kappa}}{2\kappa}\right), \tag{3}$$

yielding $\Delta Q = (0.49 \pm 0.1)\,\mathrm{J\,cm}^{-3}$ for the latent heat of the transition discussed above, and a similar good agreement with experimental findings for the latent heat at other compositions.

The thermal properties of the emulsification failure are peculiar in several aspects. After crossing the phase boundary with constant heating, oscillations in the specific heat occur. A square root dependence $\Delta T \sim (v_s/v_0)^{0.5 \pm 0.05}$ K with $v_0 = 1$ K h^{-1} of the period ΔT on scan speed v_s has been found [7], indicating that the period vanishes for quasistatic heating, whereas it takes infinitely long to heat across the first oscillation. The oscillations indicate that the dynamics of this phase separation differs from nucleation theory and spinodal decomposition. Since there is overheating there is an energy barrier against the formation of a phase-separated state from the overheated solution, but no nuclei in the classical sense exists, because it is always unfavorable for a large droplet to form and grow. Furthermore, the water rich phase being built by formation and segregation of large droplets can never become a single phase, due to conservation of the volume fraction of all components. As a consequence of the conservation laws the dynamics at the emulsification failure cannot be understood by local considerations, but it envolves a global optimization; letting a large number of droplets take their optimum size on expense of a small number of energetically unfavorable large droplets which carry away surplus volume. To clarify the origin of the oscillations we point out that the droplet free energy (1) can only be lowered when a large number N of droplets form a single, significantly larger excess droplet, where N decreases with increasing overheating. The formation of the large droplets acts as an energy barrier which can only be overcome for sufficient overheating. After the energy barrier is passed, the smaller droplets quickly hand over their surplus water to the large droplets, and relax to a close to equilibrium state. When increasing temperature further, again a similar barrier has to be passed, leading to successive bursts of precipitation and relaxation to a close to equilibrium state, followed by heating to cross another threshold for formation of large particles.

In conclusion, we have demonstrated that differential scanning microcalorimetry is a technique that allows to probe thermal properties of water–oil–surfactant mixtures. In particular, it allows to study the latent heat of first-order phase transitions and of their dynamics. The measurements indicate that only bending contributions to the free energy are needed to explain phase behavior (cf. [2]) and the latent heat related to the transitions, as well as new types of instabilities encountered in the dynamics of transitions involving droplets.

Acknowledgment This research has been supported by the Deutsche Forschungsgemeinschaft.

References

1. Gompper G, Schick M (1994) In: Domb C, Lebowitz JL (eds) Phase Transitions and Critical Phenomena. Vol 16. Academic Press, London, p 1
2. Vollmer D, Strey R (1995) Europhys Lett 32:693–698; Vollmer D, Vollmer J, Strey R (1996) Phys Rev E 54:3028
3. Kahlweit M, Strey R, Firman P (1986) J Phys Chem 90:671
4. Vollmer D, Ganz P (1995) J Chem Phys 103:4697
5. Helfrich W (1973) Z Naturforsch C 28:693
6. Strey R (1994) Colloid Polym Sci 272:1005
7. Vollmer D, Strey R, Vollmer J (1996) J Chem Phys, submitted

Progr Colloid Polym Sci (1997) 104:201–202
© Steinkopff Verlag 1997

MIX
Papier aus verantwortungsvollen Quellen
Paper from responsible sources
FSC® C105338